Curious About Nature

A Passion for Fieldwork

Notwithstanding the importance of modern technology, fieldwork remains vital, not least through helping to inspire and educate the next generation. Fieldwork has the ingredients of intellectual curiosity, passion, rigour and engagement with the outdoor world – to name just a few. You may be simply noting what you see around you, making detailed records or carrying out an experiment; all of this and much more amounts to fieldwork. Being curious, you think about the world around you, and through patient observation develop and test ideas. Fifty-one contributors capture the excitement and importance of fieldwork through a wide variety of examples, from urban graffiti to the Great Barrier Reef. Outdoor learning is for life: people have the greatest respect and care for their world when they have first-hand experience of it.

TIM BURT retired as Master of Hatfield College and Professor of Geography at Durham University in 2017. His research focuses on catchment hydrology, water quality and climate history. Tim has run the two oldest university weather stations in the UK: the Radcliffe Observatory in Oxford (dating from 1767) and the Durham Observatory (from 1850). President of the Field Studies Council and editor of its journal *Field Studies*, Tim was awarded the Linton Medal by the British Society for Geomorphology in 2017. He is an elected Fellow of the American Geophysical Union and the British Society for Geomorphology. Tim is Emeritus Professor at Durham University and Visiting Professor at the University of Bristol.

DES THOMPSON is the Principal Adviser on Science and Biodiversity with Scottish Natural Heritage. With particular interests in field ecology, his books cover a broad range of interests including birds of prey, shorebirds, alpine and upland habitats, and the Cairngorms and other mountain areas. Des chairs the Technical Advisory Group advising the UN Convention on Migratory Species on the conservation of migratory raptors in Africa and Eurasia. Awarded the Chartered Institute of Ecology and Environmental Management's (CIEEM) Medal in 2019, Des is Chairman of the Field Studies Council, and is an elected Fellow of the CIEEM and the Royal Society of Edinburgh.

ECOLOGY, BIODIVERSITY AND CONSERVATION

The world's biological diversity faces unprecedented threats. The urgent challenge facing the concerned biologist is to understand ecological processes well enough to maintain their functioning in the face of the pressures resulting from human population growth. Those concerned with the conservation of biodiversity and with restoration also need to be acquainted with the political, social, historical, economic and legal frameworks within which ecological and conservation practice must be developed. The new Ecology, Biodiversity, and Conservation series will present balanced, comprehensive, up-to-date, and critical reviews of selected topics within the sciences of ecology and conservation biology, both botanical and zoological, and both 'pure' and 'applied'. It is aimed at advanced final-year undergraduates, graduate students, researchers, and university teachers, as well as ecologists and conservationists in industry, government and the voluntary sectors. The series encompasses a wide range of approaches and scales (spatial, temporal, and taxonomic), including quantitative, theoretical, population, community, ecosystem, landscape, historical, experimental, behavioural and evolutionary studies. The emphasis is on science related to the real world of plants and animals rather than on purely theoretical abstractions and mathematical models. Books in this series will, wherever possible, consider issues from a broad perspective. Some books will challenge existing paradigms and present new ecological concepts, empirical or theoretical models, and testable hypotheses. Other books will explore new approaches and present syntheses on topics of ecological importance.

Ecology and Control of Introduced Plants
Judith H. Myers and Dawn Bazely

Invertebrate Conservation and Agricultural Ecosystems
T. R. New

Risks and Decisions for Conservation and Environmental Management
Mark Burgman

Ecology of Populations
Esa Ranta, Per Lundberg, and Veijo Kaitala

Nonequilibrium Ecology
Klaus Rohde

The Ecology of Phytoplankton
C. S. Reynolds

Systematic Conservation Planning
Chris Margules and Sahotra Sarkar

Large-Scale Landscape Experiments: Lessons from Tumut
David B. Lindenmayer

Assessing the Conservation Value of Freshwaters: An International Perspective
Philip J. Boon and Catherine M. Pringle

Insect Species Conservation
T. R. New

Bird Conservation and Agriculture
Jeremy D. Wilson, Andrew D. Evans, and Philip V. Grice

Cave Biology: Life in darkness
Aldemaro Romero

Biodiversity in Environmental Assessment: Enhancing Ecosystem Services for Human Well-Being
Roel Slootweg, Asha Rajvanshi, Vinod B. Mathur, and Arend Kolhoff

Mapping Species Distributions: Spatial Inference and Prediction
Janet Franklin

Decline and Recovery of the Island Fox: A Case Study for Population Recovery
Timothy J. Coonan, Catherin A. Schwemm, and David K. Garcelon

Ecosystem Functioning
Kurt Jax

Spatio-Temporal Heterogeneity: Concepts and Analyses
Pierre R. L. Dutilleul

Parasites in Ecological Communities: From Interactions to Ecosystems
Melanie J. Hatcher and Alison M. Dunn

Zoo Conservation Biology
John E. Fa, Stephan M. Funk, and Donnamarie O'Connell

Marine Protected Areas: A Multidisciplinary Approach
Joachim Claudet

Biodiversity in Dead Wood
Jogeir N. Stokland, Juha Siitonen, and Bengt Gunnar Jonsson

Landslide Ecology
Lawrence R. Walker and Aaron B. Shiels

Nature's Wealth: The Economics of Ecosystem Services and Poverty
Pieter J.H. van Beukering, Elissaios Papyrakis, Jetske Bouma, and Roy Brouwer

Birds and Climate Change: Impacts and Conservation Responses
James W. Pearce-Higgins and Rhys E. Green

Marine Ecosystems: Human Impacts on Biodiversity, Functioning and Services
Tasman P. Crowe and Christopher L. J. Frid

Wood Ant Ecology and Conservation
Jenni A. Stockan and Elva J. H. Robinson

Detecting and Responding to Alien Plant Incursions
John R. Wilson, F. Dane Panetta and Cory Lindgren

Conserving Africa's Mega-Diversity in the Anthropocene: The Hluhluwe-iMfolozi Park Story
Joris P. G. M. Cromsigt, Sally Archibald and Norman Owen-Smith

National Park Science: A Century of Research in South Africa
Jane Carruthers

Plant Conservation Science and Practice: The Role of Botanic Gardens
Stephen Blackmore and Sara Oldfield

Habitat Suitability and Distribution Models: With Applications in R
Antoine Guisan, Wilfried Thuiller and Niklaus E. Zimmermann

Ecology and Conservation of Forest Birds
Grzegorz Mikusiński, Jean-Michel Roberge and Robert J. Fuller

Species Conservation: Lessons from Islands
Jamieson A. Copsey, Simon A. Black, Jim J. Groombridge and Carl G. Jones

Soil Fauna Assemblages: Global to Local Scales
Uffe N. Nielsen

Curious About Nature: A Passion for Fieldwork
Tim Burt and Des Thompson

Curious About Nature

A Passion for Fieldwork

Edited by

TIM BURT

Durham University and the University of Bristol

DES THOMPSON

Scottish Natural Heritage

CAMBRIDGE
UNIVERSITY PRESS

CAMBRIDGE
UNIVERSITY PRESS

University Printing House, Cambridge CB2 8BS, United Kingdom

One Liberty Plaza, 20th Floor, New York, NY 10006, USA

477 Williamstown Road, Port Melbourne, VIC 3207, Australia

314–321, 3rd Floor, Plot 3, Splendor Forum, Jasola District Centre, New Delhi – 110025, India

79 Anson Road, #06–04/06, Singapore 079906

Cambridge University Press is part of the University of Cambridge.

It furthers the University's mission by disseminating knowledge in the pursuit of education, learning, and research at the highest international levels of excellence.

www.cambridge.org
Information on this title: www.cambridge.org/9781108428040
DOI: 10.1017/9781108552172

First published 2020

A catalogue record for this publication is available from the British Library.

Library of Congress Cataloging-in-Publication Data
Names: Burt, Tim, 1951– editor. | Thompson, D. B. A., 1958– editor.
Title: Curious about nature : a passion for fieldwork / Edited by Tim Burt, Durham University, Des Thompson, Scottish Natural Heritage.
Description: New York, NY : Cambridge University Press, 2020. | Includes bibliographical references and index.
Identifiers: LCCN 2019036957 (print) | LCCN 2019036958 (ebook) | ISBN 9781108428040 (hardback) | ISBN 9781108448642 (trade paperback) | ISBN 9781108552172 (epub)
Subjects: LCSH: Fieldwork (Educational method) | Nature study. | Environmental education–Study and teaching. | Outdoor life. | Technological innovations.
Classification: LCC LB2394 .C87 2020 (print) | LCC LB2394 (ebook) | DDC 371.3/84–dc23
LC record available at https://lccn.loc.gov/2019036957
LC ebook record available at https://lccn.loc.gov/2019036958

ISBN 978-1-108-42804-0 Hardback
ISBN 978-1-108-44864-2 Paperback

Dedication

In memory of Ian Mercer and Sam Berry

Stalwarts of the Field Studies Council, Ian and Sam encouraged large numbers of people to become passionate fieldworkers and curious about nature

Contents

The plate section can be found between pp. 202 and 203

Preface

In the primeval of our time, something happened to signal the giant step taken by *Homo erectus* to emerge as *Homo sapiens*. Scientific debate fundamentally rages over this, but there is little doubt now that fieldwork marks the evolutionary transition as well as the unravelling. Quite possibly, half a million years ago, the earliest record of a 'human'-related imprint reflects the most rudimentary of fieldwork and was revealed by a brilliant piece of field research.

In the journal *Nature*, on 3 December 2014, a large team of archaeologists, led by Josephine Joordens at Leiden University, reported a remarkable finding. But first, we have to go back to 1891, to Trinil, in east Java, Indonesia. There, the Dutch anatomist Eugène Dubois and his group discovered in a riverbank the first early remains of what is now popularly called 'Java Man' – they found a skullcap, tooth and thighbone. This was the first evidence of our ancestor *H. erectus*.

Re-working material in the *Hauptknochenschicht* ('main bone layer') of Trinil, Joordens and colleagues found a fossil shellfish assemblage. That in itself was not so unusual, for the shellfish would have been consumed and the shells used as tools. But, intriguingly, one of the shells had a geometric engraving, and in the words of the authors this suggested: 'Although it is at present not possible to assess the function or meaning of the engraved shell, this discovery suggests that engraving abstract patterns was in the realm of Asian *Homo erectus* cognition and neuromotor control.' In other words, quite possibly these markings gave a primitive sense of place, of an action, or even something more telling. Of course, there are cave paintings dating as far back as 40,000 years ago, essentially giving us the first detailed field observations made by people. But even the dates of the earliest paintings are being contested, and it has recently been argued that in the caves of three sites in Spain, some paintings may be 65,000 years old and of Neandertal origin (Hoffmann *et al.*, 2018, whose paper featured on the cover of *Science*, with the caption 'Minds at work'). But how extraordinary to think that markings made half a million years ago may be the very first field notes recorded. While we know that tool use by human ancestors dates back to something like 3.3 million years ago, we have little evidence of observations being made so far back in time.

Field studies unravelling discoveries such as these have fired our passion and curiosity for fieldwork. Our endeavour is simple – to inspire and celebrate the fantastic range of fieldwork devoted to understanding our environment, nature and the plethora of processes. Archaeologists, earth scientists, geographers, ecologists, geneticists, land and water managers and students of soils, rocks, microbes, fungi, lichens, plants, animals and any other facet of the natural world all rely on fieldwork, and we have drawn heavily on examples of their studies.

People who observe, carry out experiments or otherwise study in the field are engaged in fieldwork. Some people are simply inspired by being in the field – they may not be making observations or carrying out experiments, but they love fieldwork, find it uplifting and draw on what they have read, heard about or seen previously to form views. Is this fieldwork? Yes it is, and often simply being in the field and being in contact with the natural world and nature is life-affirming and stimulating for senses, far beyond drawing inferences on what you see, hear, touch, smell or even remember. Curiosity is the great driving force, with passion frequently ensuing as we immerse ourselves in the art and science of discovering more about our world.

Our principal aim is to inspire you to carry out fieldwork, and to enjoy it. Part I has five overview chapters. Chapter 1 is an introduction to the great range of studies and techniques employed in the field. In Chapter 2, Mike Church places field studies in the wider context of environmental science. The next two chapters provide biographies of some of our heroes in the geophysical and life sciences, with Andrew Goudie and Steve Trudgill giving vent to their enthusiasm for their areas of interest. Outdoor learning is growing in appeal, as much for educationalists as part of core curricular approaches to learning, as for students enjoying learning. Learning is a gift we should treasure, and if we can we should benefit from it every day. So, why confine it to a stuffy, soulless classroom, when we can be outside, moving, seeing, hearing and thinking as we learn. Michael Reiss traces the educational benefits of such learning in Chapter 5.

In Part II, we have a fantastic array of essays by no less than 45 fieldworkers. Our ask was simple – write about a person, topic, place, technique, publication, finding or a simple moment, which derives from fieldwork. The response has been terrific, and we believe the collection amounts to a truly enjoyable and stimulating read. In the final part of the book we try to draw some threads together on what has motivated fieldworkers and the discoveries made. We look ahead to emerging approaches and techniques and try to anticipate developments that should help embolden and strengthen the foundations of fieldwork and fieldworkers.

We are very grateful to a large number of people who have worked with us during the book's gestation and emergence. We are both trustees of the Field Studies Council (FSC) and thank the staff and fellow trustees for helping to inspire the book's contents; some are contributors. On a more practical note, we thank Rebecca Farley-Brown at FSC Publications for providing a number of photographs. We thank all our authors, not least Steve Trudgill, who stepped into the breach when the late Sam Berry passed away before he could start work on his chapter; Sam was one of those who inspired this book in the first place, and we miss his wisdom. Our authors have been marvellously patient and tolerant of our questions, and we thank them all.

Des is very grateful to Hatfield College, Durham University, for a Visiting Fellowship in Easter Term 2016, which provided a marvellous opportunity to develop with Tim the ideas underpinning this book. Janet Raine, the Master's Secretary at Hatfield College, has provided prodigious support over the years. We remain grateful for continued encouragement from the present Master, Professor Ann MacLarnon. We thank Chris Orton, cartographer in the Geography Department at Durham University, for drawing a diagram. Durham University, the FSC and both Scottish Natural Heritage and the Joint Nature Conservation Committee have provided a core influence for us, as reflected in the authorship.

At CUP, Michael Usher has been a wonderfully enthusiastic and encouraging editor, offering us a fund of ideas and suggestions. Dominic Lewis, Aleksandra Serocka and Annie Toynbee at CUP have also been very supportive; we thank Aleksandra and Annie in particular for dealing with our many questions and queries with good humour and prompt responses. We are very grateful to Zoë Lewin for her very careful copy-editing. We thank Ian Newton, doyen of field studies of raptors and bird migration, for writing the Foreword, and giving support and advice in other ways. Several colleagues have offered us frequent advice as we have worked on the book and others have inspired us through their writings, and it is a pleasure to acknowledge them: John and Hilary Birks,

Terry Burke, Ingvar Byrkjedal, Colin Campbell, Mike Church, Mark Cocker, Nick Davies, Miguel Ferrer, Alan Fielding, Bob Furness, Colin Galbraith, Andrew Goudie, Jeremy Greenwood, Bill Heal, John Atle Kålås, Ron Macdonald, David O'Brien, Robert and Stuart Rae, Steve Redpath, Chris Spray, David Stroud, Bill Sutherland, Pat Thompson, Steve Trudgill, Paul Walton, Phil Whitfield, Jeremy Wilson, and the late Chris Barnard, Richard Evans, Paul Haworth, Aubrey Manning, Simon Thirgood, Desmond Nethersole-Thompson, Dick Potts, Derek Ratcliffe and Adam Watson.

Both of us have had lives full of fieldwork! The two photos show us in our youth, already curious about nature. The sooner, the better: we are both convinced that fieldwork should start at a young age, which is why we are both strong supporters of the FSC, which has given young people the opportunity to get into the field for the past 75 years.

Finally, we thank our wives, Elizabeth and Dawn, for their strong support, tolerance and advice throughout this project.

To the late Ian Mercer, former President of the Field Studies Council, and Sam Berry, former trustee and Honorary Vice-President, we dedicate this book.

References

Hoffmann, D. L., Standish, C. D., Garcia-Diez, M, *et al.* (2018). U–Th dating of carbonate crusts reveals Neandertal origin of Iberian cave art. *Science* 359, 912–915.

Joordens, J. C. A., d'Errico, F., Wesselingh, F. P., *et al.* (2015). *Homo erectus* at Trinil on Java used shells for tool production and engraving. *Nature* 518, 228–231.

Eleven-year-old Des Thompson (on right) with his brother Pat Thompson and father Desmond Nethersole-Thompson inspecting a greenshank's nest, found by the photographer. June 1969 (Derek Ratcliffe).

Tim Burt on a school expedition to Portugal and Spain. Seen here in a valley above Cain de Valdeón in the Picos de Europa mountains, 1 August 1969.

Foreword

It is a great pleasure to introduce a book on field studies. Every branch of science has its manuals of methodology, but this welcome volume is unusual in pulling together information on field studies from an exceptionally wide base, embracing both the physical and biological sciences, and beyond. No less than 45 fieldworkers have contributed to the text, working under the capable editorship of Professors Tim Burt and Des Thompson.

But what is it that compels us all to study nature in the great outdoors? Most importantly, we are all driven by a passion and curiosity about the natural world, a deep-rooted desire to understand the underlying workings of nature. What we see, hear and record during our work is crucial. But in practice, it is often the unexpected, the puzzling or the surprising that fires our imagination, adding interest to every day in the field, and leading us on to new ideas and new discoveries.

Fieldwork has been central in my own life for as long as I can remember, beginning as a small boy with an interest in natural history and continuing through my professional life as an ornithologist. Without the possibility of working outdoors, I would probably have never become a biologist, but taken up some more applied subject, such as agriculture or forestry which, in those days, did not involve a life in an air-conditioned cab.

For biologists like me, the outdoor world is the only place where wild plants and animals can be studied in their natural settings, behaving as they should, interacting with other plants and animals around them. Without field studies, biology would be little more than laboratory-based anatomy and physiology – for the scientist, life in a white coat. But for me fieldwork involves much more than this. It gives the sheer pleasure of being outdoors, escaping the claustrophobic and pressurised human world for a while, feeling the warmth of the sun or the chill of the wind, and facing the physical challenges which some types of fieldwork present. My own work on birds of prey involved countless hours of walking and searching through interesting landscapes, and hundreds of tree-climbs every year (the sort of work that would now be constrained by health and safety measures). In addition, this type of fieldwork provides endless surprises and rare glimpses of nature. It also satisfies hunting instincts, not so easily achieved in today's world.

For this book, the editors have clearly struck a chord in securing so many enthusiastic and revealing contributions from scientists in various disciplines, hooked on field studies. Like me, both editors were ensnared by fieldwork in their early lives, and both were inspired by key people in their formative years. Tim Burt was brought up in Somerset and fell under the spell of a charismatic geography teacher called Jim Hanwell, from whom he developed a deep curiosity about landforms, including the valleys and quarries in the eastern Mendips. So began a life-long journey as an academic physical geographer and meteorologist, leading Tim to co-author a fascinating book recently published on *Oxford Weather and Climate Since 1767* (2019).

The second editor, Des Thompson, I have known since he was a teenager in Sutherland. Both he and I were heavily influenced by two exceptional field naturalists. One was Des's father, ornithologist Desmond Nethersole-Thompson and the other was botanist-ornithologist Derek Ratcliffe. Both were renowned naturalists whose field craft was legendary; they had a deep knowledge of the natural world, and both had a profound influence on the two of us. For Des as a youngster, escaping schoolwork allowed him to spend time nest hunting and observing waders in the hill country. In due course, he became a professional ecologist and went on to write and edit a number of key research volumes on upland birds and ecology, the most notable in his early career being a special publication of the British Ecological Society, edited with Michael Usher, *Ecological Change in the Uplands*, in 1988. Almost this entire volume was based on fieldwork of one sort or another.

In recent years, we have been increasingly witnessing the effects of human population growth and activities on the natural world, bringing enormous detrimental changes to flora, fauna and climate. Nowadays, these issues are continually in the news, making depressing reading for all who care about life on Earth. In May 2019, we saw the declaration of a 'Climate Emergency' in several countries, in part in response to the UN 'Summary for policymakers of the global assessment report on biodiversity and ecosystem services'. That document, based on around 15,000 scientific and government sources, points to an impending environmental calamity, largely or entirely of our own making.

The underlying evidence for most of these conclusions stems from field studies. The observers, experimentalists, modellers, analysts, synthesisers and advocates involved deserve massive recognition for their efforts. In conversation, many of them will point to their concern for the natural world and their abiding enjoyment in connecting with nature as their main motivation. Passion, curiosity and an ethos of care will all be needed as we face the challenges we have created for the future of our planet, with its much-loved lands, waters and diverse life forms. We will again rely on field studies and fieldworkers to monitor further change, whatever it may bring. In this endeavour, no doubt many future fieldworkers will gain inspiration from the contents of this book.

Ian Newton OBE FRS FRSE

Part I

Getting Curious About Nature

1 · *Fieldwork and Nature: Observing, Experimenting and Thinking*

TIM BURT AND DES THOMPSON

> A love for this subject began to revive; fresh ideas were awakened in men's minds, and a new spirit of exploration and enquiry was abroad.

At the beginning of the first chapter of the first of the Collins New Naturalist Library books, *Butterflies* (1945), E. B. Ford reflects on the history of British butterfly collecting. Ford is referring to the spread of the Renaissance, and thereby our access to the science of the Greeks, including their work on natural history.

1.1 Introduction

Fieldwork has the ingredients of intellectual curiosity, passion, rigour and engagement with the outdoor world – to name just a few. We may be simply noting what we see around us, making detailed records, employing sophisticated techniques, carrying out an experiment or, quite possibly, collaborating with a large international group of scientists. All of this and much more amounts to fieldwork.

We adopt a wide definition of what we consider to be fieldwork. We are concerned mainly with the environmental spheres of study, but we do stray into the social and human sciences where subjects like human geography and archaeology also rely on 'fieldwork', broadly defined as the collection of data beyond the laboratory, library or workplace, i.e. in the *field*. We include 'marine' fieldwork; intertidal, near-shore and further out into the deep ocean; we retain the title 'fieldwork' for these studies although there might be a case (Michael Usher, personal communication) for inventing 'seawork' to cover these studies! Some of the basic components of fieldwork are shown in Figure 1.1.

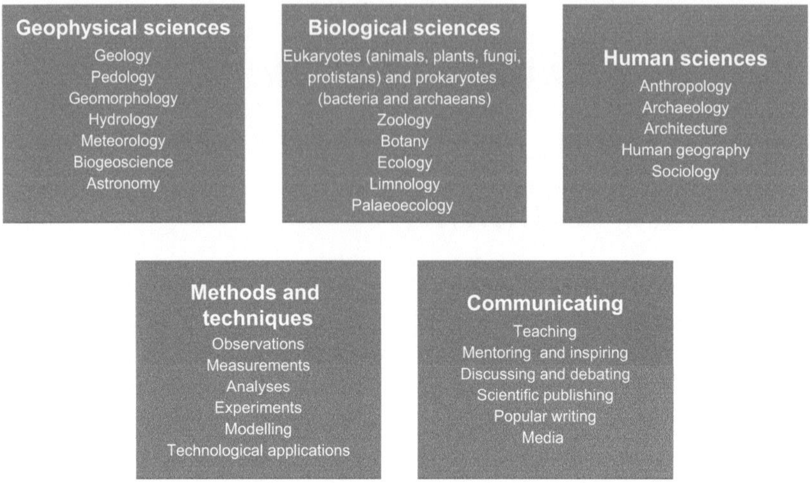

Figure 1.1 The basic components of fieldwork

One of the most important books setting out the rationale for environmental fieldwork, and its wider importance, is an obscure, 84-page book, *The Global Challenge for Field Science* (1990), published by Earthwatch and based on a seminar hosted by Earthwatch Europe and the Royal Society on 17 February 1989. What is so remarkable is the calibre of contributors and the clear sense of emerging importance attached to carrying out fieldwork. The President of the Royal Society, Sir George Porter, opened proceedings remarking: 'Field sciences and the earth sciences in general are undergoing explosive development at present . . . this is important because, more than ever before, we need to understand our environment better, so that we can take some action about it. It is an interdisciplinary subject and it has another advantage I think, in that so many people can participate – even people who are not professional scientists in any way.'

In his preface to the book, Max Nicholson was typically prescient, commenting: 'International concern over ozone depletion, global climate change, environmental degradation, and the sustainable use of natural resources foreshadowed a vastly expanded demand for programmed field research. Such international initiatives as the International Geosphere–Biosphere Programme, impending major changes in the European Community linked to the year 1992, and the pressing environmental concerns of eastern Europe, pointed to the advisability of starting without further delay to create a working programme of field science dedicated to global environmental monitoring and the sustainable use of biological and cultural resources.' If that were not enough, leading ecologist Sir Richard Southwood wrote a short essay on the role and contribution of field science, remarking: 'The point is that in physical, as well as biological sciences, fieldwork is important in the formats of hypothesis that help us to understand the natural world. There is very little in modern biology that cannot be traced back to some field observations.' In detailing the importance of experimental work, long-term field studies and field stations, he urged further that: '. . . students must see how biology functions in the field'.

There are hundreds of thousands of books and scientific papers employing fieldwork, and in each of these it is likely that at least one of the contributors will have had his or her curiosity aroused over the workings of nature. That is what places fieldwork apart from other methods or settings for research, and it is, we believe, what is at the heart of the enduring appeal of fieldwork.

In this introduction, we explore aspects of the curious mind of the fieldworker, provide some background information on the basic equipment used, reflect on some field studies which, for us, reflect the diversity of outstanding studies and the outstanding fieldworkers who carried out those studies. Finally, we offer some perspectives on what it is to be curious about the world around us.

1.2 Curiosity

> Curiosity is the very basis of education and if you tell me that curiosity killed the cat, I say only the cat died nobly.
>
> Arnold Edinborough (Canadian writer and broadcaster)

> Curiosity (noun): an eager wish to know or learn about something.
>
> Cambridge English Dictionary

In the Introduction to his book *Curiosity*, Alberto Manguel starts by reminding us that one of the first words we learn as a child is 'Why?' (Manguel, 2015, p. 1). Initially, we want to understand what the boundaries to our behaviour are, but soon we develop an interest in how we fit into the world around us and what the nature of this mysterious world is

Figure 1.2 Why have we brought you here? Andrew Goudie and Tim Burt always used to start their Dorset coast field trip with this simple question. On this photo, Andrew is talking to students at Durdle Dor (Tim Burt). A black and white version of this figure will appear in some formats. For the colour version, please refer to the plate section.

all about. Manguel adds that we feel an ancestral need to engage with other inhabitants of this world, giving us the ability not only to understand the here and now but to speculate about the there and then, of understanding the uncharted territory ahead, as Manguel puts it. Having started to ask questions, we never stop. We soon find out that curiosity is seldom rewarded with satisfying answers but rather the starting point for asking more questions.

According to Richard Dawkins (quoted in Manguel, 2015, p. 3), human imagination is a survival mechanism. In order to stay alive, *Homo sapiens* developed the ability to reconstruct reality in the mind and to conceive situations that it might confront before actually doing so. We imagine in order to exist and we are curious in order to feed our imaginative desire (Manguel, 2015, p. 3). What we need to know and what we can imagine are therefore two sides of the same coin; in relation to the natural world, this has immediate resonance with a combination of awe and wonder, perhaps the sight of a ferocious animal in close proximity or a forbidding landscape as we travel towards it. Questions are more important than answers; answers are rarely satisfactory, more questions inevitably follow. This is curiosity's inherent paradox (Manguel, 2015, p. 42).

Curious humans want to know and understand the world around them. However this knowledge is acquired, the process is likely to be hesitant and faltering. This is not the place for a detailed elaboration of scientific method, but a brief digression into this subject area allows us to see how curiosity, conjecture, observation and explanation go together. Being curious, we speculate about the world around us, and it is observation that allows us to test our ideas. When we ask 'Why?', we look to see if what we see accords with what we think. Very often, our ideas prove imperfect, leading to more questions. Manguel describes this as 'enlightened failure', quoting Samuel Beckett's aphorism: 'Fail. Try again. Fail better' (Manguel, 2015, p. 3). There are two conflicting models of scientific endeavour: induction and deduction; in both schemes the role of observation is critical.

In the inductive method, empirical generalisations are derived from observation of reality: recording of facts leads to hypothesis formulation; confident understanding grows via the accumulation of further observations. While induction is largely out of favour these days, new hypotheses have to come from somewhere, and the refinement of existing theory often relies on an essentially inductive process whereby fresh observations raise questions about existing ideas, leading on to an improved explanation: 'enlightened failure', once again. In the deductive scientific method, theory precedes observation, data are collected which can then be used to corroborate theory. The data collection is mainly achieved through formal experimentation but less-formal observational programmes often characterise the early stages of any scientific investigation. Through a circular process of conjecture and falsification, hypotheses are gradually improved and false ideas eliminated: what Karl Popper would have regarded as a process of trial and error elimination. No theory is ever proved 'true': as Popper wrote, the degree of corroboration is a guide to the preference between two theories at a certain stage of discussion with respect to their apparent approximation to the truth. But it tells us only that one of the theories offered seems – in the light of discussion – the one nearer the truth (Popper, 1972, quoted in Haines-Young and Petch, 1986).

A brief discussion of scientific method is relevant here because of the centrality of observation in fieldwork. In Part II, a wide range of fieldwork is described, but in every case, whether casual surveillance or

Figure 1.3 How come there are periglacial deposits on top of this hill? Oxford students carefully examine stony subsoil material at Stonebarrow on the Dorset coast near Charmouth (Tim Burt).

formal experiment, there is an attempt to better understand the world around us through the medium of observation. Setting aside the myriad issues associated with philosophical perspectives on scientific observation, we are content here to take a simple view: (field) observations are crucial if we are better to understand the world around us. Our perspective is largely about the natural world, but the approach is no different where the human-dominated landscape is concerned (see, for example, the essays by Carrick and Corbridge in Part II).

A brief look at perhaps the greatest fieldworker of them all, Charles Darwin, illustrates the difficulty in separating the inductive and deductive routes to scientific explanation. Darwin was naturally curious and asked the question 'Why?', wherever he travelled. Much of his initial work was necessarily inductive and he began to develop ideas from what he had observed in the field, such as: silicified trees and marine shells in rock high up in the Andean mountains; the distribution of animals on the different Galapagos Islands; coral reefs and atolls. Ayala (2009) argues that the inductive process fails to account for the actual methodology of science since no scientist works without any preconceived plan as to what kinds of phenomena to observe. But Darwin had no preconceived plan beyond being curious. No doubt some questions were of more interest to him than others but his scope was enormously wide-ranging. Partly, he claimed to work inductively to avoid being accused of subjective bias in the evaluation of empirical evidence (Ayala, 2009). However, it is clear that his major achievements were all deductive, theory-led, to which he marshalled his observations to test and refine his ideas. For example, in the case of coral reefs:

The theory which I would offer, is simply, that as the land with the attached reefs subsides very gradually from the action of subterranean causes, the coral-building polypi soon raise again their solid masses to the level of the water: but not so with the land; each inch lost is irreclaimably gone; as the whole gradually sinks, the water gains foot by foot on the shore, till the last and highest peak is finally submerged. [Darwin (1839), ch. XXll, p. 557]

Darwin goes on to set out his ideas on the formation of the Cocos (Keeling) Islands. These consist of two atolls made up of 27 coral islands, and they were discovered in 1609 by Captain William Keeling of the East India Company.

Hence we must consider this island [Keeling Island] as the summit of a lofty mountain; to how great a depth or thickness the work of the Coral animal extends is quite uncertain. If the opinion that the rock-making Polypi continue to build upwards, as the foundation of the island from volcanic agency, after intervals gradually subsides, is granted to be true; then probably the coral limestone must be of great thickness ... Hence if we imagine such an island to subside a few feet, in a manner similar, but with a movement opposite to the continent of S. America; the coral would be continued upwards, rising from the foundation of the encircling reef. In time the central land would sink beneath the level of the sea and disappear, but the coral would have completed its circular wall. [Charles Darwin's Beagle Diary. Edited by R. D. Keynes, 2001, p. 418. Entry for 12 April 1835]

Darwin distinguished between what he described as the three great classes: atolls, barrier and fringing reefs. His theory enabled him to speculate about coral reefs, as yet unseen, and to explain why there were not reefs on the South American coast or atolls in the West Indies. He was unable to provide direct evidence of ocean-floor subsidence but it is clear that his observations of uplift in the Andes were influential in convincing him that Earth's crust was anything but stable. The very long timescales involved, and the very gradual rate of formation, he drew directly from Charles Lyell's doctrine of uniformitarianism.

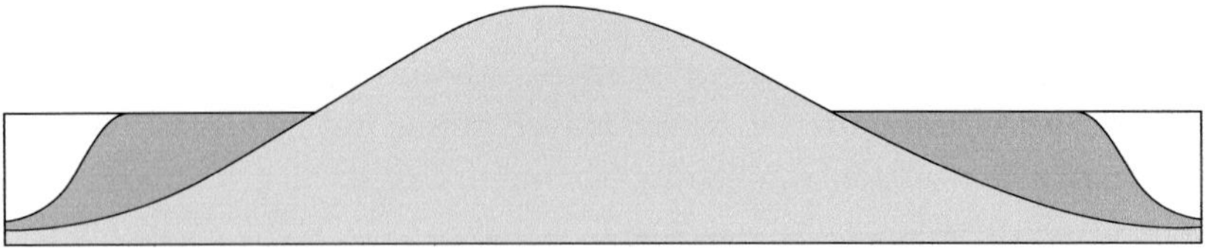

(i) A fringing reef

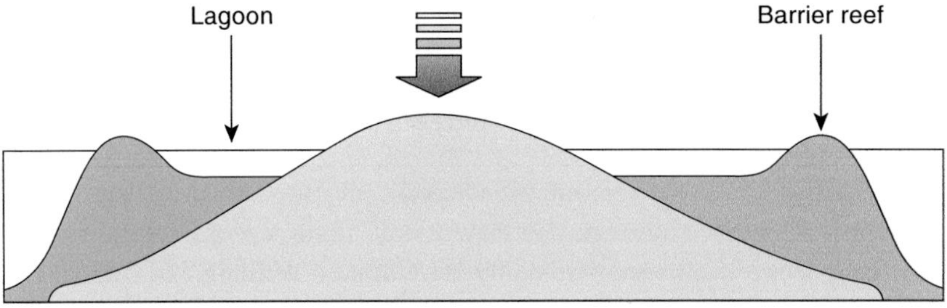

(ii) Subsidence of the island leads to a barrier reef developing

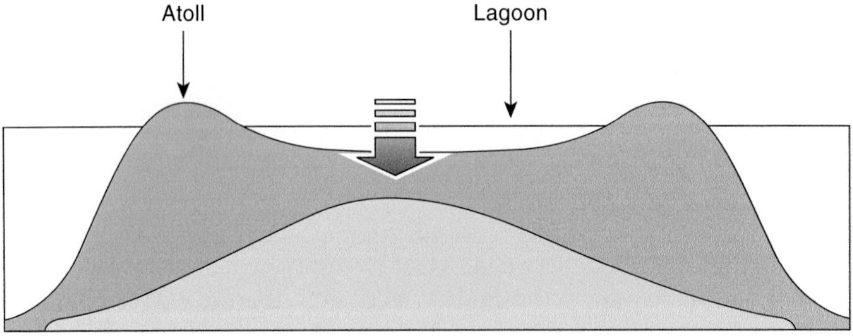

(iii) Further subsidence matched by the growth of coral leads to an atoll

Figure 1.4 Charles Darwin's theory of the transition from barrier reef to atoll (redrawn from Darwin, 1839)

Today, Darwin is best known for his observations of flora and fauna, in the Galapagos especially, but he was equally a geologist when he set off on the *Beagle* voyage, and his curiosity about rocks and relief is no less impressive. Darwin's fieldwork gave him ideas, which he then developed and refined through further observation.

It is too easy to consider fieldwork as 'terrestrial'; there is an important and growing interest in marine fieldwork too. Darwin's voyage on the *Beagle* came towards the end of a long sequence of voyages of discovery, but these were essentially still terrestrial in focus, investigating and surveying newly found land. There have long been studies of shoreline ecosystems, intertidal rocky shores being of particular interest, easily accessible to students and staff, with novel flora and fauna, usually arranged in relation to tidal limits and submergence. Good examples of such studies appear in *Field Studies*, for coastal field centres such as Dale Fort (http://fsj.field-studies-council.org/browse-by-field-centre/dale-fort.aspx; see also John Archer-Thomson's essay in Part II): initially, zonation of intertidal species was the focus (e.g. Moyes and Nelson-Smith, 1963), but more recently the impact of pollution incidents (e.g. Archer-Thomson, 1999) and invasive non-native species (Day, 2018) have been described. Deep-sea field research largely remains the preserve of well-funded research programmes, given the cost of boats and associated facilities required. Modern oceanography began with the *Challenger* expedition (1872–1876), which was specifically organised to collect a wide range of data, from sea-water temperature to marine fauna. A British Navy vessel, *HMS Challenger*, was specially converted with laboratories and equipment like winches and sea-floor sediment samplers. Among the *Challenger* expedition's discoveries was the Marianas Trench in the western Pacific, one of the deepest places on the ocean floor (8,200 metres). In other survey work, what we now know as the Mid-Atlantic Ridge was identified, but only in very broad terms. It needed much more sophisticated survey methods, nearly a century later, to map ocean-floor topography and so establish the basis for plate tectonics. Interestingly, the breakthrough was made by an American geology professor, Harry Hess, who served in the US Navy in World War II. Hess carefully tracked his travel routes to Pacific Ocean landings on the Marianas, Philippines and Iwo Jima, continuously using his ship's echo sounder. This unplanned wartime scientific surveying enabled Hess to collect ocean-floor profiles across the North Pacific Ocean, surely one of the most unlikely pieces of field data collection ever! Since the *Challenger* expedition, there have been many expeditions to all the world's seas and oceans, supported today by increasingly sophisticated remote sensing equipment. Two of the chapters in Part II (Byrkjedal, Husband) describe modern marine fieldwork, one in the deep ocean and the other in shallow waters around the Great Barrier Reef.

1.3 How Do We Carry Out Fieldwork?

In what follows, we have listed some of the most the widely used items of field equipment, ordered approximately chronologically, beginning with the earliest in use.

Field 'notebooks'. The most important of all items, the 'notebook', records what you see and hear, feel and otherwise sense – the 'when', the 'where' and the 'what' observed. As an original record of observations, field notes are essential and invaluable. Of course, the word *field* is an old one, derived from the Old Saxon *feld* – open land as opposed to woodland. The *Oxford English Dictionary* (OED) then follows this first definition with 'the country as opposed to a town or village', a definition described as obsolete except in Middle English. This sense of difference or otherness may have seemed obsolete to the compilers of the OED but it is clear that those who head to the field, as Michael Canfield (2011a) puts it, have an innate feeling for the different location and character of the 'field', whether near or far, but always apart from the classroom or laboratory, 'opposed to', as the OED puts it. As noted below, Gilbert White may have been the first to be recognised as a *field naturalist* but the contrast between town and open country is a much older sense.

Our primitive ancestors were the first observers of nature, noting what they saw in cave drawings and paintings. Scandinavian rock art goes far back – *viedekust* (hunters' art) of the Northern Tradition hunters and gatherers of the Stone Age dates to 9000–2000 BC, and the *jordbruksristninger* (agricultural petroglyphs) of the Southern Tradition is associated with people of the Bronze and early Iron ages (1800 BC–AD 400) (Lødøen and Mandt, 2010). Depicting animals, hunting and various practices, these are some of the earliest 'field notes'. Going much further back, the 600 or so Palaeolithic cave paintings of Lascaux in the Dordogne of south-western France may be up to 20,000 years old, and they marvellously depict large animals. Remarkably, the Caves of Monte Castillo in Cantabria, Spain, have cave art dated at more than 40,800 years (Callaway, 2012), and recent discoveries point to the earliest cave paintings of animals (hog deer) in Pettakere Cave on the island of South

Sulawesi, Indonesia, possibly being as old as 40,000 years (Aubert *et al.*, 2014). Aboriginal art forms the world's oldest continuous cultural tradition, dating back 30,000 years, with a rock painting in Kimberley, Western Australia, depicting the long-extinct marsupial lion (*Thylacoleo carnifex*) (Akerman and Willing, 2009).

Some of the most famous and earliest field notes on nature are by Gilbert White (1789), writing about the parish of Selborne, some 50 miles southwest of London. These notes comprised 'letters' (many not actually sent) to other naturalists (Thomas Pennant and Daines Barrington), calendars (with important phenological data) and many observations. In his introduction to *Field Notes on Science and Nature*, Michael Canfield (2011b) cites Gilbert White as the first 'field naturalist', a designation given to him by Lieutenant-Colonel Montagu in a letter he wrote to White, after the publication of White's famous book, *The Natural History and Antiquities of Selborne*.

Now, while many naturalists still use traditional notebooks, there is a wide range of apps used for biological recording, with records going to the Biological Records Centre or National Biodiversity Network, via phones and other devices. How fieldworkers maintain field notebooks in our modern era is a matter of choice. Some (usually older people) regret the loss of pen and paper but the young will no doubt embrace the computer age, armed in the field with iPad or Surface, and no doubt some of these 'field notebooks' will survive into the future in some sort of cloud-based archive.

Maps. The earliest maps of the land and sea date from the fifth century BC, depicting Earth as flat (the stars were depicted far earlier, and feature in some of the earliest cave paintings of 16,500 BC). With the advent of the Hellenistic period of the ancient Greeks, maps showing Earth as a sphere appeared (attributed in particular to Eratosthenes and Posidonius). Later, Ptolemy's world map (150 AD) was the first to use lines of latitude and longitude, as well as specifying (terrestrial) locations based on celestial observations, and these maps persisted well into the Middle Ages. Wolodtschenko and Forner (2007) detail some historical maps.

The Ordnance Survey maps have their origins in military history, with the Scottish Highlands mapped early on, following the 1745 Jacobean Rebellion (led by 21-year-old William Roy in 1747, producing the Great Map, eight years later). On the south coast of England, with fears of invasion from France, there was an impetus for a detailed southern map. Jesse Ramsden's Great Theodolite (three years in the making) enabled detailed triangulation of landmarks. On 21 June 1791, the Board

Figure 1.5 Binoculars can be used for spying vegetation on cliffs as well as birdlife (Des Thompson)

of the Ordnance bought a second (improved) Ramsden theodolite, and that gave us the advent of the Ordnance Survey (OS). In 2001, the OS Master Map was launched, with a geospatial database referencing more than 460 million man-made and natural landscape features in Britain. Today, all 243,241 square kilometres of Britain are surveyed, with 10,000 changes put into the database daily.

Compasses and clinometers. The first compasses were made of lodestone, a naturally magnetised ore of iron, in the Han dynasty of China, 300–200 BC. The dry compass, with an encased needle pivoting on a pin, was in use as early as 1269. The clinometer, used for measuring slope, elevation or depression of an object, took several early forms. The Well's in-clinometer was one of the earliest for measuring altitude; the Abney level was developed in the 1870s for measuring slopes; but the most famous one was the Rieker Inc P-1057 Degree Inclinometer, and was used by Charles Lindbergh on the 'Spirit of St. Louis' flight on 20–21 May 1927, from Long Island, New York, to Paris.

Optics. There are references to magnifying glasses/hand lenses used to light fires as early as 424 BC. The earliest lenses are typically dated to around 750 BC; the ancient Romans and Greeks filled glass spheres with water for magnification (Emperor Nero is said to have watched gladiatorial combats through an emerald lens). Glass lenses were made in the Middle Ages (monks made 'reading stones' from cut glass during the eleventh to thirteenth centuries). Telescopes appeared in 1608 (invented by Hans Lippershey), with binoculars also made around the same time (using Galilean optics – a convex objective and a concave eyepiece); later, Porro prism binoculars were devised, in 1854, and the roof prism design was used from the 1870s. Early compound microscopes appeared around 1620 (possibly invented by Galileo Galilei, later improving on the design of telescopes). The philosopher and friar, Roger Bacon, described magnifying glasses in the thirteenth century, and was a great, early observer of nature. The first photographic images produced using a 'camera' are attributed to the French inventor, Nicéphore Niépce, in 1816.

Meteorological instruments. There is a great range of these used for measuring and logging temperature, humidity, rainfall, wind speed, solar radiation and other variables, with year-on-year developments in applications. We have gone a very long way from the alcohol thermometer of 1709 and the mercury thermometer of 1714, both invented by Daniel Gabriel Fahrenheit, and the anemometer invented by John Thomas Romney Robinson in 1846. Long before then, the Ancient Greeks kept records of rainfall; and in India, record keeping began around 400 BC (with crude rain gauges used to classify land for taxation purposes in relation to grain production). The first true rain gauge, the Cheugugi, was invented by the Joseon Dynasty of Korea and used widely in the early fifteenth century. The official record of the rainfall by the Cheugugi, from King Jeongjo's reign (1776–1800) to Emperor Gojong's (1863–1907), is preserved. In the UK, early observations were often by individual enthusiasts or by astronomers (see below: Radcliffe Observatory) and the development of a 'meteorological office' came later with the need for accurate weather forecasting, especially for shipping. Equipment design evolved rapidly through the nineteenth century, moving steadily towards standardisation of design and exposure. For example, the Stevenson Screen is a box to shield meteorological instruments against precipitation and direct solar radiation, while still allowing air to circulate freely around them. It was designed by Thomas Stevenson (1818–1887), a civil engineer who designed many lighthouses (the father of author Robert Louis Stevenson). Stevenson's design of a small thermometer screen with louvred walls on all sides was reported in 1864. The Campbell–Stokes sunshine recorder was invented by John Francis Campbell in 1853 and modified in 1879 by Sir George Gabriel Stokes. The glass sphere, 10 cm in diameter, focuses the sun's rays on to a card

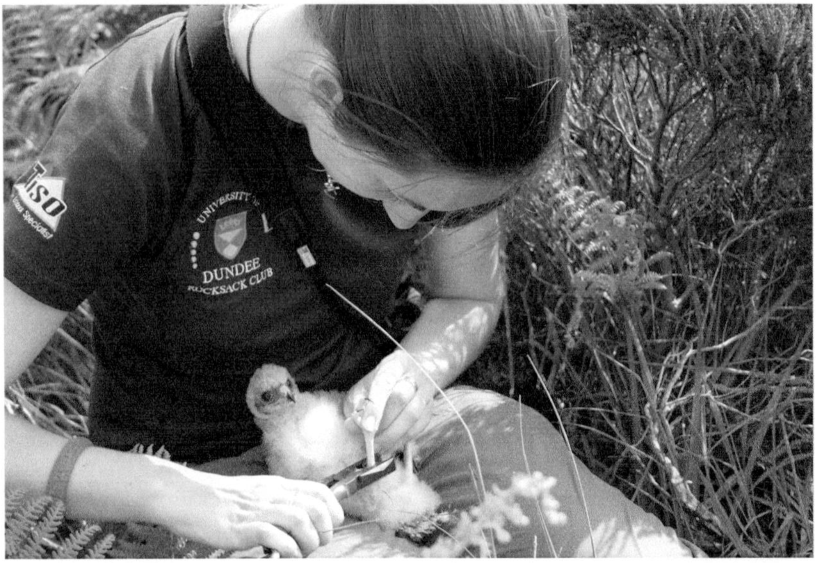

Figure 1.6 Eilidh McNab ringing a hen harrier chick on Colonsay, under the licensed supervision of David Jardine. Eilidh was a trainee with the Scottish Wildlife Trust, and is currently working for a PhD at Stirling University on the ecology of new native woods in the uplands (Des Thompson).

mounted at the back of the sphere. Burn marks on the card show when bright sunshine has occurred. The recorder is less accurate when the sun's rays are faint but, even so, many still remain in use. Today, solar radiometers connected to a data logger provide an accurate record of the timing and intensity of solar insolation and the record can be converted to 'hours of bright sunshine' to provide continuity of records. While most weather observations are taken at fixed locations, field research has included weather stations in the most remote and inhospitable locations, wherever weather observations are needed as part of the research programme.

Vasculum. Used by botanists to transport plants, cylindrical and providing cool and humid conditions, Linnaeus referred to it as a *vasculum dilletanum* – a small container (*vasculum*) named after his friend, the German botanist, J. J. D. Dillenius, active in the mid-1700s. Probably derived from tin candle-boxes used in the previous century, it was popularised by the polymath naturalist and scientist William Withering, famed for discovering digitalis, and who also made much use of the **plant press** for drying plants under pressure (the Italian physician and botanist

Luca Ghini is credited with devising the press, having established Europe's first herbarium in Pisa, in 1544).

Bird-ringing equipment. There are early records of Roman soldiers attaching threads to crows' legs to relay messages, and falconers in the Middle Ages had small plates depicting their owners' seals, attached to birds. The exceptional artist and ornithologist John James Audubon, and one of the founding pioneers of the Boy Scouts in America, Ernest Thompson Seton, lay claim to the modern method of marking birds, with Audubon tying silver threads to the legs of young eastern phoebes *(Sayornis phoebe)* in 1803, and Seton ink-marking snow buntings in Manitoba in 1882. The Danish teacher and ornithologist, Hans Christian Cornelius Mortensen, was the first to use bird-ringing for science, working first with starlings in 1899.

The formal Ringing Scheme in Britain, administered by the British Trust for Ornithology (BTO), began in 1909, with the editor of *British Birds*, H. F. Witherby (later co-author of the five-volume *The Handbook of British Birds*, 1938–41), instigating one scheme, and Sir Arthur Landsborough Thomson (editor of *A New Dictionary of Birds*, 1964) in Aberdeen beginning the other. They merged in the late 1930s, with the BTO taking responsibility in 1937. Curiosity about migration routes and destinations, and survival of species, drove much of this work. The scientific work based on bird-ringing is now some of the most important for population and conservation biology.

Field guides. Indispensable in the field, guides to rocks, soils, landforms, plants, fungi, lichens and animals abound internationally. Typically published as well-illustrated small books, 'local' guides on geology, birds and plants were published in the eighteenth century, and possibly even earlier.

The first formal natural history guide may be attributed to the American naturalist Frances T. S. D. Parsons (who published early books as Mrs William Starr Dana, whose husband, a US Commander in the US Navy, died in the flu epidemic of 1890, resulting in her spending years walking and observing nature). She published *How to Know the Wildflowers* (1893), which sold out in five days, and was warmly endorsed by family friend President Theodore Roosevelt (under whom her second husband served as Consul General in Mexico City) and Rudyard Kipling's *Plants and Their Children* (1896) and *How to Know the Ferns* (1899) followed later. Another early field guide was *Birds Through an Opera-Glass* (1890) by American ornithologist Florence A. Merriam Bailey, who went on to make substantial contributions to *Handbook of Birds of the Western United*

States (1902) and *The Birds of New Mexico* (1928). She also published *A-birding on a Bronco* (1896) and *Birds of a Village and Field: A Bird Book for Beginners* (1898). Yet another early field guide is *Birdcraft: A Field Book of Two Hundred Song, Game and Water Birds* (1895) by Mabel O. Wright, with colour plates by Louis Agassiz Fuertes. Away from birds, *American Concology: Descriptions of the Shells of North America* (1830) by Thomas Say and T. A. Conrad is an early specialist field guide. Anne Pratt's *The British Grasses and Sedges* (1858) and Mordecai Cooke's and Olive Tonge's *British Edible Fungi: How to Distinguish and How to Cook Them* (1891) are distinctively early.

Nest boxes for birds and bats. The nest box was invented by the British naturalist and explorer, Charles Waterton, in the early nineteenth century, to encourage more birdlife and wildfowl on the nature reserve he set up on his estate, Walton Hall. Some classic studies of bird population ecology have relied heavily on the deployment of nest boxes and intensive monitoring, notably in woodland and urban landscapes. Bat boxes differ from bird nest boxes, with openings on the underside of the box; Stebbings and Walsh (1997) provide an insightful guide to the history of their development and use.

Sound recorders for a wide range of animals. The Wildlife Sound Recording Society, formed in 1968, is the world's oldest and largest body dedicated to this field. The society has produced an excellent introduction to the huge range of equipment used (see www.wildlife-sound.org). Early wildlife recordings date from the early 1930s. One of the earliest pioneers was Ludwig Koch, who began recording sounds in the 1880s as a boy. In 1936, he fled Nazi Germany, and became a popular nature broadcaster; his recordings were later acquired by the BBC, establishing the BBC's library of natural history sounds.

Quadrats and point frames. These variably sized squares used for standardised sampling are first described by Raunkier (1909, 1912), and feature in classic papers by Clapham (1932) and Tinney *et al.* (1937). Tansley (1923) provides an early overview, covering transects, early field apparatus and recording-species occurrence. Stewart and Hutchings (1936) provided a key description of the point-observation-plot in the USA. The earliest systematic and critical review is provided in Greig-Smith's *Quantitative Plant Ecology* (1957). Surprisingly, we cannot find early references to their use.

Corers. The best known is the Russian 'chamber' corer or 'peat borer', but others include Glew corers (for lake sediment surfaces), brown corers, liquid nitrogen-filled 'frozen finger' tubes and Livingstone piston corers for lake sediments. Grab samplers (actually named after Van Veen Grab!) are typically clamshell buckets made of steel for sampling river and lake upper sediments. The 'D-section' Russian corer is used for sampling peat to great depth and is described by Belokopytov and Beresnevich (1955); it is also called the 'Belarussian', 'Byelorussian' or 'Jowsey' sampler. Shallower cores are typically taken with the stainless steel Wardenaar peat corer (Wardenaar, 1987). Soil augers, used for boring holes to collect samples near the surface, have a varied history, steeped in early agriculture, carpentry, ice fishing, maple syrup extraction and the creation of deep foundation piles – on land and into the seabed.

Much of the coring is to inform the field of palaeoecology, which relies heavily on continuous records of pollen in peat and lake sediments giving a continuous record of vegetation change. Sweden's Lennart von Post, in 1916, presented the technique of pollen analysis at the 16th Scandinavian meeting of natural scientists in Kristiana (now Oslo). Birks and Birks (1980) and West (2014) give overviews. Pine stumps were the first Quaternary plant fossils ever to be studied in about 1780.

V-notch weir. In the 1960s, field research in fluvial geomorphology became keenly concerned with process measurement, with a focus on small, headwater river basins. A whole range of field equipment was used, some borrowed (e.g. soil-erosion plots) and some new (e.g. automatic tensiometer systems to measure soil moisture). This was still the era of clockwork chart recorders, with unreliable clockwork mechanisms and poor ink pens to record on the paper charts; early electric chart recorders, run from car batteries, appeared in the 1970s. Gregory and Walling (1973) provide a wide-ranging summary of the types of equipment being used at that time. Today, data loggers have replaced the chart recorders and electronic probes have (to a large extent) negated the need for physical collection of water samples using pump water samplers. The V-notch weir is emblematic of those small-catchment field experiments, measuring water-level height (stage) behind the weir as a surrogate for stream discharge.

Longworth mammal trap. The most popular small-mammal trap, designed to minimise discomfort, it consists of a tunnel, with a door-tripping mechanism, and a nest box attached to the back of the tunnel. Invented in 1949 by two Oxford University zoologists, Dennis Chitty and Robert Kempson (1949), the trap was manufactured by the Longworth Scientific Instrument Company (founded by another group of Oxford academics). Several studies have contrasted their use with earlier pitfall and other traps (e.g. Williams and Braun 1983; Anthony *et al.*, 2005). Chitty is famous for devising the 'Chitty hypothesis of population

Figure 1.7 V-notch weirs have been used in all sorts of places to gauge streamflow. This one may have the hottest water of any of them, measuring outflow from hot springs at Rotorua, North Island, New Zealand. Wherever they are, the principle remains the same: a direct relation between the height of water in the stilling pond and the discharge coming over the weir. A water-level recorder is housed in the large grey box to the left (Tim Burt).

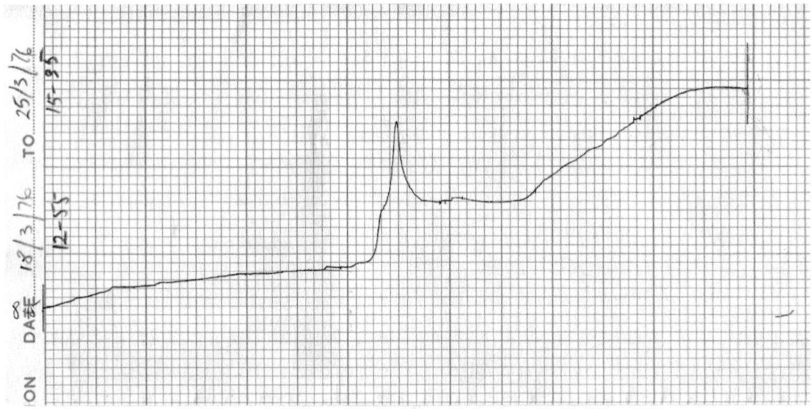

Figure 1.8 Stage recorder chart showing water level (or 'stage') behind a V-notch weir: Bicknoller Combe, March 1976 (Tim Burt). Clockwork stage recorders were unreliable – the clock would stop or the mechanism moving the pen would jam, but at least they produced an instant image of recent events. These days, a Wi-Fi link from a data logger and an app can provide the same immediacy, out in the field or back in the office!

regulation', stating that animal population density is limited by spacing behaviour, with changes in aggressive behaviour and physiology preventing unlimited population growth.

Moth trap and malaise trap. Used by entomologists to capture moths, moth traps use a light source to attract moths. These take various names – Robinson, Skinner and Heath traps. Paul Waring and Martin Townsend (2003) give a lot of detail on moths of Britain and Ireland. The tent-like Malaise trap catches flying insects, especially hymenopterans and dipterans. It was invented by the Swedish entomologist and explorer René Malaise in 1934, who in Burmese Yangon collected some 100,000 insects, many of which were unknown to science.

Hypodermic needle. Used for collecting blood samples (notably for DNA fingerprinting, but also health screening, and used under stringent licensing), this has an intriguing history. Early 'needles' devised by the Greeks and Romans were used for administering poisons and opiates. Sir Christopher Wren performed early experiments, injecting dogs with 'medicines' (in 1656, using animal bladders and goose quills), but such were the problems with injecting humans and other animals, it was only much later that needles came to be used. Arguably first to do so was the Irish physician Dr Francis Rynd, who made a successful injection on 3 June 1844, reporting his work the following year in *Dublin Medical Press*, and referring to his patient, 59-year-old Margaret Cox, as 'of spare habit' (he injected a portion of morphia into the supra-orbital nerve, evidently with a successful outcome for the lady's face pain). We have not found formal records of the earliest use of needles to collect blood from animals.

Genetic markers. These (e.g. eDNA) are now being used to identify species and individuals, and nanotechnological advances mean that minute devices can be used to track the tiniest of organisms. A remarkable paper by Peñalver *et al.* (2017) reports evidence in 99-million-year-old Cretaceous amber showing that hard ticks, and ticks of the extinct new family *Deinocrotonidae*, fed on blood from feathered dinosaurs. See Heidi Hauffe's essay in Part II.

GPS technology and telemetry. Global positioning system (GPS) is a space-based radio navigation system, owned by the US government, and it originally used 24 satellites, becoming operational in 1995. It has now got huge capabilities for tracking movements of wildlife and pinpointing recording locations with extreme accuracy. In 1956, acoustic telemetry was developed to study fish in the US Bureau of Commercial Fisheries and the Minneapolis–Honeywell Regulator Corporation. Now it is widely used in freshwater and marine environments. In the 1960s, very

Figure 1.9 Dr Virginia Morandini and Professor Miguel Ferrer radio tracking a Spanish imperial eagle in southern Spain. Virginia is a postdoctoral researcher at Oregon State University, and studied for her PhD at the Biological Station of Doñana. Miguel is a member of the Biodiversity and Global Change Council of Europe's experts, Research Professor in the Spanish National Research Council, and the world expert on these eagles (Des Thompson).

high-frequency (VHF) telemetry collars were attached to mammals to track movements, with transmitted signals triangulated to determine locations.

With rapid technological advances, passive integrated transponder (PIT) tags (integrated circuit chip, capacitor and antenna coil, encased in glass) were used from the mid-1980s, first to study fish movements, but were soon deployed to study even insects, with tagged animals simply passing an automated 'reading system', providing data on movements. First described in 1992, tiny geolocators periodically record ambient light, and are invaluable for bird migration studies. Most recently, the Global System for Mobile Communications (GSM) using mobile 'phone technology, first described in 1998, provides exceptionally detailed data on animal movements. Pop-up satellite archival tags (PSAT), surgically attached to marine mammals, are providing important data on spatial and vertical movements of large sea fish and mammals.

Data loggers. Older readers will remember clockwork chart recorders, used for a variety of purposes, from recording water level, to barometric pressure and rainfall intensity. These days, a data logger provides high-frequency data storage, limited only by the capacity of the solid-state memory. In all but the remotest locations, a Wi-Fi link can provide immediate data receipt back in the office.

Photography. Cameras get a mixed press from fieldworkers, to judge by some authors in Canfield (2011a). For good reason, many still favour field sketching, not for its artistic quality per se but because it forces the fieldworker to *look*, to identify the key elements, to ignore what is irrelevant to the question in hand, to focus. Of course, these days almost everyone has a digital camera and can happily snap wherever they go. In many ways this is good because we can record our fieldwork in great detail at effectively zero cost and, given the wonders of the electronic age, the photographs may well make it home before we do! We would make two simple points here, since good photos are there for the taking. First, it is worth taking time to choose the best vantage point, to include the most important features of what is being studied. Then think about lighting and do not shoot into the sun if you can help it. If the sun is behind a cloud, be patient! Depending on the subject, include people in the photo or not, but be aware of cultural sensitivities and, if in doubt, ask before taking any photos. The field sketcher always thinks carefully about what is significant and what is not; do the same with photographs. 'Why am I taking this photo?' should always be the question. Thus, it is important to remain curious with camera in hand. What ideas come to mind? What are the important features to include? The fieldworker is not a snap-happy tourist and should think carefully about what they *see*. That said, the camera can be an invaluable aid to fieldwork and we are rarely to be found without ours in hand!

Field sketching. In parallel with the use of field notebooks and as an antidote to photography, field sketching is still advocated by some fieldworkers as the best means to *look* at the landscape and thereby to identify the important points of interest. Keller (2011) provides a detailed assessment of the technique. 'The creation of an accurate drawing requires a systematic approach, patient observation, an openness to unforeseen possibilities, an ability to regard a topic from a variety of perspectives, a willingness to pay attention to both the exciting and the mundane, and the deliberate setting aside of preconceived ideas' (Keller, 2011, pp. 184–185). In particular, the notion of 'patient observation' deserves special emphasis, a remedy to the unthinking use of the camera.

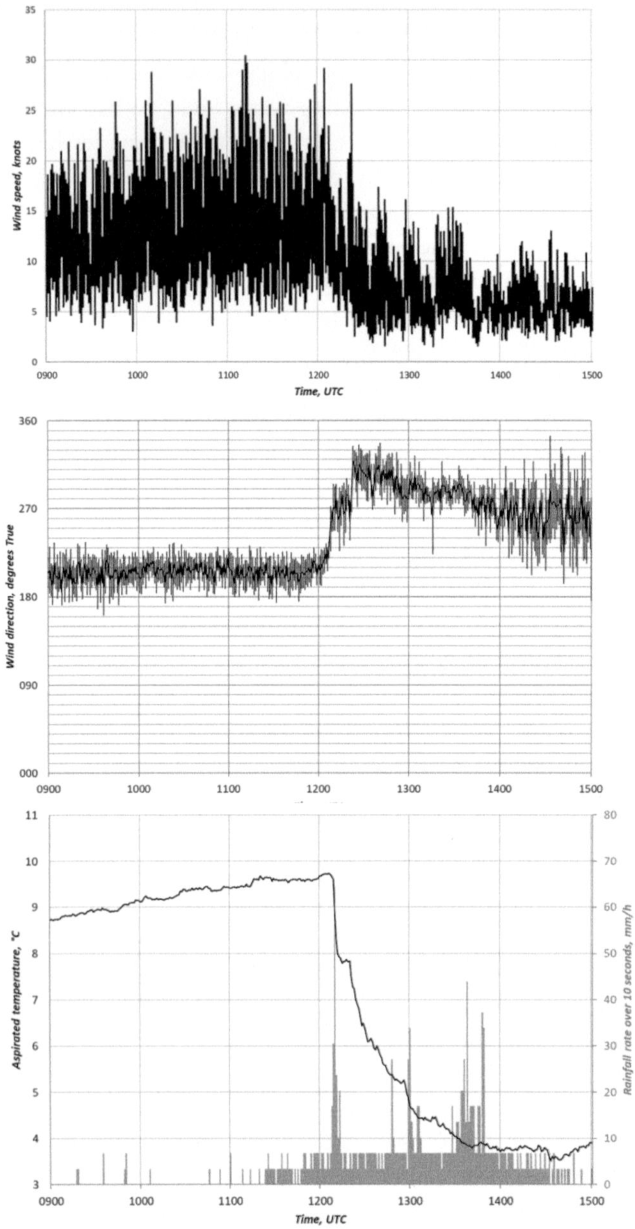

Figure 1.10 Hurricane Gareth cold front moving through Mortimer, 12 March 2019. Tim Burt and Stephen Burt (see Chapter 12) sat in Devon and watched this classic cold front move east over the Reading area, some 100 miles to the east, courtesy of a data logger, Wi-Fi and the internet. Such data do not diminish the need for fieldwork, quite the opposite: they encourage us outdoors to observe. Note the sharp decrease in wind speed and temperature and the shift in wind direction as the front passes (Stephen Burt).

Figure 1.11 Worth a second look. Repeat photographs can raise interesting questions about landscape change. Shiny Brook on the Southern Pennine peat moorlands near Huddersfield, 1981 and 2002 (Tim Burt).

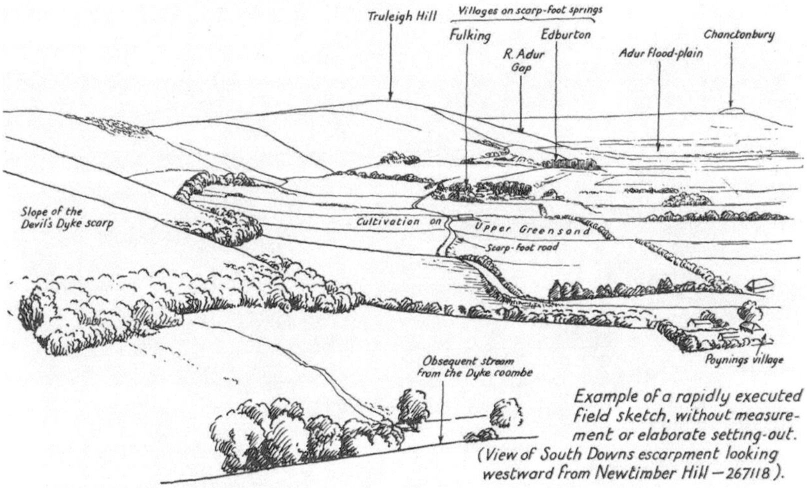

Figure 1.12 A classic field sketch by Geoffrey Hutchings, the first warden of Juniper Hall Field Centre. Courtesy of the Field Studies Council – see Hutchings (1955).

Sketching certainly does not require artistic skill, just a willingness to look and identify the salient features in front of us. See Stephen Mott's chapter in Part II for contemporary examples.

1.4 Some Classic Field Studies

A remarkable assortment of people have contributed to the advancement of field studies. Box 1.1 lists some of the pioneers, and distinguishes between the historical figures and those with us today. It is an eclectic selection but gives an impression of the wealth of talented people working in the field.

We have highlighted a number of field studies, in different parts of the world, which have inspired us. Each reader would no doubt pick a different batch of studies, but these are some of our favourites.

Red deer on Rum. The world's longest-running study of deer, begun in 1972, and now led jointly by Cambridge and Edinburgh universities. In the North Block of the Scottish Natural Heritage (SNH)-owned National Nature Reserve (NNR), all animals are individually known and followed throughout their lives. Since 1972, when culling ceased, studies have concentrated on a wide range of natural factors that are

Box 1.1 *Some of the heroes of environmental field studies, listed alphabetically.*

Early field naturalists

Aristotle – Greek philosopher and scientist, and the most famous of Plato's students, identified 500 species of mammals, birds and fish. Arguably, the founder of modern science.

John James Audubon – American ornithologist, naturalist and painter.

Comte de Buffon - French naturalist, mathematician, cosmologist and prolific author.

Rachel Carson – American marine biologist and conservationist, wrote *Silent Spring*, and pioneered government action on synthetic pesticides, banning DDT, and giving rise to the formation of the US Environmental Protection Agency.

Frederic Clements – American plant ecologist and pioneer in the study of vegetation succession.

Georges Cuvier – French naturalist and zoologist, established palaeontology.

Charles Darwin and Alfred Russel Wallace – unravelling evolution through natural selection.

Ernst Haeckel – German biologist, naturalist, philosopher, physician and artist, who discovered, described and named thousands of new species, mapped a genealogical tree, relating all life forms, and coined many terms in biology, including 'ecology'.

Victor Hensen – German zoologist, who coined the term 'plankton' and laid the foundation for biological oceanography.

Heroditus – Greek, and often called 'The Father of History', was an important early writer on environmental matters.

G. Evelyn Hutchinson – American ecologist, sometimes described as the 'Father of Modern Ecology', and expert in limnology, systems ecology, radiation ecology, entomology, genetics and biogeochemistry.

James Hutton – Scottish geologist, physician, chemical manufacturer and naturalist, who devised the theory of 'uniformitarianism', the fundamental principle of geology. Known as the 'Father of Modern Geology'. The world-respected James Hutton Institute is named after him.

T. H. Huxley – English biologist, known as 'Darwin's bulldog', largely self-taught zoologist, and brilliant educator of the public on evolution and biology.

Lamarck – French naturalist, pioneer of invertebrate classification, and widely remembered for his theory on 'the inheritance of acquired characteristics'.

Carl Linnaeus – Swedish botanist, physician and zoologist, who laid the foundations for the modern biological naming scheme of binomial nomenclature.

Robert MacArthur – Canadian-born American ecologist, who pioneered our understanding of community and population ecology, island biogeography, and who devised the term 'niche partitioning'.

Ernst Walter Mayr – one of the twentieth century's leading evolutionary biologists, renowned taxonomist, tropical explorer, ornithologist and historian of science.

Nicholas Miklouho-Maclay – Russian explorer, ethnologist, anthropologist and biologist, who researched in the Middle East, Australia, New Guinea, Melanesia and Polynesia. Established first biological research station in southern hemisphere (Australia).

Max Nicholson – English pioneering environmentalist, ornithologist and internationalist. Key founder of the BTO, in 1932, and the World Wide Fund for Nature, in 1961.

Ptolemy of Alexandria – Greek pioneer who, in his eight books *Geography*, described the entire Earth as known in around AD 100.

Emil Racoviță – Romanian biologist, zoologist, speleologist, explorer of Antarctica and the first biologist to study the arctic life.

Derek Ratcliffe – botanist, ornithologist and nature conservationist, arguably the finest all-round naturalist since Darwin, he reported breakthrough science linking eggshell thinning in peregrines with DDT prevalence in the countryside.

Richard Southwood – English academic ecologist, who pioneered descriptions of ecological methods and headed major government commissions and Working Parties, drawing on his ecological expertise.

Strabo of Amasia – Greek founder of geography.

Arthur George Tansley – English botanist and a pioneer in the science of ecology. His *The British Islands and Their Vegetation* (1939) is a *tour de force*, he was the key post-war nature conservation philosopher in the UK.

Marie Tharp – American geologist and oceanic cartographer, who, with Bruce Heezen, first mapped the Atlantic Ocean floor.

Vladimir Vernadsky – Ukrainian, Russian and Soviet mineralogist and geochemist who is considered to be one of the founders of geochemistry and biogeochemistry. Main populariser of the term 'biosphere', and has a research station, library, mountain range, lunar crater and asteroid named after him.

Alexander von Humboldt – Prussian geographer, naturalist and explorer, laid the foundation for the field of biogeography.

Eugenius Warming – Danish botanist and a main founding figure of the scientific discipline of ecology, writing first textbook on plant ecology

	and teaching the first university course in ecology. Adam Watson – Scottish advocate of the conservation importance of the Cairngorms, and made long-term studies of mountain birds, mammals and snow-patches. Gilbert White – pioneering English naturalist and ornithologist, best known for the remarkable chronicle of wildlife, *The Natural History and Antiquities of Selborne* (1789). Robert Whittaker – American plant community ecologist, who was the first to propose the five kingdom taxonomic classification.
Current advocates	David Attenborough – globally revered and gifted communicator on nature, and advocate on perils of climate change. Steve Backshall – one of the most popular naturalists and presenters on TV, renowned for getting to grips with 'deadly' animals (very much in the mould of the late Steve Irwin). BBC Natural History Unit – producer of outstanding wildlife programmes, was the initiator of 'personal views' programmes, ultimately giving rise to celebrated individual-led radio and TV programmes. John and Hilary Birks – based in Bergen, Norway, Cambridge-educated botanical, palaeoecological and geoscientific experts with a considerable academic following. Paul Ehrlich – American biologist and educator, published major studies of evolution of plants and insects, but best known for warnings on human population growth. Bernd Heinrich – author of many books on natural history, specialising in insect physiology and behaviour, and bird behaviour. Was a world class athlete, and wrote *Why We Run. A Natural History*.

Daniel H. Janzen – evolutionary ecologist, biologist, conservationist. Has led internationally renowned work in Costa Rica on ecosystem restoration.

John Lawton – eminent British ecologist, with special expertise in advocating the importance of biodiversity, and the measures needed to make it more resilient.

James Lovelock – environmentalist, best known for proposing the Gaia hypothesis, which postulates that the biosphere is a self-regulating entity with the capacity to keep our planet healthy by controlling the interconnections of the chemical and physical environment.

National Geographic – a world leader in communicating geography, cartography, exploration and environmentalism.

Ian Newton – British ornithologist, with a world-class reputation for long-term studies of raptors, and wider excellence as a lecturer and writer on ecology, migration and landscapes.

Alan Robert Rabinowitz – American zoologist, conservationist, field biologist, film maker and the CEO of Panthera (devoted to protecting the world's 37 wild cat species).

John A. Wiens – leading American landscape ecologist, whose research has focused on birds and insects in semi-arid environments on several continents, emphasising community ecology and spatial relationships.

E. O. Wilson – world leading researcher on ants, and the 'Father of Biodiversity' and deviser of sociobiology.

driving population change, and individual and sex differences in survival, ranging behaviour and productivity. Around 150 scientific publications and three books have arisen from this study. Research leaders: Tim Clutton-Brock, Josephine Pemberton, Steve Albon, Loeske Kruuk, Mick Crawley and Fiona Guinness. *Reference*: Pemberton and Kruuk (2015).

Peregrines – pesticides breakthrough. Rachel Carson's book *Silent Spring* (1962) drew attention to the pernicious impacts of pesticides in the environment. In the UK, pigeon fanciers had complained to the government's Home Office that peregrines were ruining their sport, and wanted legal protection to remove them. A survey by the BTO was commissioned to assess the current status of peregrines. Led by Derek Ratcliffe, the survey revealed that the peregrine population was crashing. He noted that peregrines and other birds of prey often had broken eggs in their nests, and that clutches and broods were evidently smaller than previously recorded. Working to an idea attributed to Desmond Nethersole-Thompson and Joe Hickey, Ratcliffe examined eggshells, comparing current with previously collected ones, and discovered that those collected after 1946 (when DDT was introduced into widespread agricultural use) were significantly thinner. His paper describing shell thinning in the peregrine and other species was published in *Nature* (Ratcliffe, 1967), followed later by a more detailed assessment of a wider range of species, in the *Journal of Applied Ecology* (Ratcliffe, 1970). This work paved the way for systematic monitoring of raptors, and indeed other birds. David Peakall made a similar discovery in America for links between DDE and shell thinning in peregrines and California condors. Research leaders: Derek Ratcliffe, Joe Hickey (organiser of the classic 1965 Madison Peregrine Conference, giving rise to the formation of the Raptor Research Foundation, in 1966), Ian Prestt, David Peakall, Norman Moore and Ian Newton. *Reference*: Newton (2015).

Rothamsted Experimental Station. The Rothamsted Experimental Station (now *Rothamsted Research*) was founded in 1843 by John Bennet Lawes, an entrepreneur and scientist who had founded one of the first artificial-fertiliser manufacturing factories in 1842, on his sixteenth-century estate, Rothamsted Manor, to investigate the impact of inorganic and organic fertilisers on crop yield. Appointing a young chemist, Joseph Henry Gilbert, as his scientific collaborator, Lawes launched the first of a series of long-term field experiments, some of which still continue.

Over 57 years, Lawes and Gilbert established the foundations of modern scientific agriculture and the principles of crop nutrition. One of the station's best-known and longest-running experiments is the Park Grass experiment. Originally set up to test the effect of fertilisers and manures on hay yields, it is notable as one of the longest-running experiments of modern science, as it has been continually monitored since 1856. Another long-term site is the Broadbalk continuous winter wheat experiment, which has received the same fertiliser treatment every year since 1852. More recently, there has been a wide range of research in areas such as entomology, biogeochemistry, meteorology and environmental statistics. Research leaders include: J. B. Lawes, J. H. Gilbert, E. John Russell, Ronald Fisher, George Cooke, John Monteith, Howard Penman and Keith Goulding. *References*: Lawes *et al.* (1881); Johnston (1991).

Darwin's finches. Also known as the **Galápagos finches** (15 species). The term 'Darwin's finches' was first applied by Percy Lowe in 1936, and popularised by David Lack in *Darwin's Finches* (1947). A long-term study over 40 years, by the Princeton University researchers Peter and Rosemary Grant, has documented micro-evolutionary changes in beak size, affected by El Niño cycles in the Pacific. Since 1973, they have shown that natural selection can operate over a single lifetime, or even within a couple of years. Darwin originally thought that natural selection was a long, drawn-out process. The Grants, however, have shown that these changes in populations can happen very quickly. Research leaders: Charles Darwin, David Lack, Peter Grant and Rosemary Grant. *Reference*: Grant and Grant (2008).

The Sir Alister Hardy Foundation for Ocean Science (SAHFOS) continuous plankton recorder – long-term ecological monitoring. This is the longest and most spatially extensive planktonic ecosystem data set currently available for the Atlantic Ocean. Collecting data from the North Atlantic and the North Sea on biogeography and ecology of plankton since 1931, the Foundation has spawned international projects with a global reach. Our understanding of climate change, eutrophication, biodiversity and non-native invasive species has been hugely advanced by this work. Currently, 698 taxa are routinely analysed, sampled along more than 6.8 million nautical miles, with detailed analyses running to 86 years. Research leaders: Sir Alister Hardy, Cyril Lucas, Martin Edwards and Darren Stevens. *Reference*: Giron-Nava *et al.* (2017).

Dung flies and sexual selection. Taking his PhD at the University of Bristol, in 1969, Geoff Parker studied the reproductive behaviour and nature of sexual selection in cow-pat-dwelling *Scathophaga stercoraria*, the yellow dung fly. His pioneering work described how natural selection operates, with much of this based on gene- and individual-centred optimality models and Game Theory, as well as careful observations in the field. In 2008, Parker was awarded the Darwin Medal of the Royal Society for his lifetime contribution to the foundations and development of behavioural ecology, in particular for understanding evolutionary adaptations and their consequences for natural populations. Geoff Parker has worked with Dan Rubenstein, Leigh Simmons, Tim Birkhead, Tommaso Pizzari and John Maynard Smith. Reference: Parker (2014).

Wytham Woods. The University of Oxford's ecological laboratory, this ancient, semi-natural woodland to the west of Oxford is one of the most researched woods in the world. There are long-running data sets on weather (as part of the UK Environmental Change Network), birds, mammals, invertebrates and vegetation. It was designated a Site of Special Scientific Interest in 1950 for its floristic richness. David Lack began the long running study of great and blue tits there in the 1940s (he was appointed Director of the Edward Grey Institute in September 1945), with nest boxes monitored since 1947 (arguably the longest-running continuous bird study in the world, with more than 1,000 nest boxes monitored in some years). Lord Krebs, who began his research career in 1966 there, commented once on the BBC that 'If there were a Nobel Prize for Ecology, and if you could award it to a place rather than a person, Wytham Woods would surely be a prime candidate'. Research leaders: Charles Elton, David Lack, Sir Richard Southwood, George Varley, G.R. Gradwell, Lord Krebs, Christopher Perrins, Mike Hassell, Andy Gosler, Peter Savill, Keith Kirby and Mike Morecroft. *Reference*: Savill *et. al.* (2011).

Radcliffe Observatory. Just a few kilometres south-east of Wytham Woods, Radcliffe Observatory, founded in 1772, provides a long-term meteorological context for the ecological research at Wytham Woods. Following a period when there were numerous if irregular observations of rainfall, temperature, winds and cloud cover, a daily record of air temperature and precipitation, which commenced in November 1813, continues to this day – the longest unbroken daily weather record at the same site in the British Isles, and one of the most enduring in the world. Regular meteorological observations were started by astronomers, who needed to correct their astronomical observations for refraction. If the first measurements were made indoors, most equipment was soon moved outside, and meteorological observation began to develop a life of its own, as the need for accurate weather forecasting was gradually acknowledged (Wallace, 1997, 2005; Burt and Burt, 2019). The Radcliffe Observatory was at the forefront of developments as meteorological instruments and their precise method of use rapidly evolved, working in close cooperation with organisations such as the Meteorological Office and the Royal Society. The Observatory building and grounds were transferred to Oxford University in 1935 when the astronomical observations were moved to the clearer skies of South Africa, since when the Radcliffe Meteorological Station, as it was renamed, has been run by the University's School of Geography and the Environment. Research leaders: Thomas Hornsby, Harold Knox-Shaw, W. G. Kendrew, Gordon Smith, Stephen Burt, Tim Burt and Richard Washington. *Reference*: Burt and Burt (2019).

Long-Term Ecological Research (LTER) Network. The US National Science Foundation-funded LTER Network supports around 240 long-term ecological experiments. These have provided unique insights into ecological patterns and processes, with many of these only evident after many years of study. The research has influenced management and policy decisions, and it has spawned research platforms supporting studies and involving investigators who were not part of the original design. *Reference*: Knapp *et al.* (2012). Two examples of the LTER follow:

Hubbard Brook Ecosystem Study. First established as an 'experimental forest' by the US Department of Agriculture (USDA) Forest Service in 1955, this has become a major centre for hydrological, earth sciences and ecological research in New England. Covered mainly by unbroken forest of northern hardwoods with spruce and fir at higher elevations, the bowl-shaped valley was set aside in 1955 for research purposes. Early pioneering research developed our understanding of the science of ecosystem processes, essentially water and chemical element cycles, soil microbial activity, soil chemical reactions and the effects of deforestation and land management practices. Since 1963, more than 1750 publications have arisen from work there. Leading researchers: Robert Pierce, Herbert Bormann, Noye Johnson, Chris Eagar, Lindsay Rustad, Richard Holmes and Gene Likens. *Reference*: Holmes and Likens (2016).

Coweeta Hydrologic Laboratory. The USDA Forest Service's Coweeta Hydrologic Laboratory (2,185 ha: originally known as the 'Coweeta

Figure 1.13 An unusual type of sharp-crested weir – a Cipoletti – is used at the main outflow to the Coweeta Hydrologic Laboratory drainage basin. Standing by the weir is Wayne T. Swank, one of the most important scientists in the history of Coweeta (Tim Burt).

Experimental Forest') lies in the Nantahala Mountains of western North Carolina, USA. Established in 1934, Coweeta is among the oldest, continuously operating environmental study sites in North America; it has been a testing ground for theoretical and applied forest hydrological research (Swank and Crossley, 1988). This is the place where hydrologist John Hewlett developed his *variable source area* model of storm runoff generation, in the early 1960s. In the 1970s, an important development was research on nutrient cycling. One outcome was the concept of nutrient spiralling, which was based on research carried out at Coweeta by Jack Webster for his PhD (Webster and Patten, 1979), complementing the river continuum concept (Vannote *et al.*, 1980). Like Hubbard Brook, Coweeta forms part of the US National Science Foundation-funded LTER Network. Leading researchers: Charles Hursh, John Hewlett and Wayne Swank. *Reference*: Swank and Crossley (1988).

The Cornell Laboratory of Ornithology. Founded in 1915, the Cornell Lab is arguably the world leader in the understanding and conservation of birds. With a community of 300 staff, 400,000 citizen science contributors and 14 million online enthusiasts following its research throughout the world, it has continuously led the field. It has a close working relationship with the Hubbard Brook centre. Leading researchers: Walter Koenig, Andre Dhondt, Irby Lovettte, Aaron Rice, Amanda Rodewald and John Fitzpatrick. *Reference*: Annual reports and an extensive website give details of wide range of research.

The Plynlimon catchments. The Plynlimon catchments in central Wales are among the most important intensively studied long-term research basins in the UK. Managed by the Centre for Ecology and Hydrology (CEH), Plynlimon is a key site in terms of catchment science – not just for the length of the records gathered (more than 40 years) but because of the breadth of measurements, spanning hydrology and including water chemistry, groundwater studies, geomorphology, ecology and atmospheric science. During the 1960s, the former Institute of Hydrology, now CEH, instigated a major hydrological study to investigate the impacts of conifer plantations on the hydrological cycle. The Plynlimon catchments were selected because they provided two, almost identical, adjacent catchments, with the exception that one was used for grazing sheep while the other was mostly under coniferous forestry. The findings are applicable to many upland areas in Britain where the land is used for water, wood and wool production. The Plynlimon research confirmed that forests 'use' more water than short vegetation, mainly because there is more evaporation of water from the tree canopy, an effect first noted by Frank Law in the Pennine hills. As a result, there is reduced total runoff (although there is little effect on flood flows). Leading researchers: Malcolm Newson, Ian Calder and Colin Neal. *Reference*: Robinson *et al.* (2012).

Great Barrier Reef Marine Park Authority. The Great Barrier Reef Marine Park, a World Heritage Site, is one of the world's greatest natural treasures. Guided by a range of plans, policies, regulations and legislation, the 'Great Barrier Reef Marine Park Act 1975' sets out the functions and responsibilities of the Park Authority. A 25-year management plan sets out work in hand. As the largest living entity on the planet, it stretches for 2,300 kilometres, has 600 types of soft and hard corals, more than 100 species of jellyfish, 3,000 varieties of molluscs, 500 species of worms, 1,625 species of fish, 133 varieties of sharks and rays and more than 30 species of whales and dolphins. *Reference*: An excellent website gives details on work undertaken, see www.gbrmpa.gov.au/. See Emi Husband's essay in Part II.

Siccar Point. Finally, we add one place to our list of classic field studies, one of our favourites, Siccar Point, on the Berwickshire coast in eastern Scotland. This is where James Hutton (1726–1797) confirmed his notion

Figure 1.14 Edinburgh students visit Siccar Point, James Hutton's famous unconformity, where a gap of some 60 million years separates the two rock formations. Unconformities are discussed in chapters 13 (Tim Burt) and 25 (Peter Higgins). A black and white version of this figure will appear in some formats. For the colour version, please refer to the plate section.

of an *unconformity*: a buried surface separating two rock strata of different ages, indicating that sediment deposition was not continuous. Hutton was a Scottish geologist, naturalist, physician, chemical engineer and experimental agriculturalist, who devised the theory of 'uniformitarianism', the fundamental principle of geology. Hutton believed that geological processes had carried on operating continuously over very long periods of time; unconformities provided clear evidence of cycles of uplift, erosion and sedimentation. Hutton reasoned that there must have been innumerable cycles, each involving deposition on the seabed, uplift, folding and faulting, erosion, then further deposition. On the belief that this was due to the same geological forces operating in the past as the very slow geological forces seen operating at the present day, the thicknesses of exposed rock layers implied to him enormous stretches of time. Hutton concluded there was 'no vestige of a beginning, no prospect of an end'. His inspiration came in part from farms he inherited from his father in Berwickshire, east of Edinburgh, where he observed soil erosion in his fields. Hutton indulged his passion for geology and he had already seen examples of unconformities when, in 1788, he famously sailed along the Berwickshire coast: having theorised that there should be an unconformity in the locality, he found this at Siccar Point. Indeed, Hutton regarded Siccar Point as conclusive evidence of his uniformitarian theory of geology. We now know that the almost vertical beds of marine Silurian sandstones (known as greywackes) underlie gently dipping rocks of Upper Devonian age (a mixture of fluvial and wind-blown sands and silts). The Silurian beds are approximately 435 million years old, while the overlying Devonian rocks are approximately 375 million years old. The unconformity therefore represents a time gap of around 60 million years.

1.5 Perspectives

What is not generally appreciated is that serious scientists experience a sense of awe. They are usually drawn to ask questions about a particular thing in the natural world. It may be flowers or stars, or it may be something that other people do not care for at all – toads, beetles, tapeworms. Whatever it might be, the study of this thing moves them to reverence. And make no mistake, this sense of reverence is real ... To the real scientist, a question that has been answered becomes not less wonderful, but more so. Increased understanding increases scientific awe. And most great scientists have named awe of this kind as their deepest reason for pursuing science at all. [Mary Midgley, 2000]

Whether we study in the humanities or the sciences, we are all curious about the world around us. Whatever drives us to be curious about one thing rather than another, inevitably there is a sense of awe: a mixed emotion of reverence, respect and wonder. The word 'wonder' itself encapsulates, in its various meanings, both the excitement and the effort of academic inquiry: to marvel, to be amazed at; to enquire, to have a wish to know something; to question, to express doubt. Whatever the virtues of laboratory or archival-based research, our purpose in this book is to acknowledge the role of out-of-classroom field study: being excited by the outdoor world, asking questions about it, doubting the views of others and ourselves. Indeed, for the core subjects covered by this book, the geophysical and biological sciences in their broadest sense, we contend that fieldwork is an essential component of the academic process, whether inductive or deductive, humanities or science.

Figure 1.15 Curiosity about the world around us leads inevitably to patient observation of whatever it is we choose to investigate (Des Thompson). A black and white version of this figure will appear in some formats. For the colour version, please refer to the plate section.

At the beginning of any inquiry, there is always a choice to be made, the choice of what shall be the things and relations forming the subject of the inquiry itself ... Whatever the focus of inquiry, choices have to be made that some things are relevant, while others are not. [Peter Gould, 1985, p. 324]

As a 2017 advertisement from the railway company Eurostar said: 'A change of scene can shift your whole perspective'. While Eurostar's focus is travel, so too in a sense is ours. To summarise Steve Trudgill (2003), fieldwork allows a spectrum of experience, from the initial sensing of the place to focusing on those aspects of the landscape that seem interesting and relevant. This leads to questions followed by focused observation, initially descriptive but inevitably quantitative. Subsequent analyses bring us back to the initial questions, helping us to make better sense of the world and to see that world as orderly rather than chaotic. As Trudgill (2003) concludes, the excitement comes not only from 'making sense of the world', but from 'being there' – the excitement of science and the excitement of experience.

Our survey of the development of fieldwork shows how subjects have evolved from purely observational, albeit with notebook in hand, through to the ultra-modern data-gathering of the electronic world, not just computers and data-loggers but drones and cameras, and satellites and their variety of sensors. In between, has come a rich flowering of instrumentation, from the first thermometers to electrical pumps and on to the world of the microchip. Measurement is at the heart of what we do, meaning that questions of experimental design, technique, validity and repeatability become central. We must choose what we measure – we cannot measure everything (Anderson and Burt, 1990). Even those who eschew the quantitative approach recognise that field investigation often requires data-gathering through case studies and site-specific examinations; they acknowledge the need for a rigorous approach so that data that have been gathered in the field are properly organised to allow for subsequent analysis in an orderly way (Caretta and Cheptum, 2017). In all quantitative fieldwork, the ability to measure successfully is central. The role of a *technique* (i.e. a measurement procedure) is to assure data reliability through the introduction of fixed rule-based procedures for obtaining information (Rhoads and Thorn, 1996). The ability to measure is essential, helping to fix one's general conception of a problem. In this sense, quantity is not the opposite of quality but rather an improvement on it (Anderson and Burt, 1990). In many field sciences, there has therefore been a steady interest in measurement technique per se underpinning the investigations being conducted. Indeed, in some areas, e.g. isotopic dating of sedimentary material, progress has sometimes been halted until a reliable technique emerges to allow headway to be made.

Given the crucial importance of fieldwork, it seems vital to us that fieldwork is supported at every stage of education, from initial out-of-classroom experiences for the youngest pupils through to field visits as part of the curriculum for high-school and university students. Too often, administrators are tempted to cut fieldwork, arguing that costs outweigh benefits. We dismiss that argument without hesitation! As we have seen, for many subjects, curiosity must involve being out there, not sitting in the classroom or library. In that sense, the money must be found, although of course it is essential that best use is then made of the investment. Fieldwork can inspire the next generation of scientists, and even if someone does not go on to study a field-based subject or to do research, field-based learning can only inspire people to become better citizens, committed to the world in which they live.

1.6 What Follows

The rest of this book falls into two halves: in the next four chapters of Part I we delve into the history and importance of fieldwork. Between them, our authors trace the development of fieldwork to the range of modern techniques available to millions of us, consider the role of different approaches and techniques adopted in the field, look at the transformative developments of taking students out of the classroom and provide biographies of some of the pioneering heroes of fieldwork in the geophysical and biological sciences. The essays in Part II provide a much more intimate and personal approach: these short essays, 'cameos' as we have taken to calling them, can be on an influential individual, a defining moment, a publication or project, a breakthrough or some other aspect of the author's field experience. Together, our essays cover the full range of emotions, from the joy of being there to the grind of field research. We round off the book with a brief epilogue, reviewing what is contained in the book and looking ahead, confident that fieldwork will remain essential notwithstanding all the other developments, which are inevitable in an increasingly electronic age. Indeed, we would expect fieldwork to take full advantage of modern technology, but the latter can never replace the former as far as we are concerned!

Acknowledgements

We are grateful to the following for providing us with initial ideas and background information: the late Sam Berry, John Birks, Andrew Goudie and Steve Trudgill

References

Akerman, K. and Willing, T. (2009). An ancient rock painting of a marsupial lion, *Thylacoleo carnifex*, from the Kimberley, Western Australia. *Antiquity* 83. See www.antiquity.ac.uk/projgall/akerman319/ (accessed 17 July 2019).

Anderson, M. G. and Burt, T. P. (1990). Geomorphological techniques – part one: introduction. In *Geomorphological Techniques*, 2nd edn, ed. A. S. Goudie. Unwin Hyman, London, pp. 1–29.

Anthony, N. M., Ribic, C. A., Bautz, R. and Garland, T., Jr (2005). Comparative effectiveness of Longworth and Sherman live traps. *Wildlife Society Bulletin* 33, 1018–1026.

Archer-Thomson, J. (1999). The *Sea Empress* incident and the limpets of Frenchman's steps. *Field Studies* 9, 531–546. See http://fsj.field-studies-council.org/media/345059/vol9.3_253.pdf (accessed 17 July 2019).

Aubert, M., Brumm, A., Ramli, M., *et al.* (2014). Late Pleistocene cave paintings from the Maros karsts of Sulawesi, Indonesia. *Nature* 514, 223–227.

Ayala, F. J. (2009). Darwin and the scientific method. *Proceedings of the National Academy of Sciences* 106 (supplement 1), 10033–10039; DOI: 10.1073/pnas.0901404106.

Belokopytov, I. E. and Beresnevich, V. V. (1955). Giktorf's peat borers. *Torfânaâ promyslennost* 8, 9–10.

Birks, H. J. B. and Birks, H. H. (1980). *Quaternary Palaeoecology*. Edward Arnold, London.

Burt, S. D. and Burt, T. P. (2019). *Oxford Weather and Climate Since 1767*. Oxford University Press, Oxford.

Callaway, E. (2012). Spain claims top spot for world's oldest cave art. *Nature, News*, 14 June.

Canfield, M. R., ed. (2011a). *Field Notes on Science and Nature*. Harvard University Press, Cambridge, MA.

Canfield, M. R. (2011b). Introduction. In *Field Notes on Science and Nature*, ed. M. R. Canfield. Harvard University Press, Cambridge, MA, pp. 1–18.

Caretta, M. A. and Cheptum, F. J. (2017). Leaving the field: (de-)linked lives of the researcher and research assistant. *Area* 49, 415–420; DOI: 10.1111/area.12342.

Chitty, D. and Kempson, D. A. (1949). Prebaiting small mammals and a new design of live trap. *Ecology* 30, 536–542.

Clapham, A. R. (1932). The form of the observational unit in quantitative ecology. *Journal of Ecology* 20, 192–197.

Darwin, C. (1839). Journal of Researches into the Natural History and Geology of the Countries Visited During the Voyage of H.M.S. Beagle Round the World: under the command of Capt. Fitz Roy, R. N. (This final volume of Fitz Roy's three-volume official account of the voyage was re-published by Darwin as a revised second edition under the title *The Voyage of the Beagle*, in 1846.)

Day, F. (2018). Monitoring the invasive/non-native species of Jetty beach, FSC Dale Fort, Pembrokeshire, Wales. *Field Studies* online paper. See http://fsj.field-studies-council.org/media/5143752/fs2018_day.pdf.

Earthwatch (1990). *The Global Challenge for Field Science. Monitoring a Changing Planet*. Proceedings of a one-day seminar hosted by Earthwatch Europe at the Royal Society on 17 February 1989. Earthwatch, Boston, MA.

Ford, E. B. (1945). *Butterflies. The New Naturalist*. Collins, London.

Giron-Nava, A., James, C. C., Johnson, A. F., *et al.* (2017). Quantitative argument for long-term ecological monitoring. *Marine Ecology Progress Series* 572, 269–274. https://doi.org/10.3354/meps12149.

Gould, P. (1985). *The Geographer at Work*. Routledge & Kegan Paul, London.

Grant, B. R. and Grant, P. R. (2008). Fission and fusion of Darwin's finches populations. *Philosophical Transactions of the Royal Society B: Biological Sciences* 363, 2821–2829.

Gregory, K. J. and Walling, D. E. (1973). *Drainage Basin Form and Process*. Edward Arnold, London.

Greig-Smith, P. (1957). *Quantitative Plant Ecology*. Blackwell Scientific Publications, Oxford (3rd edn, 1983).

Haines-Young, R. H. and Petch, J. R. (1986). *Physical Geography: Its Nature and Methods*. Harper and Row, London.

Holmes, R. T. and Likens, G. E. (2016). *Hubbard Brook. The Story of a Forest Ecosystem*. Yale University Press, London.

Hutchings, G. (1955). *An Introduction to Geographical Landscape Drawing (Lowland Country)*. Field Studies Council, Telford.

Johnston, A. E. (1991). Benefits from long-term ecosystem research: some examples from Rothamsted. In *Long-Term Ecological Research: An International Perspective*, ed. P.G., Risser. Chichester: Wiley, pp. 89–114.

Keller, J. (2011). Why sketch? In *Field Notes on Science and Nature*, ed. M. R. Canfield. Harvard University Press, Cambridge, MA, pp. 161–185.

Keynes, R. D., ed. (2001). *Charles Darwin's Beagle Diary*. Cambridge University Press, paperback edition.

Knapp, A. K., Smith, M. D., Hobbie, S. E., *et al.* (2012). Past, present, and future roles of long-term experiments in the LTER Network. *BioScience* 62, 377–389.

Lawes, J. B., Gilbert, J. H and Warrington, R. A. (1881). On the amount and composition of rain and drainage water collected at Rothamsted. Parts I, II and III. *Journal of the Royal Agricultural Society of England* 17, 241–279; 17, 311–335; 18, 1–70.

Lødøen, T. and Mandt, G. (2010). *The Rock Art of Norway*. Oxbow Books, Oxford.

Manguel, A. (2015). *Curiosity*. Yale University Press, New Haven, CT.

Midgley, M. (2000). The need for wonder. In *God for the 21st Century,* ed. R. Stannard. Templeton Press, West Conshohocken, PA, pp. 186–188.

Moyes, J. and Nelson-Smith, A. (1963). Zonation of plants and animals on rocky shores around Dale Fort. *Field Studies* 2, 1–31. See http://fsj.field-studies-council.org/media/344333/vol1.5_28.pdf (accessed 17 July 2019).

Newton, I. (2015). Pesticides and birds of prey – the breakthrough. In *Nature's Conscience: The Life and Legacy of Derek Ratcliffe*, eds. D. B. A. Thompson, H. H. Birks and H. J. B. Birks. Langford, Peterborough, pp. 281–299.

Parker, G. A. (2014). The sexual cascade and the rise of pre-ejaculatory (Darwinian) sexual selection, sex roles, and sexual conflict. *Cold Spring Harbour Perspectives Biology* 6, a017509.

Pemberton, J. M. and Kruuk, L. E. B. (2015). *Red Deer Research on the Isle of Rum NNR: Management Implications*. SNH, Battleby.

Peñalver, E., Arillo, A., Delclòs, X., *et al.* (2017). Ticks parasitised feathered dinosaurs as revealed by Cretaceous amber assemblages. *Nature Communications* 8, a1924. See www.nature.com/articles/s41467-017-01550-z.

Ratcliffe, D. A. (1967). Decrease in eggshell weight in certain birds of prey. *Nature* 215, 208–210.

Ratcliffe, D. A. (1970). Changes attributable to pesticides in egg breakage frequency and eggshell thickness in some British birds. *Journal of Applied Ecology* 7, 67–115.

Raunkier, C. C. (1909). Formnationsundersogelse og' Formationsstatistik. *Botanisk tidsskrift* 30, 110 pp.

Raunkier, C. C. (1912). Measuring apparatus for statistical investigations of plant-formations. *Botanisk tidsskrift* 33, 45–48.

Rhoads, B. L. and Thorn, C. E. (1996). Observation in geomorphology. In *The Scientific Nature of Geomorphology*, eds. B.L. Rhoads and C.E. Thorn. Wiley, Chichester, pp. 21–56.

Robinson, M., Rodda, J. C. and Sutcliffe, J. V. (2012). Long-term environmental monitoring in the UK: origins and achievements of the Plynlimon catchment study. *Transactions of the Institute of British Geographers* 38, 451–463.

Savill, P., Perrins, C. M., Kirby, K. and Fisher, N. (2011). *Wytham Wood's Ecological Laboratory*. Oxford University Press, Oxford.

Stebbings, R. E. and Walsh, S. T. (1997). *Bat Boxes: A Guide to the History, Function, Construction and Use in the Conservation of Bats.* Bat Conservation Trust, London.

Stewart, G and Hutchings, S. S. (1936). The point-observation-plot (square-foot density) method of vegetation survey. *Journal of the American Society of Agronomy* 28, 714–722.

Swank, W. T. and Crossley, D. A. (1988). *Forest Hydrology and Ecology at Coweeta.* Springer, New York, NY.

Tansley, A. G. (1923). *Practical Plant Ecology*. Allen and Unwin, London (republished in 1946 as *Introduction to Plant Ecology*).

Tinney, F. W., Asmodt, O. S. and Ahlgren, H. L. (1937). Preliminary report of a study on methods in botanical analysis of pasture swards. *Journal of American Society of Agronomy* 29, 835–846.

Trudgill, S. T. (2003). Meaning, knowledge, constructs and fieldwork in physical geography. In *Contemporary Meanings in Physical Geography*, eds. S. T. Trudgill and A. Roy. Arnold, London, pp. 25–46.

Vannote, R. L., Minshall, G. W., Cummins, K. W., Sedell, J. R. and Cushing, C. E. (1980). The river continuum concept. *Canadian Journal of Fisheries and Aquatic Science* 37, 130–137.

Wallace, G. (2005). Meteorological observation at the Radcliffe Observatory. In A History of the Radcliffe Observatory Oxford: The Biography of a Building, eds. J. Burley and K. Plenderleith. Green College at the Radcliffe Observatory, Oxford, pp. 102–128.

Wallace, J. G. (1997). *Meteorological Observations at the Radcliffe Observatory, Oxford: 1815–1995.* School of Geography, University of Oxford, Oxford.

Wardenaar, E. P. C. (1987). A new hand tool for cutting peat profiles. *Canadian Journal of Botany* 65, 1772–1773.

Waring, P. and Townsend, M. (2003). *Field Guide to the Moths of Great Britain and Ireland.* Bloomsbury, London.

Webster, J. R. and Patten, B. C. (1979). Effects of watershed perturbation on stream potassium and calcium dynamics. *Ecological Monographs* 49, 51–72.

West, R. (2014). *Quaternary Research in Britain and Ireland.* Sidestone Press, Leiden.

White, Gilbert (1789). *The Natural History and Antiquities of Selborne*. Cassell & Company, London (at least 300 editions published).

Williams, D. F. and Braun, S. E. (1983). Comparison of pitfall and conventional traps for sampling small mammal populations. *Journal of Wildlife Management* 47, 841–845.

Wolodtschenko, A. and Forner, T. (2007). Prehistoric and Early historic maps in Europe: conception of Cd-Atlas. *e-Perimetron* 2, 114–116.

2 · *The Place of Field Studies in Environmental Science*

MICHAEL CHURCH

2.1 Beginnings

Modern environmental science has grown out of natural history. Natural history is as old as the interest of humans in the world around them – indeed, it must have been of overriding importance to the earliest humans, and for most since, whose very survival depended on a close and reasonably accurate knowledge of the resources and hazards of the natural environment around them. This is, of course, the case for any animal: that knowledge derives directly from what is seen and experienced in the field. Field observation has always been and remains the signal feature of natural history.

Natural history began to be recorded in the West by the Greeks and Romans (Figure 2.1). Later scholars have attributed to Aristotle the formal founding of 'natural history' as the disciplined description of the world, and of 'natural philosophy' as the logical consideration of antecedent causes of the present world. The great surviving documents that informed the medieval and Renaissance West are the *Historia naturalis* (written ca. AD 72–79) of Pliny the Elder, a Roman naval commander and naturalist – whose title names natural history – and *De Materia Medica* (ca. AD 50–79) of Pedanius Dioscorides, a Roman physician. The knowledge contained in these books remained the foundation for knowledge of nature in the West until the time of the European Renaissance.

The Renaissance heralded a renewed interest in contributing fresh observations, best exemplified in the notebooks of Leonardo da Vinci (1452–1519). Aspects of his observations on running water, for example, retained interest into the twentieth century. The following

Figure 2.1 Founders of natural history in the West: (a) Aristotle (384–322 BC) - the image is of a Roman copy in marble of an original bronze bust by Lyssipus, from life, ca. 330 BC (the mantle is a modern addition); (b) Pliny the Elder (AD 23-79), from a nineteenth century painting (no contemporary likeness exists) (images from Wikipedia).

three centuries witnessed a steadily developing interest in natural history, stunningly illustrated in the botanical and entomological paintings of Maria Sibylla Meriam (Figure 2.2a), a remarkable German–Dutch naturalist and artist of precociously analytical mind (Etheridge, 2011). The second year before Meriam's death – 1715, the year of the death of Louis XIV – conventionally marks the advent of the eighteenth-century European Enlightenment, the culmination of the foregoing period of early natural history. The Enlightenment witnessed the appearance of landmark works as varied as Carl Linnaeus's *Systema Naturae* (1735), the defining work of modern biological taxonomy, and Gilbert White's *Natural History and Antiquities of Selborne* (1789), the work that established the importance of phenological observations in natural history. Encyclopaedic summaries of natural phenomena reached their zenith in the monumental *Histoire Naturelle* (in 44 volumes) of Georges-Louis Leclerc (Comte de Buffon; 1717–1788). The period also saw the rise of formal learned societies (the Royal Society of London, 1660; the Académie Royale des Sciences, Paris, 1666 (after 1790, the Académie des Sciences); followed by St. Petersburg, Berlin and Bologna). With these institutions came the establishment of more or less regularly issued journals, the early subject matter of which included increasingly specialised reports on aspects of natural history.

Figure 2.2 Early pioneers of modern environmental science. (a) Maria Sibylla Meriam (1647–1717), pioneer naturalist, entomologist and taxonomic illustrator, portrait by Jacob Marrel (1679). (b) Charles Marie de La Condamine (1701–1774), an engraving by Pierre-Philippe Choffard (1759) (images from Wikipedia).

Natural history in the age of the Enlightenment encompassed observation and recording of all aspects of the external world, including humankind and human society. Accordingly, it included what have become the modern subjects of astronomy, physics, chemistry, geology (mainly recognised then as mineralogy), geography, botany, zoology, anthropology, sociology and even medicine. But, as knowledge expanded, this very range of interests presaged the development, from the late eighteenth century on, of these more specialised and distinct modern subjects. In the organisation of knowledge recognised by the great *Encyclopédie* of Denis Diderot and Jean le Rond d'Alembert, and by the pre-eminent philosopher of the Enlightenment, Immanuel Kant, natural history encompassed the observable, descriptive knowledge of the world. In particular, and not unlike today, it carried no significant connotation of temporal history. History was the province of natural philosophy, which considered – mainly speculatively – the causal history of the world.

The dichotomy of natural history and natural philosophy was problematic when considering the question of the history of Earth itself, especially for those studying the physical geography of the landscape and the rocks beneath. Largely constrained, well into the eighteenth century, by a biblical timescale for Earth history, natural philosophy consisted largely of speculative attempts to explain the current landscape in terms of the Noachian deluge. But in mid-eighteenth-century France,

Nicholas Desmarest, a French public official and avocational physical geographer, was engaged in an extended study of the landscape in the volcanic province of the Auvergne, where he recognised the sequential history of the successive lava flows. He identified erosional landforms in the older flows and attributed them to the action of running water. Working on the south side of the Massif Central, Jean-Louis Giraud-Soulavie extended these observations to the claim that valleys were formed by the streams that flow in them. They both recognised the extended timescale that this would require. In 1757, Desmarest contributed an article on physical geography to the *Encyclopédie* in which he emphasised the need to base knowledge of the landscape on detailed field observation. His claim was taken up by the influential Swiss surveyor–geographer, Horace-Bénédict de Saussure, who, in 1796, published an *Agenda* for physiographic studies in the Parisian *Journal des Mines* in which he asserted the primacy of fieldwork. The claim was forcefully echoed by Georges Cuvier, the most influential naturalist of his day (and the founder of modern palaeontology) in his 1810 report to the Emperor Napoleon on the status of the natural sciences. Thus was the primacy of fieldwork in the study of Earth and its environment formally institutionalised at the turn of the nineteenth century.

2.2. The Hegemony of Fieldwork

With the rapid development of the individual modern sciences from ca. 1800, 'natural history' as a recognisable pursuit has increasingly come to represent the continuation of descriptive field study by more- or less-informed amateur observers. 'Amateurs' have possessed all levels of ability and insight right up to Alexander von Humboldt and Charles Darwin, observers whose insights changed the fundamental concepts associated with nature. But while exceptional individual efforts have remained important even to the present, field study in modern science has passed through a recognisable sequence of increasingly institutionalised stages, beginning in the mid-eighteenth century with the inception of expeditions sponsored by European governments, wholly or in part for scientific ends. Barely disguised in most cases, there was, variously, exploration for commercial opportunity, in pursuit of imperial ambitions, to secure territorial claims or to seek military advantage.

One of the earliest such expeditions, apparently in sole pursuit of scientific ends, was that of Charles Marie de La Condamine (Figure 2.2b) to equatorial South America, between 1735 and 1745. Initially tasked with

Figure 2.3 Influential figures in the evolution of European exploratory natural science, whose attitudes were formed on voyages with Cook. (a) Sir Joseph Banks (1743–1820; portrait by Joshua Reynolds, 1773, shortly after Banks's return from the first Cook voyage), botanist and long-term president of the Royal Society, who strongly influenced British exploration at the turn of the nineteenth century. (b) Georg Forster (1754–1794; portrait by J. H. W. Tischbein, 1781, six years after the second Cook voyage), who strongly influenced Alexander von Humboldt (both images from Wikipedia).

geodetic measurements to test Isaac Newton's claim that Earth was an oblate spheroid (a parallel expedition was sent to subarctic Lapland), de La Condamine returned to Europe with knowledge of rubber, quinine (a natural antimalarial drug), an account of the first scientific descent of the Amazon and proof of Newton's conjecture – a striking early example of field observation being the arbiter of the validity of theory.

The expeditionary surveying voyages of the late eighteenth and early nineteenth centuries, notably into the Pacific by the Russians (Bering and Chirikov, 1742), French (Bougainville, 1766–1769; La Pérouse, 1785–1786), English (Cook, 1770s; Vancouver, 1791–1795) and Spanish (many exploring voyages between 1774 and 1795), mostly carried naturalists, and these expeditions began to return systematic reports of the natural history of exotic foreign lands. Accompanying Cook on his first voyage – ostensibly to observe a transit of Venus in the South Pacific for the Royal Society – were the botanist, Joseph Banks, and Daniel Solander, a student of Linneaus. Banks (Figure 2.3a) subsequently became a long-time president of the Royal

Society and an influential voice for field exploration. On his second voyage, Cook carried the German naturalist–geographers Johann Reinhold (father) and Georg (son) Forster (Figure 2.3b). Only 21 years old when he completed the voyage in 1775, the younger Forster's book, *A Voyage Round the World*, contains significant ethnological information about the people of Polynesia and is considered one of the founding examples of modern scientific travel reports. Georg later dictated that observations must not only be systematic, they must be focused, so that they would contribute toward the realisation of systematically defined objectives. This is a specification that we would make today.

On land, the travels into Siberia of Peter Simon Pallas represented a similar hybrid of scientific and commercial purposes and, in the New World, the transcontinental journey of Lewis and Clark at the behest of US President Thomas Jefferson, in search of a possible route to the Pacific and the oriental trade, was of parallel character. The most influential of the overland expeditions was the five-year (1799–1804) private expedition of Alexander von Humboldt and Aimé Bonpland to Central and northern South America, while the apogee of voyaging explorations was the second *Beagle* voyage (1831–1836) under Captain Robert Fitzroy, a part of the task of the Royal Navy to map the coastal waters of the world in support of commercial navigation. The last such oceanic expedition with wholly scientific purpose was the *Challenger* expedition of 1873–1876, while on land (or, rather, on ice), polar exploration continued into the twentieth century, culminating in Alfred Wegener's Greenland expeditions.

These overtly expeditionary forays progressively gave way to more systematic surveys of land and resources, conducted by agencies of governments specifically established for that purpose. The shift occurred most definitively in North America where, after 1800, survey offices were established by various American states, to map and appraise the resource potentials of the continent that was now becoming settled by Europeans. Agricultural land capability and mineral resources being of principal interest, land survey initiatives were established in New York (1820) and North Carolina (1823), followed by the establishment of formal geological surveys. The 'Ordnance Geological Survey' of Great Britain was established in 1835 and the Geological Survey of Canada in 1842. Similar surveys followed in other states and territories. Those of the American West began with the early surveys of F. V. Hayden and J. S. Newberry, before the Civil War. After the war, their surveys continued (e.g. Figure 2.4) and they were joined by additional survey groups.

Figure 2.4 Members of the Hayden survey expedition to Idaho and Montana (1871) in camp. (United States Geological Survey photo by W. H. Jackson.)

Eventually, the work was consolidated into the newly established (1879) United States Geological Survey. Their field seasons were long and arduous; their tools were basic surveying instruments and rock and soil sampling apparatus; their method consisted of field observation summarised in narrative and analytical reports, including the pioneering use of photography to record the landscape. The focus was the entire landscape.

There was no parallel institutionalisation of botanical or zoological survey in this period, though some of the geological surveys (notably the Geological Survey of Canada) formally included such surveys. These were, however, topics better suited to the efforts of individual collectors, notably Humboldt, Darwin, Alfred Russel Wallace and Henry Walter Bates, and a host of colonial administrators. But their collections required systematic curation and study. Thus arose the first great institutions of science with official support, the national natural history museums (Figure 2.5). These establishments, first rising to prominence in the latter half of the nineteenth century, presaged an important shift in the nature of what, in the nineteenth century, fast became natural science:[1] the removal to a study or laboratory of a significant part of the scientific activity associated with exploration of the environment.

Two other aspects of environmental science that required formal institutional support emerged as state-sponsored activities in the Victorian period; the offices of meteorological/climate and hydrological surveys.

1 'Science' and scientists were formally invented by William Whewell, in 1833.

Figure 2.5 The Natural History Museum, London, formally founded in 1881 when the natural history collections of the British Museum were transferred to the new establishment. (Photo by Christine Matthews, 2011. Reproduced under collective commons licence CC BY-SA2.0.)

Again, while the primary observations were obtained in the field (often, in the case of meteorology, just outside the door), interpretive and analytical activity occurred in an office. Hence, fieldwork in environmental science increasingly shared the allocation of time with office and laboratory investigations that were designed to amplify the understanding, and hence the value, of the field-derived observations. The trend was strongly reinforced by the emergence of the natural sciences in the universities in the latter half of the nineteenth century. While this served to re-emphasise in substantial degree the importance of individual investigators acting on their own initiative in the field, it also significantly changed the relative effort devoted to fieldwork. Anchored to the institutional home for much of the year by the need to lecture and mentor the students, professors increasingly engaged both in laboratory work and in theorising the bases of their science.

2.3 The Field in Eclipse

The most important influence on fieldwork in the twentieth century has been the effect of technology on the conduct of environmental science.

The earliest radical change involved transport. A significant part of the earlier need for long and arduous field seasons was the means of travel to and in the field. Until the twentieth century one travelled overland on foot, on horseback or in horse-drawn carriages. Where there were roads, they were mainly dirt tracks. At sea, one travelled in sailing ships at the mercy of the winds. Time passed. In part for these reasons, some of the most influential field research was conducted in one's own backyard (consider, for example, Darwin's many experiments at Down). As the twentieth century advanced, one increasingly could access even exotic field areas in days or weeks, rather than months, using automobiles (on much-improved roads), railways, steamships or aircraft. R. A. Bagnold and others used automobiles to explore the Western (also called Libyan) Desert in the 1930s (Goudie, 2008; see also Chapter 3). By late in the century, almost any site in the world could be accessed in hours.

Another signal technological contribution has been the development of automatic recording instruments, beginning with the likes of water-level recorders, recording thermographs and similar contraptions, using clockwork timing mechanisms. Today, electronically governed sensors, many of them able to transmit their readings to the laboratory by radio, can record almost any environmental signal of interest, from microscale atmospheric turbulence to continental drift (Figure 2.6). Such instruments record a single measurement or suite of closely related measurements, such as a panel of water-quality parameters or atmospheric energy fluxes, on a continuous or near-continuous basis for a substantial period of time. For the purposes of many increasingly specialised subjects within the umbrella of environmental science, these measurements may for many purposes be nearly all that is required from the field. Armed with such tools, one might do 'fieldwork' – that is, record field observations – without the need to be in the field at all, save for site reconnaissance and for installing and periodically maintaining the equipment (and senior scientists usually have technicians to do that).

The most radical technological effect of all on the conduct of environmental science has been the development of methods of remote sensing. Beginning with aerial photography in the 1920s, Earth's surface is now observed from space at frequencies varying from hours to seasons by a panoply of sensors, mounted on space satellites. Today, satellites can observe environmental processes and features, varying from Earth's physiography and tectonic movements, through land-cover condition, vegetative processes and aspects of environmental chemistry, to weather and ocean state, with a spatial resolution down to ca. 1 metre (and the

Figure 2.6 Modern fieldwork: raft equipped with acoustic Doppler current profilers and single-beam sonar. With this equipment, continuous data of velocity profile and water depth were obtained in an eight-day traverse covering more than 500 km of Fraser River, British Columbia, including passage through the Fraser Canyon (Rennie *et al.*, 2018). Data were supplemented by channel widths derived from aerial photography (Michael Church). A black and white version of this figure will appear in some formats. For the colour version, please refer to the plate section.

world's spooks can no doubt do much better than that) (Figure 2.7). Airborne survey captures spatially and temporally more highly resolved information for more limited but still substantial areas (Figure 2.8).

There have been, as well, important developments in the character and objectives of science itself that have influenced the role of fieldwork in environmental science. It is fair to say that, as modern science began to take form at the turn of the nineteenth century, the great underlying questions in environmental science – which were perhaps first coherently defined by Georges Cuvier – were to understand the history of Earth and the history of life on Earth. These questions characterised the mainstream divisions of the environmental sciences, which individually defined themselves throughout the course of the nineteenth century. Such questions lead to a descriptive and explanatory kind of science that attempts to comprehend the world in an integrated way, which fully

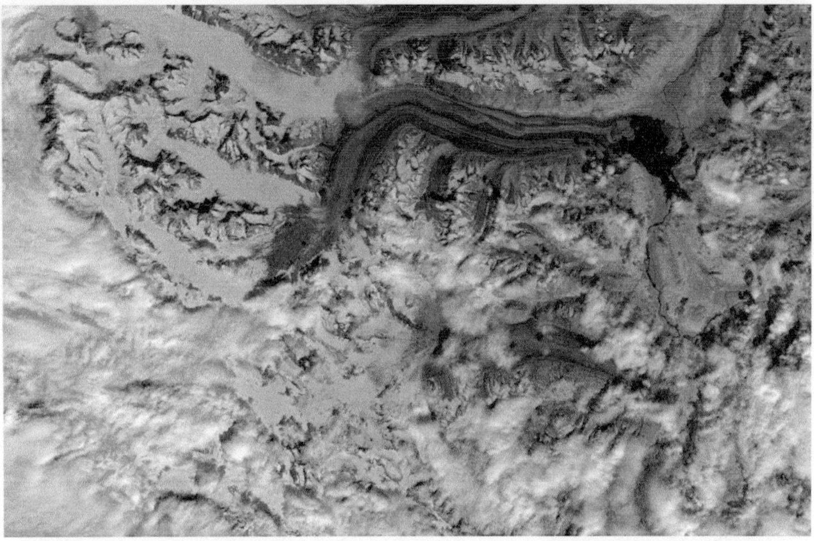

Figure 2.7 Image from space: Landsat 8 image of Lowell Glacier, Yukon Territory, Canada, 26 July 2018, showing extensive rapid snowmelt (dark grey) covering 40 km^2 of the glacier surface, and continued snow cover on the upper glacier (pale grey). Interpretation by Mauri Pelto, Nichols College, USA (NASA image).

recognises the complexity of the environment, and hence must make frequent reference to primary field observations. 'Theory' was mainly a generalised temporal narrative.

Two important developments have changed this situation. The larger endeavour of science was (and still is) to summarise its findings in a compact way in theory, preferably in a closed-form theory of mathematical precision that permits exact and reliable prediction. By the mid-twentieth century (but beginning in the nineteenth century in physical science), the environmental sciences had accordingly fallen under the influence of a positivist, reductionist approach whereby – in imitation of the classical bench sciences of physics and chemistry – problems were broken down into simple steps that focused on the processes of development, both of organisms and of environments. This served to focus attention on individual elements of the environment and their immediate connections. The grand historical issues faded from prominence. Fieldwork for the most part no longer considered whole landscapes, and it no longer necessarily dominated the activities of the environmental scientist.

Figure 2.8 Modern air photography: a portion of Cinderford, Gloucestershire, and the Forest of Dean, centred at 51° 50′ 00″ N, 02° 30′ 24″ W on 12 July 2013. Land uses are clearly demarcated in such imagery (Google Earth image).

In the new perspective data analysis and generalisation about environmental processes, achieved in the laboratory, the office and seminars became more prominent.

But it has proven difficult for the environmental sciences to construct theory in this mould, hence some continued preoccupation with the traditional questions and with fieldwork. In the later twentieth century, however, the concept of 'system science' emerged; the idea that some systems – especially environmental systems – are inherently complex constructs that entail the interaction of many elements and processes (of energy and mass exchange). Such sciences exhibit the generic characteristics of contingency (Simpson, 1963), complexity (Lorenz, 1963) and emergence (Anderson, 1972). They are largely intractable to closed-form theory, but might yield to generalisations embodied in 'system models', combinations of statements derived from theories about sufficiently simple parts of the system and empirical statements that bridge

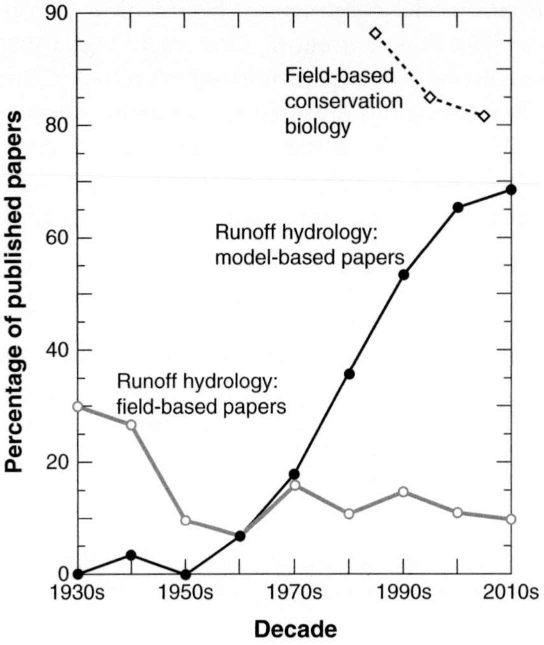

Figure 2.9 Two examples of the change in relative prominence of fieldwork and model exercises in particular environmental sciences: streamflow generation studies from 1930 to 2014 (after Figure 1 of Burt and McDonnell, 2015); and conservation literature from 1980 to 2014 (after Figure 1 of Rios-Saldina *et al.*, 2018), field studies only. Data displayed by decade, except for the most recent four-year period.

theoretical lacunae. Computers were now available to operationalise model constructs. This development has led to a major preoccupation with the construction of system models in the environmental sciences, an exercise that promises significant predictability in an age when environmental change, and the ability to predict it, have become critical socio-political concerns. Modelling has largely replaced fieldwork as an attempt to understand nature (Figure 2.9).

The consequence of these technological developments has been an orders of magnitude increase in the volumes of data that are ultimately derived, in one way or another, from the field. But traditional fieldwork – observe and describe – delivers only a small minority of the data; by far most derives from autonomous instruments. Fieldwork has ceased to be the principal activity in the environmental sciences. It is now only one part of a triumvirate consisting of data acquisition (from the field or laboratory), data analysis and model construction. Fieldwork is usually the activity absorbing the least time and effort. One might argue that instrument deployments and remote sensing substitute for fieldwork. But do the arrayed products of autonomous instruments and remote sensing constitute fieldwork at all? How do we ascertain that instrument readings are accurate? How do we confirm that the interpretation of a remote image is correct? We continue to need calibration to the field situation. Further, there remains the question how to recognise phenomena that are not expected or not previously understood.

A powerful incentive to prioritise modelling activity is the circumstance that it is less expensive than modern fieldwork while promising more comprehensive answers to key questions about the environment. Perhaps, then, it deserves its contemporary prominence in the scientific enterprise. Even the basic, state-operated meteorological and hydrological networks of observing stations have contracted in favour of forecast models in most countries in recent decades (e.g. Shiklomanov *et al.*, 2002). But we know that models are at best an approximation of the environment (e.g. Carslaw *et al.*, 2018) and, at worst, a dangerously biased approximation. How, then, do we hold models of environmental processes up to critical evaluation? To achieve assurance, direct observations in the field remain imperative – observations trump theory. In particular, contingent, complex and emergent properties of environmental systems need be discovered by observation before their appearance (or not) in a model can be verified. In environmental science, observations ultimately mean field observations.

2.4 Revaluing Fieldwork

Not only do expense and organisational complexity (including obtaining the right, in many cases, to access field areas to conduct work) serve to discourage fieldwork, but science has passed through a phase of evaluating modelling work as somehow a superior undertaking, making it relatively more difficult to publish field results; at least, field results that do not immediately support model derivation or test. There are, however, indications that this situation may be changing. While preparing this discussion I have encountered a number of articles advocating the need for renewed field investigation. One cogent reason for this is the need to improve predictive models. Yet 'making models work for the right reasons is wholly dependent on our ability to characterize and understand mechanistically such processes in the field' (Burt and McDonnell, 2015).

Accordingly, Lowenstern *et al.* (2018) urge the need for new geophysical investigations of crustal magma reservoirs and scientific drilling into magma chambers in order to improve the 'rudimentary nature' of models for the prediction of volcanic activity. Ward *et al.* (2018) advocate the need for observations of currents and water chemistry in the mixing zone where the Amazon River flows into the Atlantic Ocean in order to improve the understanding of the ecosystem of the coastal mixing zone – at present, there simply are no observations in this region. Oschlies *et al.* (2018) emphasise the need for additional observations of ocean oxygen content with the intention to search for additional deoxygenation mechanisms, the knowledge of which will correct currently biased models of ocean oxygen exchange. And, in a rapidly changing environment, fresh observations of the geographical range of all organisms are always necessary (Delibes-Mateos and Ferreira, 2018).

More significantly, perhaps, the participation of ordinary citizens in the gathering of field data has never withered. Natural history pursuits were a logical extension of the early studies and, in the nineteenth century, became institutionalised themselves with the establishment of natural history societies. In Great Britain, the Natural History Society of Northumbria was founded in 1829, and the London Natural History Society in 1858. Most such organisations pursue interests in living organisms such as birds (ornithology), insects (entomology), animals (mammalogy, herpetology), plants (botany) and fungi (mycology). But there are also societies for the amateur pursuit of geology, mineralogy and astronomy. Some of the findings of society members are decidedly professional in character. In particular, 'amateur' observers are well known for important contributions to astronomical discovery, and they return the bulk of field observations of bird distribution, nesting success and wintering survival. For example, the annual Christmas Bird Count that has been organised by the Audubon Society of America has recorded wintering bird populations in North America since 1900. Some societies' activities have evolved to the point that they can support significant professional science (for example, the British Royal Society for the Protection of Birds).

With the advent of convenient mass communication amongst individuals via computers and, now, smartphones, the notion of 'citizen scientists' has taken root (*Nature*, 2015; Irwin, 2018). Citizen science is, in one sense, a fusion of natural history and environmental science. In this sense, professional scientists design programmes in which amateur volunteers conduct set analyses (often procedures that must be massively repeated and must be done by humans) or report field observations that must be repeated often or over wide areas. Many observations in the environmental sciences are of the latter type (Figure 2.10). Kampf *et al.* (2018) describe two programmes in which citizen scientists regularly observe and report stream channel conditions at sites they frequent; recruitment, training and reporting are achieved through smartphone apps. The programme is particularly oriented towards gaining observations on ephemeral streams that are almost never gauged but nonetheless have significant effects on water quality and riparian ecology.

Citizen science can also be citizen-motivated, or it can arise from the activities of the natural history clubs. Citizens concerned with environmental features or change in their neighbourhood might institute a systematic programme of observations to document their concern. For some, the systematic study of some aspect of nature is a satisfying hobby that leads to important results. Remarkably, the members of the Krefeld Entomological Society (Germany) have been observing and collecting insects from their region since 1905. By the 1980s, they became aware that their records were showing significant declines in insect numbers and the local disappearance of some species. They have since established stations of professional standard for systematic collections, which have confirmed remarkable declines. The effect reverberates up the food chain to affect insectivorous animals, principally birds, which are also in decline. The Krefeld club data are now the world's major field data of this crucial trend (Vogel, 2017).

The rise of citizen science has presented novel problems for the scientific enterprise (Guerrini *et al.*, 2018), especially in the context of amateurs who are recruited by professional scientists to participate as observers in professionally planned and sponsored projects. The question of intellectual property arises – who 'owns' the observations, the citizen observer or the project leader(s)? In most cases, this issue is specifically addressed and, usually, the arrangement is for joint ownership and free access for the citizen to the project's entire database. Another issue in cases when some formal body (a chartered society; a company; a university or government laboratory) has initiated the project is the question of liability in the event of an accident in the field. A critical issue in all cases is data quality, an issue that arises amongst professional scientists themselves as well as with amateur cooperators. In most cases of volunteer cooperators, the issue does not arise – indeed, in some areas of natural history, the amateurs are the leading experts in field observation. But in

Figure 2.10 Citizen science. (a) Members of the Cascades Butterfly Project, a long-term survey in the North Cascades and Mount Rainier National Parks, Washington State, USA. The objective is to understand the effects of environmental change in alpine meadows, for which butterflies are important indicator species. Photo by National Park Ranger. A black and white version of this figure will appear in some formats. For the colour version, please refer to the plate section. (b) Birdwatchers in Sea Island Conservation Area, adjacent to the airport in Vancouver, British Columbia. The objective is to survey the long-term effect of airport operations on birdlife in the adjacent area, which has been established as a compensation habitat for airport development (J. M. Ryder).

some cases, citizen observers undertake observations because they have in mind some specific issue related to environmental management (pesticide use, for example, or a decision to allow development in some valued natural area); then, one must ask whether the observations have been structured with some particular outcome in mind – or not. A careful construction of observing protocols is necessary to obviate this problem.

2.5 Prospect

Field study in the environmental sciences appears to have two overriding purposes: (i) to ascertain the history and current status of the elements of the planetary environment, from the rocks of the planetary crust through to the distribution and actions of human society; and (ii) to understand the processes of change in individual elements of the environment and the structure and dynamics of interactions amongst them. The latter undoubtedly requires the field, experimental and analytical activity of professional scientists – issuing in model constructs – to be fully developed. The former, increasingly dominated by remote sensing methods, nevertheless requires ongoing and widespread observation of Earth's surface for many factors that are not accessible by remote means, and to calibrate remote and instrumental observations. It appears that some combination of professional scientific guidance and amateur (citizen) observers will increasingly take on these tasks. Without the fruits of such endeavours, the necessary understanding of our planetary home cannot be reached. There certainly is an important future in the environmental sciences for fieldwork.

Acknowledgement

Thanks are due to Eric Leinberger, who prepared the illustrations.

References

Anderson, P. W. (1972). More is different. *Science* 177, 393–396; DOI: 10.1126/science.177.4047.393.

Burt, T. P. and McDonnell, J. J. (2015). Whither field hydrology? The need for discovery science and outrageous hydrological hypotheses. *Water Resources Research* 51, 5919–5928; DOI: 10.1002/2014WR016839.

Carslaw, K. S., Lee, L. A., Regayre, L. A. and Johnson, J. S. (2018). Climate models are uncertain, but we can do something about it. *Eos* 99, 15–16.

Delibes-Mateos, A. M. and Ferreira, C. C. (2018). Are fieldwork studies being relegated to second place in conservation science? *Global Ecology and Conservation* 14, doi.org/10.1016/j.gecco.2018.e00389.

Etheridge, K. (2011). Maria Sibylla Meriam and the metamorphosis of natural history. *Endeavour* 35, 17–21.

Goudie, A. (2008). *Wheels across the Desert*. London, Silphium Books.

Guerrini, C. J., Majumder, M. A., Lewellyn, M. J. and McGuire, A. L. (2018). Citizen science, public policy. *Science* 361, 134–136; DOI: 10.1126/science.aar8379.

Irwin, A. (2018). Citizen science comes of age. *Nature* 562, 480–482.

Kampf, S., Strobl, B., Hammond, J., *et al.* (2018). Testing the waters: mobile apps for crowdsourced streamflow data. *Eos* 99, 31–34.

Lorenz, E. (1963). Deterministic nonperiodic flow. *Journal of the Atmospheric Sciences* 20, 130–141.

Lowenstern, J. B., Sisson, T. W. and Hurwitz, S. (2018). Probing magma reservoirs to improve volcano forecasts. *Eos* 99, 16–21.

Nature (2015). Rise of the citizen scientist (Editorial). *Nature* 524, 265; DOI: 10.1038/524265a.

Oschlies, A., Brandt, P., Stramma, L. and Schmidtko, S. (2018). Drivers and mechanisms of ocean deoxygenation. *Nature Geoscience* 11, 467–473; DOI: 10.1038/s41561-018-0152-2.

Rennie, C. D., Church, M. and Venditti, J. G. (2018). Rock control of river geometry: the Fraser canyons. *Journal of Geophysical Research – Earth Surface* 123, 1860–1878; DOI: 10.1029/2017JF004494.

Shiklomanov, A. I., Lammers, R. B. and Vorosmarty, C. J. (2002). Widespread decline in hydrological monitoring threatens Pan-Arctic research. *Eos* 83, 13–17; DOI: 10.1029/2002EO000007.

Simpson, G. G. (1963). Historical science. In *The Fabric of Geology*, ed. C. C. Albritton. Addison-Wesley, New York, NY, pp. 24-48.

Vogel, G. (2017). Where have all the insects gone? *Science* 356, 576–579.

Ward, N. D., Sawakuchi, H. O. and Richy, J. E. (2018). The Amazon River's ecosystem where land meets the sea. *Eos* 99, 29–33.

3 · *The History of Fieldwork in the Geosciences*

ANDREW S. GOUDIE

3.1 Introduction

For much of its history, geography has been very largely a field science, and while today there is an increasing weight placed upon such modern technologies as computer modelling and remote sensing, fieldwork continues to be a major approach to the subject and a fundamental source of many types of inspiration and of data. It is not an exaggeration to say that many of the most significant and influential themes in physical geography and across the geophysical sciences more broadly resulted from the labours of scientists who undertook sustained programmes of work in the field. Among the themes to which one can point with pride are the ideas of Alexander von Humboldt on the indivisibility of nature, Louis Agassiz's discovery of the Ice Age, Charles Darwin's work on both the evolution of species and the development of coral reefs, the work of American geologists in the opening up of the American West and the development of ideas on erosion and slope evolution, Baron von Richthofen's discovery of the true origin of loess, William Morris Davis's elaboration of the concept of the cycle of erosion, Ellsworth Huntington's discovery of evidence of climate change in central Asia and its role in driving human history (which led on to environmental determinism), Roy Chapman Andrews's discovery of dinosaur eggs in Mongolia and Ralph Bagnold's work on the fundamentals of wind movement of sand and the formation of sand dunes in deserts.

3.2 Some Nineteenth-Century Heroes

Field research in the earth sciences has a long history. In 1571, in his *Idea Medecinae Philosophicae*, a Danish scientist, Peter Severinus, gave advice to his students 'Go my Sons, buy stout shoes, climb the mountains, search the valleys, the deserts, the sea shores, and the deep recesses of the earth . . . in no other way will you arrive at a knowledge of the nature and property of things' (quoted in Adams, 1938, p. 210). Archibald Geikie (1905) drew attention to two particularly important fieldworkers in the eighteenth century: Pierre Simon Pallas (1741–1811), who led a six-year-long great scientific expedition through the Russian Empire, and Horace-Benedict de Saussure (1746–1799), who pioneered mountain science and urged people to 'quit the beaten tracts, and climb to the lofty summits' (Geikie, 1905, p. 184). It was also a time that saw the development of geological mapping (Ireland, 1943), as pioneered in France by the gruff Jean-Étienne Guettard (1715–1786). Jackson (2007) provides a series of papers on early geological mappers and travellers who visited the world 'on foot, bicycle, sledge and camel'.

In the nineteenth century, observations and recording were greatly valued by bodies like the Royal Geographical Society (RGS), as is exemplified by Colonel Julian Jackson's *What to Observe or, the Traveller's Remembrancer* (Jackson, 1841; see also Goudie, 1978). Jackson (1841, p. 4) 'endeavoured to excite a desire for useful knowledge by awakening curiosity'. He hoped that the intending traveller would discover through the use of his book 'what an immense field of physical and moral research lies open to his investigation, and be encouraged to exertion by the assurance that, without being what is termed a philosopher, he may not only do much to enlarge the sphere of his own ideas, but acquire the means of communicating to others a great mass of valuable or interesting information'. Sir Francis Galton's *The Art of Travel or Shifts and Contrivances Available in Wild Countries* (Galton, 1872) was to play a similar role, some decades later. It provided a *vade mecum* of how to conduct expeditions, giving details, inter alia, of equipment, food, medicine, surveying and the 'management of savages'. Galton (1822–1911) was a biostatistician, geneticist and eugenist, who led an expedition from the coast of the Namib Desert to the interior, in the early 1850s (Goudie, 2016).

Alexander von Humboldt

The development of a fieldwork tradition in the USA can be traced back to Alexander von Humboldt (1769–1859) (Figure 3.1), whose tropical American travels with his partner Aimé Bonpland, between 1799 and 1804, showed the potential value, scale and scope of scientific fieldwork (Mathewson, 2001). His influence was, however, worldwide (Wulf, 2015)

Figure 3.1 Portrait of Alexander von Humboldt by Friedrich Georg Weitsch, 1806 (image from Wikipedia)

and included an influence on the work of Powell and others in the exploration of the American West (Mathewson, 2006). He formulated ideas about the indivisibility of nature, the web of life and so was an inventor of ecology and environmentalism. Driven by a sense of wonder, whether in the mines of Germany, the high Andes, the Venezuelan rainforest or in the steppes of Russia, he undertook researches and observations of an unprecedented quality and range. Charles Darwin was hugely stimulated by his voluminous and detailed writings, as was George Perkins Marsh, the writer of *Man and Nature* (Marsh, 1864), who was himself a great traveller.

Louis Agassiz

In the early nineteenth century, the diluvial theory, arising from a belief in the biblical flood, was usually invoked to explain many geomorphological phenomena in Europe. However, in the 1820s, various brave scientists started to suggest that glaciers had once been considerably more extensive than today and could account for much of what was then called 'drift'. The most famous exponent of the glacial theory was Louis Agassiz (1807–1873), the Swiss natural historian. In 1836, he was guided around his home country by Ignaz Venetz and Jean de Charpentier and as a result of their inspiration became an enthusiast for the idea that, in the past, glaciers had been much more extensive than now – the so-called 'Glacial Theory'. Agassiz wrote it up as his famous *Discours de Neuchâtel: Etudes sur les Glaciers* (Agassiz, 1840). The Glacial Theory, and the discovery of the Ice Age, had an impact on the earth sciences in the 1830s that was as striking as the impact of plate tectonics was to be in the 1960s (Imbrie and Imbrie, 1979). In 1840, Agassiz visited Scotland and recognised evidence for former glaciations there as well. On long walks, he managed to persuade Oxford's William Buckland of the veracity of his views, even though Buckland had been an arch-diluvialist and ardent catastrophist (Oldroyd, 1999).

When he went to Harvard College in America, Agassiz championed fieldwork and his dictum was 'Study nature, not books'. He built a seaside laboratory on Penikese Island off Cape Cod. His heir, Nathaniel S. Shaler, continued this tradition (Gladfelter, 2002). Shaler (1841–1906) was an American palaeontologist and geologist. At Harvard University, he promoted the geography programme, which produced the first cohort of academic geographers under the direction of William Morris Davis.

Charles Darwin

Charles Darwin (1809–1882) was both a geologist and a biologist and his twin skills came together in his work on coral reefs. On the voyage of the *Beagle* he saw many coral reefs. In 1842, he summarised his subsidence theory to explain the sequence of fringing reefs, barrier reefs and atolls (Darwin, 1842). This was 'a simple and luminous explanation', which received rapid acceptance (Geikie, 1882, pp. 466–467). It explained how large-scale structures arising from the deep ocean floor might be built by light-dependent organisms, restricted to the upper 100 metres, or less, of the water column (Spencer *et al.*, 2008). As Davis (1913, p. 173) remarked: '. . . for forty years the scientific world accepted it as demonstrated. Darwin's diagram of a subsiding island and an up growing reef have been reproduced over and over again on countless blackboards, as representing one of the great discoveries of geological science'.

John Wesley Powell

Fieldwork was especially valued as the exploration and development of the American West took place. After the American Civil War, Major John Wesley Powell (Figure 3.2) and Professor Nathaniel S. Shaler played complementary roles. Powell (1834–1902) got various government agencies to sponsor the scientific exploration of the American West, and himself made heroic journeys into the Colorado Plateau. This great geologist, anthropologist and explorer was born in 1834, in Mount Morris, New York. After a spell of teaching, museum work and fighting on the side of the Union in the American Civil War, where he lost part of one arm at the Battle of Shiloh, he became a professor of geology at Illinois Wesleyan University in Bloomington. In May 1869, having obtained funds and equipment from various sources, he set out on his first journey – the Rocky Mountain Scientific Exploring Expedition – to navigate the largely unknown Green and Colorado rivers. He made a second river trip in 1871–1872, in which he retraced part of the 1869 route, travelling the Colorado River from Green River, Wyoming, to Kanab Creek in the Grand Canyon (Figure 3.3). Powell (1875) in his *The Exploration of the Colorado River and its Canyons* wrote thus of the impact upon his senses of the Grand Canyon:

Figure 3.2 The bust of John Wesley Powell from a lookout point over the Grand Canyon (Andrew Goudie)

Figure 3.3 The Grand Canyon in Arizona and neighbouring areas of the American South-west were the scene of major geological surveys by American geologists, such as J. W. Powell and G. K. Gilbert, which revolutionised geomorphology in the late nineteenth century (Andrew Goudie)

> The Grand Canyon of the Colorado is a canyon composed of many canyons. It is a composite of thousands of gorges. In like manner, each wall of the canyon is a composite structure, a wall composed of many walls, but never a repetition. Every one of these almost innumerable gorges is a world of beauty in itself. In the Grand Canyon there are thousands of gorges like that below Niagara Falls, and there are a thousand Yosemites. Yet all the canyons unite to form one Grand Canyon, the most sublime spectacle on the Earth. [Powell, 1875, p. 390]

Powell's experiences in the Colorado Plateau and the Uinta Basin were crucial for the development of ideas in geomorphology. He appreciated the sheer power of river erosion, and could see from the unconformities present in the area's starkly exposed stratigraphic columns that rivers could cut down to a 'base-level of erosion', and continue to cut down through deep layers of slowly uplifting strata.

In 1870 Congress granted Powell funds for what came to be designated the 'Geographical and Geological Survey of the Rocky Mountain Region, J. W. Powell in charge'. In the event the Rockies hardly figured at all in his programme, and Worster (2001), in *A River Running West*, has suggested that a better name might have been 'The survey of the high plateau country of Utah and Arizona through which the Colorado River flows'. Within six years of the first river expedition and three years of the second, the work of Powell's surveyors and geologists in the Colorado Plateau had become hugely appreciated for their results and their professionalism (Sack, 2013). In this role he employed and encouraged some other great investigators of the arid west, including Grove Karl Gilbert and Clarence Dutton. Their researches were vital in the development of modern geomorphology as is made evident in the pages of the magisterial survey of Chorley *et al.* (1964), contributing to the acceptance of fluvialism, helping to formulate ideas about slope evolution, and recognising the importance of landscape planation. From this flowed the concepts of the cycle of erosion and of denudation chronology, which were to dominate the subject until the 1950s.

Grove Karl Gilbert

Grove Karl Gilbert (1843–1918) was a remarkable American geomorphologist who, in many respects, was ahead of his time (Baker and Pyne, 1978). His name is most often associated with what is often termed dynamic geomorphology. As Slaymaker (2004, p. 307) remarked, 'the work of G. K. Gilbert is the first seminal antecedent of the study of geomorphic process or dynamic geomorphology'. This is exemplified in Gilbert's report on the geology of the Henry Mountains (Gilbert, 1877), his study of the deltaic deposits and old shorelines of pluvial Lake Bonneville, his study of the convexity of hill tops, his examination of the effects of gold mining debris in California and his work on the transportation of debris by running water. He was a champion of the value of observation in geomorphology (Rhoads and Thorn, 1996).

Baron Ferdinand von Richthofen

Loess, a mostly non-stratified and non-consolidated silt, containing some clay, sand and carbonate, is a widespread deposit of geomorphic significance. It was Baron Ferdinand von Richthofen (1823–1905) who cogently argued, following fieldwork in inner China, where he gave the Silk Road its name, that these intriguing deposits probably had an aeolian origin and that they were produced by dust storms, transporting silts from deserts and depositing them on desert margins:

In regions where the rains are equally distributed through the year, little dust is formed, and the rate of growth of the soil covered with vegetation will be exceedingly small. But where a dry season alternates with a rainy season, the amount of dust which is put in motion and distributed through atmospheric agency can reach enormous proportions, as witnessed by the dust storms which in Central Asia and Northern China eclipse the sun for days in succession. A fine, yellow sediment of measurable thickness is deposited after every storm over large extents of country. Where this dust falls on barren ground, it is carried away by the next wind; but where it falls on vegetation, its migration is stopped. [von Richthofen, 1882, pp. 297–298]

Among von Richthofen's students was the great Swedish explorer and archaeologist, Sven Hedin (1865–1952), who travelled in central Asia and Mongolia and introduced the term 'yardang' into the geomorphological literature (Goudie, 2016).

William Morris Davis

William Morris Davis (1850–1934) (Chorley *et al.*, 1973) was arguably the leading American geomorphologist of the late nineteenth and early twentieth centuries. His great contribution was to produce a deductive model of landscape evolution, called the Cycle of Erosion or the Geographical Cycle (Davis, 1899). This was developed during the 1880s and 1890s (Orme, 2004). Davis believed that landscapes were the product of three factors: structure (geological setting, rock character, etc.), process (weathering, erosion, etc.) and time (stage) in an evolutionary sequence. Stage was what most interested him. He suggested that the starting point of the cycle was the uplift of a broadly flat, low-lying, surface. This is followed by a phase he termed *youth*, when streams become established and start to cut down and to develop networks. Much of the original flat surface remains. In the phase he termed *maturity*, the valleys have widened so that the original flat surface has been largely eroded away and streams drain the entire landscape. The streams begin to meander across wide floodplains and the hillslopes become gradually less steep. In *old age*, the landscape becomes so denuded that a low relief surface, close to sea level, develops, with only low hills rising above it. This surface is then called a peneplain.

Davis promoted field observations as one of the subject's organising principles, and he travelled the world, basing his theory on personal examination of many landscapes in many different settings (Rhoads and Thorn, 1996). Chorley *et al.* (1973, pp. 392–394) record that as a fieldworker 'he was remarkably indefatigable and loved to exert himself', and was able to sustain himself 'long after most of the students had had enough'. His fieldwork had four imperatives: Go, see, think, and draw.

Ellsworth Huntington

Ellsworth Huntington was born in Illinois, USA, in 1876, and, like another great explorer of central Asia, Roy Chapman Andrews, attended Beloit College in Wisconsin. He was not only one of the greatest, most well-known and controversial academic geographers of his generation, but he also made major contributions to our knowledge of the geography and history of large tracts of dry Asia. After Beloit, Huntington obtained a missionary position in present-day Turkey. He took the opportunity to travel in Anatolia and, in 1901, to travel down the Euphrates on a raft made of saplings, buoyed up by inflated sheepskins. On his journey back to America, he read James Geikie's *The Great Ice Age and Its Relations to the Antiquity of Man* (Geikie, 1874), which stimulated his life-long effort to link climate change to the history of human societies. It sowed the seeds of the environmental determinism for which Huntington became famous; or, depending on your point of view, notorious. After that, he studied physiography at Harvard under Davis.

Davis took Huntington on a field trip to the arid south-west of the USA in 1902, and in 1903, accompanied again by Davis, Huntington participated in a central Asian expedition led by R. W. Pumpelly and supported by the Carnegie Institution. This expedition took him to Kashgar at the western end of the Tarim Basin. During the winter of 1903–1904 he worked in the Seistan Basin (which he described as 'a hell'), at the borders of Persia and Afghanistan. He then made a trip in 1905–1906, which was sponsored by the newly formed Association of American Geographers (AAG). This expedition took him first to Kashmir and Peshawar (where he met another great fieldworker, Aurel Stein) and then over the Karakorams to the Lop Nor basin in East Turkestan and on to Turfan. Upon his return, Huntington wrote *The Pulse of Asia* (1907), in which he argued that the climatic factor was crucial for understanding the human history of the regions he had been through. He speculated that pulsations of climate had served as a driving force in the history of Eurasia, impelling nomadic invaders in dry periods to overrun the civilisations that surrounded them. Huntington was especially struck by the information about past climates that could be provided by lake deposits and old strandlines in arid regions, noting the relic shorelines of Lop Nor, which rose to up to around 180 m above the present lake and indicated to him a mega-lake, which was formerly 480–640 km long and 160 km wide.

Roy Chapman Andrews

Roy Chapman Andrews (1884–1960) was an American explorer of the Gobi Desert, known for his pioneering use of motor transport and for his collections of dinosaur remains. He was born in 1884, in Wisconsin. On graduation from Beloit College (also attended by Huntington), he moved to New York to seek work at the American Museum of Natural History. He started as a janitor and floor scrubber in the taxidermy department, but while working there also earned an MA in mammalogy from Columbia University. Andrews is best known for his expeditions to the Gobi in Outer Mongolia, made between 1922 and 1930 (Figure 3.4). One feature of the expeditions was that they were orchestrated, multi-disciplinary affairs, in which an increasing range of scientists

Figure 3.4 An armed Roy Chapman Andrews searching for dinosaur remains in Mongolia (image from Wikipedia)

were brought together to tackle specific problems in the field. Another was that the expeditions were made possible by successfully raising large sums of money, which in turn was facilitated by courting publicity and approaching financiers such as John D. Rockefeller, Jr, Cleveland Dodge and J. P. Morgan, Jr. A third characteristic was that these large, complex expeditions required a new style of efficient management, while a fourth was that Andrews made use of motor vehicles. Finally, Andrews's camps were relatively comfortable. He believed that scientists cannot work unless well fed and argued that hardships should be avoided, for they lessen effectiveness. In other words, he modernised the approach to field research. By 1925, the expedition involved 40 people, 7 vehicles and 125 camels. The American flag flew proudly and ostentatiously from the vehicles as they sped across the desert, and Andrews himself, an Indiana Jones type of figure, was a striking and charismatic individual, with his large hat, field glasses and guns. The greatest achievement of the expeditions was the discovery, in 1923, of remarkable dinosaur bones and eggs in Cretaceous beds in the snake-infested 'Flaming Cliffs of Shabarrakh Usu'. These were the first dinosaur eggs ever to have been found. Andrews described his philosophy of exploration to the RGS in 1926:

> I realized that we could obtain satisfactory results only by bringing to bear upon the problem every branch of science which could assist in its solution. A group of highly trained specialists must be taken into the field *together*. Thus the work of each man could assist all the others.
>
> The value of such correlated work was demonstrated continually. As we sat in the mess tent at night discussing the day's work, it was most interesting to see how puzzling situations in geology would be clarified by the palaeontologist; how the topographer brought out important features which gave the key to physiographic difficulties; and how the palaeontologist would be assisted by the palaeobotanist or geologist solving stratigraphic problems. [Andrews, 1927, p. 1]

Andrews believed that this was the first expedition of such magnitude to employ these methods and predicted that it would in the future become the recognised type of scientific exploration. The various projects launched by the RGS and other bodies in the decades after World War II demonstrated that his prediction was correct.

Ralph Alger Bagnold

Ralph Alger Bagnold (1896–1990) (Figure 3.5) was a professional soldier who used Ford cars to explore the Western (Libyan) Desert of North

Figure 3.5 The Bagnold Medal of the European Geophysical Union (Andrew Goudie)

Africa and commanded the Long Range Desert Group in World War II. Remarkably for a soldier and explorer, he was made a fellow of the Royal Society. He was 'one of the founders of modern geomorphology despite having no formal academic affiliation, no cadre of students or postdocs under his command, no steady financial support, and no scientific training beyond a degree in engineering. What he did have, and used to great effect, were a deep curiosity about natural phenomena, a powerful physical intellect, a talent for clever experimentation, extensive opportunities to observe geomorphic processes at work in the field, and – perhaps most important of all – the time and freedom to focus his energies on significant scientific challenges' (Kirchner, 2013).

In 1926, he was posted to Egypt, and it was while on leave there that he developed a passion for desert exploration by motor car, building upon the experience that Dr John Ball and others had gained in using Model T Fords for desert surveys and patrols during the Great War (Goudie, 2008b). After a couple of years in Egypt he was sent to India, but in 1931 was posted back to Catterick Garrison in the north of England. However, the army granted him permission to conduct an expedition in the Libyan Desert in 1932 (building upon those he had done in 1929 and 1930). Thereafter, he served in the Far East but was taken ill with 'tropical sprue' (a digestive complaint) and was discharged from the army as 'a permanent invalid'. However, Bagnold was durable. In 1938, he decided that he needed to validate the work he had been doing on sand movement in his pioneering wind-tunnel experiments against real-world conditions. Thus, he returned to the desert accompanied by a team of archaeologists and surveyors, very much in the spirit of Andrews. One of the team was R. F. Peel, who became a leading desert geomorphologist at Leeds and then Bristol. The journey took them to Uweinat, the Gilf Kebir and the Selima Sand Sheet in southern Egypt. For much of the time Bagnold's fieldwork involved waiting for a sandstorm to occur so that he could witness and measure sand grains in action, but he also wanted to do some real exploration. One feature of Bagnold's expeditions was that besides exploring new regions and developing new techniques for desert travel, he also had time for science. Not only did he use the 1938 expedition to validate his own wind-tunnel experiments, his team made remarkable discoveries that included a profusion of archaeological sites and evidence of profound changes in past climates. Bagnold wrote his classic book on dunes, *The Physics of Blown Sand and Desert Dunes*' (Bagnold, 1941), which is still much cited today. He attributed his fascination for dunes directly to the experience of motoring into the interior of an exceptionally dry desert. There were, of course, many others, including French and Germans, who contributed to dune studies in the twentieth century (see Goudie, 1999, 2008a), but Bagnold was paramount.

3.3 Student Fieldwork in the First Half of the Twentieth Century

In the USA, during the first half of the twentieth century, there were movements to establish the role of student fieldwork (Platt, 1959), and as early as 1915, two products of Rollin Salisbury's Chicago School, Carl Sauer and Wellington Jones, produced a detailed outline of how to organise fieldwork and what to observe (Jones and Sauer, 1915). Sauer, who was initially no mean geomorphologist (Stoddart, 1997), progressed to Michigan and to Berkeley, where he stressed the importance of overseas fieldwork. This became a feature of the Berkeley School, while the role of fieldwork in Chicago has been described by Martin (2005, pp. 372–374).

In Britain, fieldwork was a major component of the meetings of Section E of the British Association for the Advancement of Science, and right from its inception, in 1933, the Institute of British Geographers thought in terms of a serious, professionally organised programme of fieldwork (Steel, 1987, p. 70). After World War II, three towering figures in British geography were major exponents of the value and importance

Figure 3.6 James Alfred Steers at a meeting of Section E of the British Association in Oxford, 1954 (the late E. W. Gilbert)

of work in the field. At Cambridge, James Alfred Steers (Figure 3.6) was head of department from 1949 to 1966, and in addition to his own fine career as a fieldworker on the Great Barrier Reef and on the Norfolk coast, he encouraged fieldwork by members of his department, partly based on a research station on Scolt Head Island, but also through a group of younger staff, which included A. T. Grove (who spearheaded research on climate change in African deserts and the East African Rift), Vaughan Lewis (who, with Jean Grove and others, conducted major field programmes on glaciers in Norway; Lewis, 1960) and David Stoddart, who worked indefatigably on coral reef problems in Belize and elsewhere (see Starrs, 2015 for a first-class summary of Stoddart's wonderfully individualistic career). Steers (1987, pp. 149–152) describes the coastal fieldwork undertaken by Cambridge investigators, and the work done, particularly in polar regions, by long vacation expeditions. Released from Cambridge, Marjorie Sweeting undertook missionary work in Oxford, stimulating generations of karst fieldworkers. Other women fieldworkers of note at this time, who acted as role models, were Jean Grove at Cambridge, Cuchlaine King at Nottingham and Hanna Bremer at Cologne (Sack, 2004; Hart, 2007).

In London University, S. W. Wooldridge, whose personal interests were mainly in southern England, was a frequent exponent of the value of fieldwork (see, for example, Wooldridge, 1955), a passion he shared with another London professor, W. G. East. As they remarked, 'The first principle which must be affirmed is that for the geographer as for other field-workers the ground, not the map, is the primary document ... anyone of scientific mind or conversant with received standards of scholarship will tend to view with suspicion work, however excellent, uninformed by knowledge of the actual ground' (Wooldridge and East, 1951, p. 162). Wooldridge, a very distinguished geomorphologist, played a formative role in the establishment of the Council for the Promotion of Field Studies (subsequently the Field Studies Council), to provide field centres for school parties. He thought that a fundamental role of fieldwork was to build up the power to read a piece of country, something he termed 'an eye for country' (Wooldridge, 1955, p. 78; Stoddart, 1986, p. 144). Also in London University was Sir L. D. Stamp, who, as early as the 1930s, brought generations of people, some 22,000 in total, into the field as part of his great Land Utilisation Survey (Board, 1965).

3.4 Geoscience Fieldwork Today

As remarked at the start of this chapter, while today there is an increasing weight placed upon such modern technologies as computer modelling, laboratory simulation and remote sensing, fieldwork continues to be a major approach to the subject and a fundamental source of many types of inspiration and of data. This is made evident when, for example, one considers the many field projects that have been organised by the RGS (with the Institute of British Geographers). Between 1977 and 2001, it ran 11 major multi-disciplinary overseas field research programmes, and those with a major geoscience component included: the International Karakoram Project (Pakistan), 1980 (Figure 3.7); the Jordan Badia Programme, 1992–1996; the Kimberley Research Project, 1988; the Kora Research Project (Kenya), 1983; the Maracá Rainforest Project (Brazil),

Figure 3.7 Professor David Jones (left) and Professor Denys Brunsden (right), both former students and colleagues of S. W. Wooldridge, undertaking a plane table survey of the Minapin Valley at ca. 2,500 m above sea level during the International Karakoram Project, 1980 (Andrew Goudie)

1987–1988; the Mulu Sarawak Expedition, 1977–1978; the Nepal Middle Hills Projects, 1991–1998; the Oman Wahiba Sands Project, 1985–1987; and the Shoals of Capricorn (Indian Ocean), 1998–2001. As examples of the work undertaken, the geoscience projects during the International Karakoram Project included studies of earthquake-resistant structures, the mapping of glacier retreat, the measurement of screes, the determination of solute and sediment loads in the Hunza River, mass movements and other hazards along the Karakoram Highway and the salt weathering of moraines (Miller, 1984). The geoscience components of the Kimberley Research Project included the study of Quaternary dune systems, the reconstruction of past flooding in the area's gorges, the analysis of karstic phenomena, the classification of tufas, the description of aligned gilgai and the study of cave deposits (Goudie *et al.*, 1990). Although such projects are currently in abeyance, the Society continues to recognise the great value of fieldwork and exploration, providing advice and many grants (www.rgs.org). Its Thesiger-Oman Fellowships, for example, have supported much research in the geosciences in arid and semi-arid lands.

Fieldwork has also been facilitated by the establishment of research stations, field universities and the like. For example, after World War II, a number of countries established research stations in their desert regions. These permitted long-term monitoring and provided bases for sustained investigations. Such stations included those at Bardai in the central Sahara, Fowlers Gap in New South Wales (Australia), Gobabeb in the Namib, Jodphur in India, the Jornada Experimental Range in New Mexico, Sidi Boqer in the Negev, the Taklamakan Desert Research Station in north-west China, the Desert Institute of Turkmenistan and the Walnut Gulch Experimental watershed in Arizona.

The value of fieldwork has been strengthened rather than weakened by the development of new techniques. Since the review by Goudie (1990), an explosion of techniques has occurred, both allowing and requiring fieldwork: better field measurements (e.g. global positioning system (GPS) and data loggers); surveying of landform distribution and morphometry (e.g. remote sensing, lidar, geographic information systems (GIS), unmanned aerial vehicles (UAVs)); geophysical techniques to allow a two- or even three-dimensional view of the materials and structures which make up the landscape (e.g. ground-penetrating radar and resistivity surveys); studies of geochemical properties (e.g. portable X-ray fluorescence (XRF)); assessment of rock hardness (e.g. by the Schmidt hammer and the Equotip); absolute dating of landforms and deposits by isotopic means, optically stimulated luminescence (OSL), cosmogenic nuclides, and thermochronology; experimentation under real field conditions (using such devices as portable wind tunnels and rainfall simulators); and environmental reconstruction (especially by means of ice, lake bog and ocean cores). Fieldwork has also been facilitated by the ready availability of flights to many parts of the world, and by the development of four-wheel drive and all-terrain vehicles. Nonetheless, as Burt and McDonnell (2015) indicated in their review of fieldwork in hydrology, there are concerns about its relative significance in comparison with modelling approaches.

Furthermore, to end on an upbeat note, notwithstanding financial stringencies, geologists and physical geographers still expect their students to undertake fieldwork (Figure 3.8) as part of their training and in order to produce dissertations, and most of the main US geology departments still run field camps (http://geology.com/field-camp.shtml). Field trips are a key aspect of undergraduate degrees in geology in the UK. Fieldwork will normally be carried out once or twice per year during undergraduate degrees for anything from a single day to over two weeks

Figure 3.8 Oxford geography students undertaking fieldwork in the Henry Mountains, Utah, USA, in 1987 (Tim Burt)

at a time (www.geolsoc.org.uk/Geology-Career-Pathways/University/During-your-degree/Fieldwork). Field trips and fieldwork are also a required component of many school geography programmes in the UK (http://geography.org.uk/resources/fieldwork/), and are a named part of the National Curriculum. The edited work of Thornbush *et al.* (2014) provides a resource for students and teachers planning geomorphological fieldwork.

References

Adams, F. D. (1938). *The Birth and Development of the Geological Sciences*. Dover Publications, New York, NY.

Agassiz, L. (1840). *Discours de Neuchâtel: Etudes sur les Glaciers*. Neuchâtel, privately published.

Andrews, R. (1927). Explorations in Mongolia: a review of the Central Asiatic Expeditions of the American Museum of Natural History. *Geographical Journal* 69, 1–19.

Bagnold, R. A. (1941). *The Physics of Blown Sand and Desert Dunes*. Methuen, London.

Baker, V. R. and Pyne, S. (1978). G. K. Gilbert and modern geomorphology. *American Journal of Science* 278, 97–123.

Board, C. (1965). Field work in geography, with particular emphasis on the role of land-use survey. In *Frontiers in Geographical Teaching*, eds. R. J. Chorley and P. E. Haggett. Methuen, London, pp. 186–214.

Burt, T. P. and McDonnell, J. J. (2015). Whither field hydrology? The need for discovery science and outrageous hydrological hypotheses. *Water Resources Research* 51, 5919–5928.

Chorley, R. J., Dunn, A. J. and Beckinsale, R. P. (1964). *The History of the Study of Landforms, Volume 1*. Methuen, London.

Chorley, R. J., Beckinsale, R. P. and Dunn, A. J. (1973). *The History of the Study of Landforms, Volume 2: The Life and Work of William Morris Davis*. Methuen, London.

Darwin, C. R. (1842). *The Structure and Distribution of Coral Islands*. Smith Elder, London.

Davis, W. M. (1899). The Geographical Cycle. *Geographical Journal* 14, 481–504.

Davis, W. M. (1913). Dana's confirmation of Darwin's theory of coral reefs. *American Journal of Science* 35, 173–188.

Galton, F. (1872). *The Art of Travel or Shifts and Contrivances Available in Wild Countries*. John Murray, London.

Geikie, A. (1882). *Geological Sketches at Home and Abroad*. Macmillan, London.

Geikie, A. (1905). *The Founders of Geology*. Macmillan, London.

Geikie, J. (1874). *The Great Ice Age and its Relations to the Antiquity of Man*. Appleton, New York, NY.

Gilbert, G. K. (1877). *Report on the Geology of the Henry Mountains*. US Geological and Geographical Survey, Washington, DC.

Gladfelter, E. H. (2002). *Agassiz's Legacy*. Oxford University Press, New York, NY.

Goudie, A. S. (1978). Colonel Julian Jackson and his contribution to geography. *Geographical Journal* 144, 264–270.

Goudie, A. S., ed. (1990). *Geomorphological Techniques*, 2nd edn. Unwin Hyman, London.

Goudie, A. S. (1999). The history of desert dune studies over the last 100 years. In *Aeolian Environments, Sediments and Landforms*, eds. A. S. Goudie, I. Livingstone and S. Stokes. Wiley, Chichester, pp. 1–13.

Goudie, A. S. (2008a). Aeolian processes and forms. In *The History of the Study of Landforms or the Development of Geomorphology, Volume 4: Quaternary and Recent Processes and Forms (1890–1965) and the Mid-Century Revolutions*, eds. T. P. Burt, R. J. Chorley, D. Brunsden, N. J. Cox and A. S. Goudie. Geological Society, London, pp. 767–804.

Goudie, A. S. (2008b). *Wheels across the Desert. Exploration of the Libyan Desert by Motor Car*. Silphium Press, London.

Goudie, A. S. (2016) *Great Desert Explorers*. Silphium Press, London.

Goudie, A., Viles, H., Allison, R., *et al.* (1990). The geomorphology of the Napier Range, Western Australia. *Transactions of the Institute of British Geographers* 15, 308–322.

Hart, J. K. (2007). The role of women in British Quaternary science. *Geological Society, London, Special Publications* 281, 83–95.

Huntington, E. (1907). *The Pulse of Asia*. Houghton Mifflin, New York, NY.

Imbrie, J. and Imbrie, K. P. (1979). *Ice Ages: Solving the Mystery*. Macmillan, London.

Ireland, H. A. (1943). History of the development of geologic maps. *Geological Society of America Bulletin* 54, 1227–1280.

Jackson, J. (1841). *What to Observe or, the Traveller's Remembrancer.* James Madden, London.

Jackson, P. W. (2007). *Four Centuries of Geological Travel: The Search for Knowledge on Foot, Bicycle, Sledge and Camel.* Geological Society of London, London.

Jones, W. D. and Sauer, C. O. (1915). Outline for field work in geography. *Bulletin of the American Geographical Society* 47, 520–525.

Kirchner, J. W. (2013). The physics and chemistry of Earth's dynamic surface (Ralph Alger Bagnold Medal Lecture). *EGU General Assembly Conference Abstracts* 15, 14258.

Lewis, W. V., ed. (1960). *Norwegian Cirque Glaciers.* Royal Geographical Society, London.

Marsh, G. P. (1864). *Man and Nature.* Sampson Low, Son and Marston, London.

Martin, G. J. (2005). *All Possible Worlds: A History of Geographical Ideas*, 4th edn. Oxford University Press, New York, NY.

Mathewson, K. (2001) Between "in camp" and "out of bounds": notes on the history of fieldwork in American geography. *Geographical Review* 91, 215–224.

Mathewson, K. (2006). Alexander von Humboldt's image and influence in North American geography, 1804–2004. *Geographical Review* 96, 416–438.

Miller, K. J., ed. (1984). *The International Karakoram Project.* Cambridge University Press, Cambridge.

Oldroyd, D. (1999). Early ideas about glaciation in the English Lake District: the problem of making sense of glaciation in a glaciated region. *Annals of Science* 56, 175–203.

Orme, A. R. (2004). American geomorphology at the dawn of the 20th century. *Physical Geography* 25, 361–381.

Platt, R. S. (1959). *Field Study in American Geography: The Development of Theory and Method Exemplified by Selections.* University of Chicago, Department of Geography, Chicago, IL.

Powell, J. W. (1875). *Exploration of the Colorado River of the West and its Tributaries.* US Government Printing Office, Washington, DC.

Rhoads, B. L. and Thorn, C. E. (1996). Observation in geomorphology. In *The Scientific Nature of Geomorphology*, eds. B. L. Rhoads and C. E. Thorn. John Wiley & Sons, Chichester, pp. 21–56.

Sack, D. (2004). Experiences and viewpoints of selected women geomorphologists from the mid-20th century. *Physical Geography* 25, 438–452.

Sack, D. (2013). Geomorphology and nineteenth century explorations of the American West. In *Treatise on Geomorphology, Volume 1*, ed. J. Shroder. Academic Press, San Diego, CA, pp. 53–63.

Slaymaker, O. (2004). Dynamic geomorphology. In *Encyclopedia of Geomorphology*, ed. A. S Goudie. Routledge, London, pp. 307–310.

Spencer, T., Stoddart, D. R. and McLean, R. F. (2008). Coral reefs. In *The History of the Study of Landforms or the Development of Geomorphology. Volume 4: Quaternary and Recent Processes and Forms (1890–1965) and the Mid-Century Revolutions*, eds. T. P. Burt, R. J. Chorley, D. Brunsden, N. J. Cox and A. S. Goudie. Geological Society, London, pp. 863–922.

Starrs, P. F. (2015). In memoriam "To Restore One's Faith in the Subject We Profess": the Geography of David R. Stoddart (1937–2014). *Geographical Review* 105, 612–619.

Steel, R. W., ed. (1987). *British Geography 1918–1945*. Cambridge University Press, Cambridge.

Steers, J. A. (1987). Physical geography in the universities, 1918–1945. *British Geography 1918–1945*, ed. R. W Steel. Cambridge University Press, Cambridge, pp. 138–155.

Stoddart, D. R. (1986). *On Geography*. Blackwell, Oxford.

Stoddart, D. R. (1997). Carl Sauer: geomorphologist. In *Process and Form in Geomorphology*, ed. D. R. Stoddart. Routledge, London, pp. 340–379.

Thornbush, M. J., Allen, C. D. and Fitzpatrick, F. A, eds. (2014). *Geomorphological Fieldwork*. Elsevier, Amsterdam.

von Richthofen, F. (1882). On the mode of origin of the loess. *Geological Magazine* 9, 293–305.

Wooldridge, S. W. (1955). The status of geography and the role of field work: address to the Geographical Association. *Geography* 40, 73–83.

Wooldridge, S. W. and East, W. G. (1951). *The Spirit and Purpose of Geography*. Hutchinson & Co, London.

Worster, D. (2001). *A River Running West: The Life of John Wesley Powell*. Oxford University Press, New York, NY.

Wulf, A. (2015). *The Invention of Nature: The Adventures of Alexander von Humboldt, the Lost Hero of Science*. John Murray, London.

4 · *Pioneering Fieldwork Heroes in the Life Sciences*

STEPHEN TRUDGILL

4.1 Fieldwork and Seeing What Is There

Fieldwork in biology is the study of organisms in their environment. It involves the observation of the organisms themselves, their interactions with each other and with their environment, together with the study of spatial patterns and changes over time. Measurement and experimentation may also be involved.

Inevitably, behind these activities there are questions of 'making sense of' what the investigator is seeing. As discussed by Trudgill (2003) in a reflection on fieldwork, these questions can include searches for pattern, explanations and meanings together with ideas about the classification of organisms, their characteristics and behaviour. However, such endeavours should always be thought about carefully. First, the methods of observation, especially the timing, frequency and distributions of observations, can influence the outcome of the investigation. Different approaches can give different answers. Second, while the type of methodology chosen is necessarily relevant to, and appropriate for, the questions posed, the ideas of the investigator can act to limit the selection of methodology. As this may then limit the range of possible answers, it is usually very useful to use more than one methodology. Third, the whole way the investigator is studying, looking, seeing, questioning and finding answers can be limited by the prior knowledge of the investigator. While it is not a biological example, it is nonetheless salutary to note that Charles Darwin (1887, p. 11) visited Cwm Idwal in the mountains of north Wales in 1831, and did not notice any glacial features at all. However, having learnt about glacial landscapes by the time of his second visit (Darwin, 1842), he found the interpretation of the area as a glacial landscape 'inescapable' (Trudgill, 2012a).

The fieldworker should always be open to new ideas, not only because knowledge can influence observation but also because the history of science shows us that progress is made when an observation challenges prior knowledge. Confirmation bias, or picking out observations which confirm our knowledge, is all too easy. We should always be thinking about what might disprove our ideas, as much as anything. Because we so readily confirm what we think we already know, we should actually be especially interested in phenomena which do not fit into our current thinking. Such phenomena may be the very things which lead us forward to new insights.

To me, then, fieldwork heroes are those who not only go out and see things for themselves but who also ask thoughtful questions about what they think they are seeing.

4.2 Early Foundations

The earliest humans were in a sense 'fieldworkers' – looking for sources of food and danger while learning from their experience. These appreciations related to animals, crop growing and human well-being, with plants having medicinal as well as food value. What might be seen as less immediate concerns did, however, develop alongside such practical approaches, with studies of patterns and process burgeoning from the seventeenth century onwards, and very often developing from some foundational ideas in antiquity.

In ancient Greece, **Aristotle** (ca. 384–322 BC) moved on from the conceptual teachings of his mentor Plato and made many practical observations of his own. These involved the close inspection of organisms and what we might call fieldwork: studying organisms as they exist in nature. Making many observations on animal form and function, as well as suggesting systems of classification, he laid the foundation for the mindset which seeks the causes of natural phenomena, as recorded in his *De Animalibus* series, notably *Historia Animalium*. Many of his primary observations in the field were made on the island of Lesvos (Lesbos), as recorded in a documentary 'Aristotle's Lagoon' by biologist Armand Leroi (BBC Four, 17 January 2011; see also Leroi, 2014).

Similarly, **Theophrastus** (ca. 372–287 BC) produced volumes on the then current knowledge of plants, coupled with his own observations. His books described the parts of plants and their reproduction together with types of plants and their useful products; many of the names he gave to plants survive in the scientific names in use today. **Hippocrates** (ca. 460–370 BC) made progress by challenging existing ideas through observation: he realised that diseases were not caused by the gods but had tangible physical causes. However, the idea championed at the time of the Roman Empire by the Greek physician **Galen** (AD 129 to ca. AD 200/ca. 216) of the 'humours' (black bile, yellow bile, blood and phlegm) went unchallenged in Western medical thought for over 1,000 years, into medieval times. Some ideas are more difficult to update than others.

It is interesting to reflect on the power of oral tradition and the written word in perpetuating ideas, such as Galen's 'humours'. The existence of such received knowledge means that later generations can learn about animals and plants without recourse to their own fieldwork. This represents both an efficiency in amassing information and contains the possibility of a more enlightened interpretation of phenomena in the field, but it also contains an element of passing on of information which may be of limited applicability. This is not so much of a problem if each generation tests the ideas of their predecessors.

Thus, I reflect on the fundamental necessity of fieldwork in its broadest sense – going out and testing existing or new ideas by observation, measurement and field experimentation. It is equally as important as the promulgation of ideas in print and laboratory work. We should not forget that in this sense learning is like a three-legged stool – learning from books and teachers, learning from laboratory work and learning from fieldwork are equal partners – and without fieldwork as an ultimate test of how things actually work in nature, we are less secure in what we think we know and understand.

Even so, we must be ever alert to what the great French naturalist **Georges-Louis Leclerc** (**Comte de Buffon**), 1707–1788) wrote:

> This is the most delicate and most important point in scientific studies, to know how to distinguish well what there is of the real in a subject and that which we add to it arbitrarily as we consider it; to recognise clearly which properties belong to the subject and which properties we only imagine it to have. [de Buffon, 1749]

In this context, it is interesting that beliefs can have a validity almost irrespective of any evidence for or against them. Here, it is useful to think about some of the early discussions of the migration of birds. The observation of swallows flying low over water in the autumn and then disappearing over winter led to a hibernation theory. This even included a search for dormant birds asleep in the mud of ponds or elsewhere in winter. Indeed, in 1769, the famous early naturalist **Gilbert White** (1720–1793) refers to this in his account of *The Natural History and Antiquities of Selborne* and he spent some considerable time in the field trying to find swallows hibernating (White, 1769; see, for example, the edition by Scott, 1962, pp. 46–48).

4.3 Early Fieldworkers: Local Study and Global Biogeography

John Ray (1627–1705), when a teacher at the University of Cambridge, undertook fieldwork to get to know the plants of his local area. He catalogued the local flora, producing his first book: a flora of Cambridgeshire (*Catalogus Plantarum circa Cantabrigiam nascentium*; Ray, 1660). He went on to become interested in anatomy and classification, and he published *Synoposis Methodical Animalum Quadripedum* (1693) and *Historia Insectorum* (Raven, 1942). **Giovanni Maria Lancisi** (1654–1720) was physician to Pope Innocent XI and made a field study in 1714 of the changes of the coastal vegetation in the vicinity of Rome (Pignatti and Ubrizsy Savoia, 1989). Regional field studies were also developed. For example, **Charles Flahault** (1852–1935) initiated a botanical and forest map of France, discussing the vegetation of the country as a whole in his 'Introduction' to Coste's *Flore descriptive et illustrée de la France* (Coste, 1901).

Berry (1987) wrote in his review of scientific natural history, published in the *Journal of the Linnaean Society*, that part of the nineteenth-century local study in the UK stemmed from the requirement of trainee doctors to have knowledge of local flora and not just the generalised medicinal knowledge of plants, as hitherto fostered by monks and apothecaries. This sent the medical students out into the field:

> A knowledge of natural history was important to doctors in past generations. Two of the key figures in the procession of naturalists drawn up by Charles Raven [1947], William Turner and John Caius, were leading physicians of their time. A major factor in introducing biology into the university curriculum was the Apothecaries' Act of 1815 which required all medical students to know the British flora, and led to the introduction of field classes in botany. [Berry, 1987, p. 20]

By contrast, for many, fieldwork was synonymous with exploration and travelling widely. One of the most significant early fieldworkers was the German-born **Maria Sybilla Merian** (1647–1717), who was also an accomplished artist, recording her first-hand observations in considerable detail. In 1699, she travelled to Dutch Surinam and observed insects directly in the field. Merian depicted the life cycle of butterflies and moths from the egg to caterpillar larva, pupa/chrysalis and adult, and her beautifully illustrated *Metamorphosis Insectorum Surinamensis (Veranderingen der Surinaamsche Insecten)* was published in 1705 (Merian, 1705).

Much of the focus of the early fieldwork was biogeographical in the sense that it charted the distributions of plants and animals in space over the surface of the Earth rather than over time. This was especially true of field observations made during sea voyages to different parts of the globe. This and subsequent work became summarised in cartographic form by **Nicholas Polunin** (1960).

Alexander von Humboldt (1769–1859) was a pioneer in the study of the distribution of plants in relation to the physical environment, as presented in his *Geography of Plants* (von Humboldt, 1807). Andrea Wulf (2015), in her book, *The Invention of Nature*, sees Humboldt as 'the Lost Hero of Science'. She relates his meetings with **Goethe** (1749–1832) where they discussed natural history intensively. She also relates his influence on **Charles Darwin** (1809–1882) who was inspired by his writings and carried them with him on the voyage of the *Beagle*. **Ludwig Karl Schmarda** (1819–1908), an Austrian naturalist at the universities of Graz and Prague, also travelled widely in 1853–1857, and published *Die Geographische Verbreitung der Thiere* (*The Geographical Distribution of Animals*; Schmarda, 1853), proposing 21 zoogeographical regions. Through his fieldwork, **Alfred Russel Wallace** (1823–1913) identified the main zoogeographic regions of the world recognised today (Wallace, 1876). **Thomas Huxley** was also on the voyage of the *Rattlesnake* to Australia, in 1822, where classification and geographical distribution became very much part of field observation.

August Grisebach (1814–1879) was at the University of Göttingen, and was Director of its Botanical Garden from 1875. He published *Die Vegetation der Erde nach ihrer klimatischen Anordnung* (*The Earth's Vegetation after its Climatic ArrangementI*; Grisebach, 1872), which again saw plant distributions in relation to the physical environment. Other key works were those of **Andreas Schimper** (1856–1901): *Botanische Mitteilungen aus den Tropen* (*Botanical Communications from the Tropics*; Schimper, 1888) and *Pflanzengeographie auf physiologischer Grundlage* (*Plant Geography on a Physiological Basis*; Schimper, 1898). He is credited by Richards (1996, p. 1) with coining the term 'tropical rainforest' (tropische Regenwald) in Schimper (1898).

4.4 Appreciations of Change and Dynamic Ecological Interrelationships

Many early fieldwork observations were seen in the context of contemporaneous ideas about evolution. The French naturalist and zoologist **Georges Cuvier** (Jean Léopold Nicolas Frédéric, Baron Cuvier, 1769–1832) concluded that the fossils he found were either extinct or no different from the animals living today, as discussed by Coleman (1964). The fieldwork of **William Buckland** (1784–1856) led him to write *Geology and Mineralogy Considered with Reference to Natural Theology* (Buckland, 1836) where he could only interpret what he saw by proposing that there were several creations. However, zoologists like the Frenchman **Jean-Baptiste de Lamarck** (1744–1829) thought that that species changed gradually in response to their environment, passing on acquired characteristics, and Georges-Louis Leclerc de Buffon, who wrote *Histoire Naturelle* (de Buffon, 1949), also held that species were mutable.

The idea of changes over time steadily became more prevalent in fieldwork, especially through the work of the geologist **Charles Lyell** (1797–1875) who was instrumental in promoting the ideas of change and dynamism. The pioneer field geologist **William Smith** (1769–1839), who produced the first geological map of Britain in 1815, had wonderful fieldwork opportunities during the cutting of the canals – there were ready-made sections exposed as a record of how past conditions had changed during geological times. Smith, in his *Stratigraphical System of Organized Fossils* (first published in 1817) showed that certain strata each have a characteristic series of fossils (Smith, 2010). The Danish geologist **Johannes Japetus Smith Steenstrup** (1813–1897) studied deposits containing postglacial fossils and developed the idea that the changes he found evidence for, such as the succession from tundra vegetation to birch, pine, oak and beech, formed a record of changing climate (Steenstrup, 1842).

As described by Worster (1994) in his *History of Ecological Ideas*, there were many key players in the refocusing of field ecology from the earlier biogeographical study of distributions in relation to environment to an understanding of changes, interrelationships and interactions between organisms.

In France, de Buffon (1742) recognised how the tree species composition changed over time as a forest developed and, later, **Dureau de la Malle** (1825; Trudgill, 2012c) also made a significant long-term study of forest changes in cut-over areas. In Sweden, **Biberg**, a student of the plant classifier **Carl Linnaeus**, made important observations on the colonisation of bare rock by plants (Biberg, 1749).

Oscar Drude (1852–1933) was Professor of Dresden Polytechnic and Director of Dresden Botanic Gardens from 1879. Tansley (1909, p. 223) considered Drude's writings on habitat in *Deutschlands Pflanzengeographie* (Drude, 1896) and his other works (Drude, 1890, 1912, 1913) as paving the way for this more dynamic study of plants.

Three people in particular, however, had the greatest influence in changing the perspective to a more dynamic one: Ernst Haeckel, Charles Darwin and Eugenius Warming (Trudgill, 2012b) as discussed further below.

It was **Ernst Haeckel** (1834–1919; Figure 4.1), a scientist much inspired by von Humboldt, who coined the word 'ecology'. Haeckel worked at the University of Jena in eastern Germany. As described in the review by Coleman (1986), he promoted the study of the interaction of organisms rather than of the simple deterministic relationship between organism and environment. Haeckel (1866) wrote:

We considered especially inorganic conditions, factors such as the influence of light, heat, moisture and inorganic nutrients. Far more important than these,

Figure 4.1 Ernst Haeckel (image from Wikipedia)

however, and exerting a much more powerful influence on the transformation and adaptation of species are organic conditions of existence, that is, the mutual relations of organisms to one another. [Coleman, 1986, p. 183]

Coleman writes how **Charles Darwin** also developed this idea:

That the interactions between all neighbouring organisms are extraordinarily important and that they exert far more influence on the change and adaptation of species than do the inorganic conditions of existence was first emphasised with due clarity by Darwin. [Coleman, 1986, p. 183]

Darwin was very aware of plant distributions, partly from his Cambridge mentor **John Henslow** (1796–1861) of Cambridge University Botanic Gardens (Walters and Stow, 2001) and partly though his friendship with the much-travelled **Joseph Hooker** (1817–1911), who became Director of the Royal Botanic Gardens, Kew, London. His ideas on ecological processes were, however, drawn much more from his first-hand observations, none less than his observations on earthworms. His book *Vegetable Mould and Earthworms* (Darwin, 1882) is a *tour de force* of field observations by him and many other naturalists. He appreciated that the soil was very much a living entity showing constant change through earthworm activity. This idea of a dynamic nature was a major contribution of Darwin, as Berry (1987, p. 27) notes: 'Darwin's great achievement was to change the focus of biologists from the "much reflection" of pattern to the processes that determine the pattern.' This Berry saw as occurring though observation and, writing on nature as a dynamic process rather than as a static, non-temporal mechanism. He wrote:

[nature as a dynamic process] is well illustrated by Charles Darwin's approach to biology. The fixed ideas of his upbringing were broken by the unplanned observations he made on the *Beagle*; his habit and passion throughout life was a natural history which could be opened up by experiment [Clark, 1984]. Evolution by natural selection was not constructed by a closet biologist or any form of theoretician, but by a naturalist who, in the words of his *Autobiography*: 'I had no great quickness of apprehension or wit which is so remarkable in some clever men, for instance, Huxley. I am therefore a poor critic; a paper or a book, when first read, generally excites my admiration, and it is only after considerable reflection that I perceive the weak points . . . On the positive side of the balance, I think that I am superior to the common run of men in noticing things which easily escape attention, and in observing them carefully. My industry has been as great as it could have been in the observation and collection of facts. What is far more important, my love of natural science has been steady and devout'. [Berry, 1987, p. 26]

Coleman shows how Darwin played a role in the early development of vegetation dynamics and also downplayed abiotic, environmental factors:

> In chapter 4 of *The Origin of Species*, Darwin [1860] provided a striking example of local ecological change. This was on a plot of Staffordshire heath where it is recorded how the enclosure and planting with larch (*Larix*) and fir (*Abies*) some 25 years earlier led to dramatic floral and faunal changes. The new forest brought in its wake several other species of plants and many new insects, and had even allowed oaks [*Quercus*] to germinate on the previously treeless moor. These changes Darwin attributed 'to the introduction of a tree [and] to no other change whatever'. [Coleman, 1986, p. 187]

Coleman felt that this provided a 'fine example of the power of biotic forces, species impinging upon species, other possible participatory factors being dealt a subsidiary role or no role at all'.

However, Coleman perhaps overstates the case and it was **Eugenius [Eugen] Warming** (1841–1924), a Dane from Copenhagen (Figure 4.2), who made a wealth of primary field observations and came up with greater insights on the dynamics of plant ecology and the mutual interaction of biotic and abiotic factors. Warming very much believed in field

Figure 4.2 Eugenius Warming. Photo ca. 1900, scanned from obituary in Botanisk Tidsskrift (Copenhagen), 1924.

Figure 4.3 A. G. Tansley, seated, right, with F. E. Clements, on the First International Phytogeography Excursion of the British Isles in 1911. (Reproduced by permission of the British Ecological Society.)

study and many of his ideas came from his own observations of the sand dunes, saltmarshes and wooded areas in western Jutland, Bornholm and Zealand. His classic studies of the development of Danish dune vegetation (Warming, 1891), and with the more widely read 1906 version (Warming, 1906), together with *Plantesamfund* (*Plant Society*; Warming, 1895), consolidated the findings from his fieldwork. The 1896 German translation of the 1895 book was widely read in England and America and played an important part in stimulating fieldwork in both countries.

The significance of Warming's work was recognised by **Sir Arthur Tansley** (1871–1955) (Figure 4.3), himself a key figure in the development of ecological ideas (Trudgill, 2007). Tansley wrote in his review of the early history of plant ecology (Tansley, 1947, p. 130) that the foundations of ecology were 'laid on the continent'. This is indeed well exemplified by Acot (1998) in the volume entitled *The European Origins of Scientific Ecology*. Tansley felt that:

In Britain, while there were a great many excellent observations on various aspects of the natural history of plants and animals, among which Darwin's work was outstanding, there was no systematic study of what we now call ecology. [Tansley, 1947, p. 130]

He recognised Warming's specific contribution to the way in which field ecology moved away from a static view of vegetation, which had become rather entrenched in Britain, to a more dynamic viewpoint. He wrote: '. . . we [in the UK] still tended, I think, to treat plant communities too statically' and 'The modern meaning of the word "ecology" was largely fixed by the publication of Warming's 1895 book' (Tansley, 1947, p. 137).

Coleman also paid tribute to Warming, who worked 30 years after Darwin, by writing:

Thirty years later Warming described a similar instance of what Darwin had called rotation, and what the Dane [Warming] and we today call plant succession. In ancient times Jutland had been covered by oak [*Quercus*] forest; the present dominant vegetation is heather (*Calluna*) and stands of beech [*Fagus*]. Only scattered groves of oak are [now] found. [Coleman, 1986, p. 187]

Clearly a change in vegetation had occurred, and Warming's fieldwork not only stressed biotic interaction but also the subtlety of the interaction with, and co-evolution of, abiotic factors involving the soil:

He attempted to articulate the causal factors behind the changes leading from oak to heath, noting the role of light and shade due to changing forest cover, the effects of humus formation and especially chemical change and loss, species replacement among low-rising plants of the forest floor, the advance of soil consolidation, and the disappearance of earthworms. Biotic factors were not neglected: Jutland beech and oak competed directly with each other, with the latter emerging at a distinct disadvantage. [Coleman, 1986, p. 187]

Thus, the chain of interaction was established with competition for light leading to species change, which in turn led to soil change, which led to further vegetation change. All of this knowledge came from first-hand fieldwork.

A contemporaneous review of the German edition of Warming's key book, which was provided by an author identified only by his initials (J. M. C., 1896), sums up the key aspects of plant dynamics in the book. The importance of changes and interactions evident from Warming's field observations are clear:

> The last section discusses the struggle between plant societies ... [including] changes in vegetation induced by slow changes in soil, changes in vegetation without changes in climate or soil. The general characteristics of vegetation occupying new soil are given as follows: the first vegetation is sparse and open; the number of species is small at first, then larger when the physiognomy is diverse, then comes an equilibrium and fewer species; annuals and biennials are common at first, but afterwards subordinated to perennials; the first species are those whose seeds are carried by winds and birds; light-trees appear before shade-trees; there is a gradual transition to former conditions, and thus there may be primal, transitional and final plant societies. [J. M. C., 1896, p. 175]

Warming's work was cited by the renowned American ecologist **Frederic Clements** (1874–1945) (Clements, 1916, pp. 23–24) and remains much cited today, for example in the book by Packham *et al.* (2001) on the ecology and geomorphology of coastal shingle (pp. 277–289). Warming's fieldwork thus marks a milestone in ecological study as he moved well beyond the spatial associations between organism and environment to the study of the importance of the functional and dynamic interactions between organisms, and between them and the environment.

Henry Chandler Cowles (1869–1939) (Figure 4.4) was a leading American figure in early ecological fieldwork, who developed the ideas of Warming. Cowles studied at Chicago University and became a populariser of the term 'ecology' in the USA (Sprugel, 1980; Cassidy, 2007). His 1898 PhD thesis on the vegetation of the sand dunes at Lake Michigan was directly inspired by the work of Eugenius Warming. His paper (Cowles, 1911) on vegetation cycles stressed the importance of Warming's pioneering work. Overall, Cowles felt that as well as Warming's work, the significance of the fieldwork of Dureau de la Malle (1825; Trudgill, 2012c) on forest change was often unacknowledged and he also wrote that the awareness of the subject of dynamism was then increased further with Treub's (1888) field study of the new vegetation of the volcanic island Krakatoa in Indonesia after its eruption in 1883.

Many other botanists contributed to changes in ideas and methods over time in history, and several of these are detailed in the admirable volume (Oliver, 1913) edited by **Francis Oliver** (1864–1951), botanist at University College London (1890–1925): *Makers of British Botany. A Collection of Biographies by Living Botanists.* Included are chapters on John Stevens Henslow (1796–1861), by George Henslow; Arthur Henfrey (1819–1859), by F. W. Oliver; William Henry Harvey

Figure 4.4 Henry Chandler Cowles in the Santa Catalina Mountains, Arizona, 1913. (Reproduced courtesy of the Peabody Museum of Natural History, Yale University, New Haven, Connecticut, USA.)

(1811–1866), by R. Lloyd Praeger; and Sir Joseph Dalton Hooker (1817–1911), by F. O. Bower.

This early pioneering work and the growing appreciation of ecological dynamics led on to the quantitative ecological work, such as seen in *Quantitative Plant Ecology* by Peter Greig-Smith (1957) and Charles Krebs's *Ecology: The Experimental Analysis of Distribution and Abundance*

(Krebs, 1972) and his *Ecological Methodology* (Krebs, 1989). Quite when the quadrat methodology for the quantitative study of vegetation was developed seems difficult to establish clearly, but it was certainly in use in the early 1900s as described by Weaver (1918) and such methods were discussed by Oliver and Tansley (1904).

4.5 Interaction and Behavioural Ecology

We have seen that the ideas of change and mutability increasingly became major topics of discussion for many field scientists. Indeed, the early search for order and classification, together with a passion for collecting and cataloguing, became superseded by the study of process, which then dominated fieldwork. Here, the studies of the interaction between organisms as well as between organism and environment burgeoned as time went on, including the study of 'phytosociology' and plant communities and, in zoology, the study of animal behaviour, or ethology.

Charles Otis Whitman (1842–1910) was an American zoologist, interested in animal behaviour; much of his practical work involved passenger pigeons. He was involved in the unsuccessful attempt to breed the last remaining individuals in Cincinnati Zoo, in 1902. **Karl Ritter von Frisch** (1886–1982) was an early ethologist from Austria, whose pioneering fieldwork initiated the understanding of the honey bees' 'waggle dance'. He observed how the abdomen is successively moved on returning to the hive, to communicate the location of nectar sources to other bees, as was described (von Frisch, 1927) in his book *Aus dem Leben der Bienen* (*The Dancing Bees*). **Oskar Heinroth** (1871–1945) was a German biologist who studied the behaviour of ducks and geese, including the phenomenon of imprinting (where hatchlings bond with the first moving object they see).

Heinroth's pupil, **Konrad Zacharias Lorenz** (1903–1989) followed this up in his well-known work with geese. He met Dutch biologist **Nikolaas Tinbergen** (1907–1988) in 1936. Tinbergen became a pivotal figure in behavioural ecology. He outlined four topics to address when studying animal behaviours: proximate causes, ontogeny, survival value and phylogeny of behaviour. In that they were to do with coming into being, growth, continued existence and evolution, these are not dissimilar to Aristotle's four causes: matter, form, agent and purpose.

A major body of Tinbergen's fieldwork focused on the 'supernormal stimulus', where he exaggerated markings, for example on eggs or models, to see which an animal would respond most to. One of his classic studies in animal behaviour was on the chicks of herring gull, *Larus argentatus*, as described in Tinbergen's (1953) *The Herring Gull's World*. He found which features of the adult beak induced chicks to peck in order to be fed by using a variety of cardboard models of adult herring gull heads in different shapes and colours. His work on the 'red patch' stimulus that elicits begging behaviour in young chicks is assessed in a modern perspective in a paper by Carel ten Cate (ten Cate, 2009) of Leiden University where Tinbergen was based. This paper shows what a debt present-day behavioural studies owe to such pioneers.

4.6 The British Natural History Tradition of Fieldwork

Present-day studies in field ecology likewise owe a considerable debt to the long tradition of natural history study in Britain. As Berry (1987) wrote:

> The British tradition of natural history is almost unique. A similar tradition is found in a somewhat weaker form in other countries of north-west Europe, and there are, of course, excellent naturalists in most countries of the world. But nowhere else than Britain is it so widespread and deep-rooted. Moreover, it is closely linked with sense of place, not a simple love of fauna and flora. [Berry, 1987, p. 17]

As my review in this chapter is necessarily selective, I refer the reader to David Elliston Allen's *The Naturalist in Britain* (Allen, 1976) and *Books and Naturalists* (Allen, 2010) for comprehensive overviews. Allen gives detailed accounts of the large number of people who have contributed to the British natural history tradition which 'influenced the present shape of field biology' (Berry, 1987, p. 18). There is, however, one person who stands out in the pantheon of British field naturalists and that is **Gilbert White** (1720–1793). Curate of Selborne, in Hampshire, he is often credited as a pioneering field ecologist. His many observations included the timing of the first appearance in the year of plants and animals together with their activities. His bird study helped to answer some questions about bird migration which were developed later in North America by **John James Audubon** (1785–1851), who devised a method of marking migratory birds by tying silver threads on to the legs of young birds, in 1803.

The county of Norfolk seemed to have been very well endowed with field naturalists, such as **Arthur Patterson** (1857–1935) and **Ted Ellis**

(1909–1986), of whom accounts are found in Patterson (1930) and Stone (1968). David Matless's book *In the Nature of Landscape* (Matless, 2014) provides a comprehensive account of them and many others. Growing up in Norfolk myself, it was Ted Ellis who taught me so much about field natural history.

William Gardner Smith (1866–1928) was a Scottish botanist and ecologist. His brother Robert had studied plant geography with Charles Flahault at the University of Montpellier and had started his own field-work on English plant geography, but he died in 1900. William took up his brother's work and published the first vegetation maps of England, with Charles Edward Moss and W. Munn Rankin, in 1903. In 1904, he, Tansley, Moss, Rankin and others established the Central Committee for the Survey and Study of British Vegetation, later renamed the British Vegetation Committee. In 1913, they organised the British Ecological Society, the first ever professional body of ecologists, and Smith served as president of the Society, in 1918 and 1919.

Edward Salisbury (1886–1978) was Director of the Royal Botanic Gardens, Kew, from 1943 to 1956. His fieldwork was focused on forest ecology, the ecology of garden weeds and dune plants.

William Pearsall (1891–1964) was Professor of Botany at University College London, from 1944 to 1957. He undertook fieldwork on aquatic plants in British lakes and published the well-known book *Mountains and Moorlands* (Pearsall, 1950).

One particular biological pioneer who stands out is **Charles Elton** (1900–1991). His fieldwork involved the study of population and community ecology, and also invasive organisms. In 1932, he established the Bureau of Animal Population at Oxford and became the first editor of the *Journal of Animal Ecology*. His book *Animal Ecology* was published in 1927, covering ecological studies of animal behaviour and life history, including food chains, ecological niches and the 'pyramid of numbers' describing feeding relationships in ecosystems (Elton, 1927). His book *The Ecology of Invasions by Animals* was first published in 1958 (Elton, 1958). After World War II, Elton started a 20-year survey of animals and their interrelationships on Oxford University's Wytham Estate, including animals in meadows, woods and water. The field ecology studies continue in Wytham Woods to this day (see also Chapter 1).

Frank Fraser Darling (1903–1979) was an ecologist and author who spent much of his life in Scotland and is known for the book *The Highlands and Islands* (Fraser Darling and Morton Boyd, 1964). His fieldwork involved studies of the social and breeding behaviour of

Figure 4.5 Frank Fraser Darling (courtesy of the Field Studies Council)

the red deer, gulls and the grey seal. Zoological fieldworkers also included **John Cloudsley-Thompson** (1921–2013), who studied desert zoology, and **Richard Southwood** (1931–2005), whose fieldwork involved the study of populations and ecosystems.

Edward Max Nicholson (1904–2003) was a significant figure in the establishment of important institutions which had the effect of fostering fieldwork by conserving habitats and wildlife. He was a founder of the World Wildlife Fund and was involved in the foundation of the British Trust for Ornithology and the International Union for the Protection of Nature (IUPN) (now International Union for Conservation of Nature (IUCN)), with Julian Huxley. He also steered The National Parks and Access to the Countryside Act 1949, which established the Nature Conservancy with National Nature Reserves and Sites of Special Scientific Interest (SSSI). He helped to set up Monks Wood Experimental Station in the early 1960s, which became a bastion and beacon of field science until 2009, when the endeavour was run down, though the woodland does remain a focus for field study.

4.7 Birds

Birds hold a special place in field natural history, as the study by Peter Bircham, *A History of Ornithology*, shows (Bircham, 2007).

Edward Jenner (1749–1823) is a fascinating character – and in many ways a classic combination of physician and field naturalist. He is well known as a pioneer of the smallpox vaccine and immunology in general and, as such, his work is often said to have saved more lives than that of any other human. He also made extensive and detailed field observations in zoology and he was the first person to describe the behaviour of the cuckoo in laying its egg in the nest of a host and the subsequent ousting of the host eggs or young by the hatchling cuckoo (Bircham, 2007, pp. 113–116).

The field observations by Jenner laid the foundation for later fieldwork, for example that by Professor Nick Davies of the University of Cambridge's Zoology Department on egg mimicry at Wicken Fen. Davies was inspired by the work on Niko Tinbergen and has a good summary of what field research means to him: 'I get most of my ideas by watching animals and simply asking "I wonder why they're doing that?" The key to research is coming up with a good question and devising an experiment to answer it' (see www.cam.ac.uk/research/features/the-reed-warbler-and-the-cuckoo-an-escalating-game-of-trickery-and-defence)

In the early days there was a lot of shooting and stuffing, with specimens being placed in museums. However, there were people like **Florence Merriam Bailey** (1863–1948), who was a field ornithologist who became one of the earliest advocates for the protection of wildlife in the USA. She studied birds in nature, focusing on their behaviours in the field. She wrote her first book *Birds through an Opera-Glass* aged 26, considered one of the first modern field guides for bird-watching since it included both notes on behaviour and illustrations (Bailey, 1899). Her later books continue to influence field guides to this day, and some people still consider them the standard because of their detailed entries. She and many other female pioneers of field natural history are celebrated by Lew (2017).

David Lack (1910–1973) also studied the living bird in the field. He became Director of the Edward Grey Institute of Field Ornithology at Oxford University (1945–1973). He visited the Galapagos Islands in 1938, and published two version of *Darwin's Finches*, the later one (Lack, 1947) with revised ideas. Another well-known book is his *The Life of the Robin* (Lack, 1943).

A person who described how to do field surveys of birds in woodlands was **William Brunsdon Yapp** (1909–1990). He was a zoologist, who lectured at the University of Birmingham. His methodology was one of walking round the perimeter of a designated area, noting the locations of individual bird songs within it, as described in his book *Birds and Woods* (Yapp, 1962).

4.8 Marine Biology

Fieldwork in marine biology holds a special fascination for many, especially the revelations of the rocky shore at low tide.

Philip Henry Gosse (1810–1888) was an innovator in the study of marine biology, well known for promoting the use of aquaria for viewing and study. His field studies resulted in the publication of *British Sea Anemones and Corals* (Gosse, 1860). He also held contentious views on creation and evolution; he is remembered in his son Edmund's account: *Father and Son* (Gosse, 1907).

Sir C. Maurice Yonge (1899–1986) was leader of the Great Barrier Reef Expedition, 1928–1929. His *The Sea Shore* (Yonge, 1949) and especially the *Collins Pocket Guide to the Sea Shore* (Barrett and Yonge, 1958) has greatly assisted many a marine biological fieldworker. **John Barrett** (1913–1999) was a naturalist and the first warden at Dale Fort field centre.

Silent Spring by **Rachel Carson** (1907–1964) has become an iconic book (Carson, 1962). Her lesser known book *The Sea around Us* (Carson, 1951) covers her observations and thoughts on marine biology.

The Woods Hole Oceanographic Institution, founded in 1930, was a signal development in marine biology, as charted by Dexter (1988) in the paper, 'History of American marine biology and marine biology institutions. Introduction: origins of American marine biology' and by Benson (2002) in his paper, 'The study of vertical zonation on rocky intertidal shores: a historical perspective'. Benson charts how many early marine biologists attempted to determine the physical features that could explain the distribution of intertidal communities, much like early ecologists had done in botany. These included **Addison E. Verrill** (1839–1926) of Yale University, who surveyed the biology of the shore near where Woods Hole came to be built, and **Victor Shelford** (1877–1968), who published many papers on marine biota (for example Shelford, 1925, 1929, 1935). However, just as we saw with botanical fieldwork above, the appreciation of the dynamic interaction between organisms prevailed, and so the same kind of poised understanding was reached that, while physical factors were significant, it was species interaction which was a key factor. Benson concluded:

> . . . it is important to understand how early notions of marine communities were initially influenced by investigators closely tied to the botanical biases of physiological ecology, especially in the work of Clements, Cowles, and Shelford. A new direction was promised with the self-conscious formation of marine

ecology within the context of the new marine laboratories, which soon called for more comprehensive natural history investigations of marine communities. These studies, especially along the US west coast in the 1930s revealed the inadequacy of both the botanical approach and laboratory-based experiments. By the Second World War, several marine ecologists suggested that physical factors, so popular with earlier studies, were important only as limiting factors for the marine fauna and flora. The intriguing zonation patterns that attracted the attention of biologists in the first place, were the product of biotic factors. [Benson, 2002, p. 779]

The pioneering marine biological fieldwork of **Robert 'Bob' Paine** (1933–2016) was foundational to the development of ecology as a whole. Working mostly from the University of Washington, Paine (1966, 1969) developed the idea of a 'keystone' species. The removal of a top keystone predator can lead to a complete change in a marine ecosystem. An oft-cited study showed how the removal of the purple sea star *Pisaster ochraceus* converted a diverse rocky shore to a uniform mussel bed. The importance of keystone species has subsequently been emotively misrepresented as causing the 'collapse' of an ecosystem but in *The Importance of Species: Perspectives on Expendability and Triage* the editors (Kareiva and Levin, 2003) express the concept which Paine developed admirably:

It is now widely appreciated that certain keystone species exert effects on communities or ecosystems that are disproportionately large relative to the species' abundance. Thus in some cases, real communities are strongly altered by the loss of a single species but in other cases not and the magnitude of the effects appears to depend upon the specific traits of the deleted species or on the nuances of how the deleted species interacted with others in the community, rather than on the changes in the numbers per se. [Kareiva and Levin, 2003, p. 158]

This description is fundamental: sometimes a species is removed and there is no perceptible difference, other times another can step in to fulfil the same role, other times there can be a marked change. The functional interaction between the species is crucial. This sums up the whole tenor of this chapter in that we have seen the way in which fieldwork has gone from the study of an association between environment and organism to elucidating the importance of the interaction between species.

4.9 Pioneers of Field Education

Louis Agassiz (1807–1873) was a former student of Cuvier. He founded the Museum of Comparative Zoology, at Harvard University, and published his pioneering *Nomenclator Zoologicus* (Agassiz, 1847). He exhorted students to 'study nature, not books'.

Anna Botsford Comstock (1854–1930) was an American pioneer of nature education in the field. According to Lew (2017):

> Anyone who enjoyed taking nature field trips in school owes a debt of gratitude to Anna Botsford Comstock (1854–1930). [She] pushed for outdoor education in public schools in New York after seeing how well her students at Cornell University – where she was the institution's first female professor – responded to spending class time seeing the subjects of their study in their natural environment.

She published several successful books, including the *Handbook of Nature Study*, which had more than 20 printings. Later, another important field manual was produced by Vinson Brown (1948): *The Amateur Naturalist's Handbook*.

The development of the Field Studies Council (FSC) by pioneering field ecologists is well described in Berry and Crothers (1987). There, John Barrett (1987) describes **Francis Butler**'s vision in setting up the FSC, at first called the Council for the Promotion of Field Studies (CPFS). He describes how Butler, an inspector of science in London schools, who lamented the lack of contact of children with the countryside, met with a local general practitioner **Eric Ennion** (1900–1981), a keen field naturalist. Ennion introduced Butler to Cambridge biologists, including **Harry Godwin** (1901–1985), field botanist at Clare College and renowned for his work on peatland history and vegetation, **J. T. Saunders**, zoologist at Christ's College, and **Sir Arthur Tansley**, recently retired from Oxford and then living in Grantchester.

Tansley galvanised the formation of the CPFS, and the first meeting was held in the British Museum (Natural History) on 10 December 1943 – note the timing, in the middle of World War II. Many pioneering field scientists rapidly became involved: **C. Maurice Yonge**, **Edward Salisbury** and **William Pearsall** we have already mentioned; **S. W. Wooldridge** is cited in the companion chapter in this book by Andrew Goudie. Also involved were: **A. S. Watt** (1892–1985), who specialised in plant community dynamics; **A. R. Clapham** (1904–1990), renowned for his work on the flora of the British Isles; **P. W. Richards** (1908–1995), renowned for his work on the tropical rainforest; **C. T. Ingold** (1905–2010), renowned for his work in mycology; and **E. J. Fritsch** (1879–1954), a biologist specialising in algae. Eric Ennion became the first warden (i.e. head) of the very first FSC field centre at Flatford Mill.

Figure 4.6 Francis Butler with his wife Eleanor, *ca.* 1967 (courtesy of the Field Studies Council)

Tansley published a pamphlet in 1951 (reproduced as Tansley, 1987) where he listed the contributions of many of the field scientists we have discussed above. Any field scientist in the UK today owes this pioneering group a great debt – and also to the like of those who came after them.

Notably, we remember **R. J. 'Sam' Berry** (1934–2018). He was Professor of Genetics at University College London between 1974 and 2000, president, from 1983 to 1986, of the Linnean Society, the British Ecological Society and the European Ecological Federation. He did such a great deal to foster fieldwork through his long association with the FSC. Alas, Sam died before he could contribute to this book by writing this very chapter.

For myself, I count it a great privilege to have been associated with the FSC for many years (www.field-studies-council.org/about/fsc-stories/1960s/1966-steve-trudgill.aspx). I so relished being surrounded by those who inspired generations of people, young and old, sharing a mutual enthusiasm for nature (see www.field-studies-council.org/about/fsc-stories.aspx). The key characteristic of all the FSC people is the vision of promoting first-hand personal experience. There is a sense of wonder which is engendered by seeing, experiencing and understanding nature

Figure 4.7 Eric Ennion, a pioneering colossus of fieldwork and the FSC (courtesy of the Field Studies Council)

Figure 4.8 FSC wardens in November 1961. Left to right: John Sankey, Charles Sinker, Ian Mercer, Jim Bingley, John Barrett, Paul Holmes, Geoffrey Hutchings. (Photo by Dierdre Williams. Reproduced by Permission of The Linnean Society and OUP.)

for yourself, a point I made when writing on 'the fieldwork experience' (Trudgill, 2003, p. 37) and as championed by Rachel Carson (1965) in her book: *The Sense of Wonder* (published posthumously from her 1956 text).

I cannot but end with such an expression quoted from the book by **John Steinbeck** (1902–1968): *The Log from the Sea of Cortez*. Here, the writer, more famed for his *Grapes of Wrath*, wrote about marine biology, stemming from his association with marine scientist **Ed Ricketts** (1897–1948) on the California Coast (see Steinbeck and Ricketts, 1941, and later Steinbeck editions). Ricketts is well known for his co-authored book, *Between Pacific Tides*, a pioneering study of intertidal ecology (Ricketts and Calvin, 1939; see also Tamm, 2004 and Lannoo, 2010). Steinbeck takes up the theme of a man who 'has strength and energy of mind' when looking at a tide pool, can see it is both the microcosm and the macrocosm: 'digs back to electrons and leaps space into the universe' in a timeless way: 'Then ecology has a synonym which is ALL (Steinbeck, 1960, p. 146)'. He continues this theme later in the book:

> What we know as life meets and enters what we think of as non-life: barnacle and rock, rock and earth, earth and tree, tree and rain and air. And the units nestle in the whole and are inseparable from it . . . And it is a strange thing that most of the feeling we call religious, most of the mystical outcrying which is one of the most prised and used and desired reactions of our species, is really the understanding and the attempt to say that man is related to the whole thing, related inextricably to all reality, known and unknowable. This is a simple thing to say, but the profound feeling of it made a Jesus, a St Augustine, a St Francis, a Roger Bacon, a Charles Darwin, and an Einstein. Each of them in his own tempo and with his own voice discovered and reaffirmed with astonishment the knowledge that all things are one thing and that one thing is all things – plankton, a shimmering phosphorescence on the sea and the spinning planets and the expanding universe, all bound together by the elastic string of time. It is advisable to look from the tide pool to the stars and then back to the tide pool again. [Steinbeck and Ricketts, 1941, p. 178]

There is much wisdom in Steinbeck's writings. He is my fieldwork hero.

References

Acot, P., ed. (1998). *The European Origins of Scientific Ecology 1800–1901*. Overseas Publishers Association, Amsterdam.

Agassiz, L. (1847). *Nomenclator zoologicus: continens nomina systematica generum animalium tam viventium quam fossilium, secundum ordinem alphabeticum disposita, adjectis auctoribus, libris, in quibus reperiuntur, anno editionis, etymologia et familiis, ad quas pertinent, in singulis classibus. Soloduri: Jent et Gassmann.* See www.biodiversitylibrary.org/bibliography/49761#/summary.

Allen, D. E. (1976). *The Naturalist in Britain. A Social History*. Princeton University Press, Princeton, NJ.

Allen, D. E. (2010). *Books and Naturalists*. HarperCollins, London.

Bailey, F. M. (1899). *Birds through an Opera-Glass*. The Chautauqua Press, New York, NY.

Barrett, J. (1987). The Field Studies Council: how it all began. In *Nature, Natural History and Ecology*, eds. R. J. Berry and J. Crothers. Linnean Society/Academic Press for the Field Studies Council, London, pp. 31–41. Reprinted from *Biological Journal of the Linnean Society* 32.

Barrett, J. H. and Yonge, C. M. (1958). *Collins Pocket Guide to the Sea Shore*. Collins, London.

Benson, K. R. (2002). The study of vertical zonation on rocky intertidal shores: a historical perspective. *Integrative and Comparative Biology* 42, 776–779; DOI: 10.1093/icb/42.4.776.

Berry, R. J. (1987). Scientific natural history: a key base to ecology. In *Nature, Natural History and Ecology*, eds. R. J. Berry and J. Crothers. Linnean Society/Academic Press for the Field Studies Council, London, pp. 17–29. Reprinted from *Biological Journal of the Linnean Society* 32.

Berry, R. J. and Crothers, J. (1987). *Nature, Natural History and Ecology*. Linnean Society/Academic Press for the Field Studies Council, London. Reprinted from *Biological Journal of the Linnean Society* 32.

Biberg, I. J. (1749). Oeconomia Naturae. *Amoenitates Academicae* 2, 1–52 (cited from Clements, 1916).

Bircham, P. (2007). *A History of Ornithology*. HarperCollins, London.

Brown, V. (1948). *Amateur Naturalist's Handbook, Illustrated by Don Greame Kelly*. Bramhall House, New York, NY.

Buckland, W. (1836). *Geology and Mineralogy Considered with Reference to Natural Theology*. W. Pickering, London.

Carson, R. (1951). *The Sea around Us*. Oxford University Press, Oxford.

Carson, R. (1962). *Silent Spring*. Houghton Mifflin. New York, NY.

Carson, R. (1965). *The Sense of Wonder*. Harper and Row, New York, NY (published posthumously from her 1956 text, published in *Woman's Home Companion*, entitled 'Help your Child to Wonder').

Cassidy, V. M. (2007). *Henry Chandler Cowles: Pioneer Ecologist*. Kedzie Sigel Press, Chicago, IL.

Clark, R. W. (1984). *The Survival of Charles Darwin*. Weidenfeld & Nicolson, London.

Clements, F. E. (1916). *Plant Succession: An Analysis of the Development of Vegetation*. Carnegie Institute, Washington, DC.

Coleman, W. (1964). *Georges Cuvier, Zoologist: A Study in the History of Evolution Theory*. Harvard University Press, Cambridge, MA.

Coleman, W. (1986). Evolution into ecology? The strategy of Warming's ecological plant geography. *Journal of the History of Biology* 19, 181–196.

Coste, H. (1901). *Flore descriptive et illustrée de la France*. Librairie des Sciences et des Arts, Paris.

Cowles, H. C. (1911). The causes of vegetational cycles. *Annals of the Association of American Geographers* 1, 3–20.

Darwin, C. (1842). Notes on the effects produced by the ancient glaciers of Caernarvonshire, and on the boulders transported by floating ice. *The London, Edinburgh and Dublin Philosophical Magazine* 21, 180–188.

Darwin, C. (1860). *On the Origin of Species by Means of Natural Selection*, 2nd edn. John Murray, London.

Darwin, C. (1882). *The Formation of Vegetable Mould through the Action of Worms with Observations on Their Habits.* John Murray, London.

Darwin, C. (1887). Autobiography. In *The Life and Letters of Charles Darwin*, ed. F. Darwin. See www.victorianweb.org/science/darwin/darwin_autobiography.html (accessed July 2019).

de Buffon, G.-L. L. (1742). *Mémoire sur la culture des forêts.* l'Académie Royale des Sciences, Paris (cited from Cowles, 1911).

de Buffon, G.-L. L. (1749). *Premier discours: de la manière d'étudier et de traiter l'Histoire Naturelle.* De L'Imprimerie Royale, Paris. (Part 1 of the series *Histoire Naturelle* 1749–1804.)

Dexter, R. W. (1988). History of American marine biology and marine biology institutions. Introduction: origins of American marine biology. *American Zoologist* 28, 3–6.

Drude, O. (1890). *Handbuch der Pflanzengeographie.* J. Engelhorn, Stuttgart.

Drude, O. (1896). *Deutschlands Pflanzengeographie. Ein geographisches Charakterbild der Flora von Deutschland und den angrenzenden Alpen sowie Karpathenländern.* J. Engelhorn, Stuttgart.

Drude, O. (1912). The international phytogeographical excursion in the British Isles. IX. The flora of Great Britain compared with that of Central Europe. *New Phytologist* 11, 236–255.

Drude, O. (1913). *Die Ökologie der Pflanzen.* Vieweg, Braunschweig.

Dureau de la Malle, A. J. C. A. (1825). Mémoire sur l'alternance ou sur ce problème: la succession alternative dans la réproduction des espèces végétales vivant en société, est-elle une loi générale de la nature? *Annales des Sciences Naturelles, Botaniques* 5, 353–381.

Elton, C. (1927). *Animal Ecology.* Macmillan, New York, NY.

Elton, C. (1958). *The Ecology of Invasions by Animals.* University of Chicago Press, Chicago, IL.

Fraser Darling, F. and Morton Boyd, J. (1964). *The Highlands and Islands.* Collins, London.

Gosse, E. (1907). *Father and Son.* Heinemann, London.

Gosse, P. (1860). *British Sea Anemones and Corals.* Van Voorst, London.

Greig-Smith, P. (1957). *Quantitative Plant Ecology.* Butterworth's Scientific Publications, London.

Grisebach, A. H. R. (1872). *Die Vegetation der Erde nach ihrer klimatischen Anordnung (*1884 second edition). Wilhelm Engelmann, Leipzig.

Haeckel, E. (1866). *Generelle Morphologie der Organismen II.* Georg Reimer, Berlin.

J. M. C. (1896). Review: Warming's plant geography. *Botanical Gazette* 22, 173–175. See www.jstor.org/stable/2464035.

Kareiva, P. and Levin, S. A., eds. (2003). *The Importance of Species: Perspectives on Expendability and Triage*. Princeton University Press, Princeton, NJ.

Krebs, C. J. (1972). *Ecology. The Experimental Analysis of Distribution and Abundance.* Harper and Row, New York, NY.

Krebs, C. J. (1989). *Ecological Methodology*. Harper and Row, New York, NY.

Lack, D. (1943). *The Life of the Robin*. Witherby, London.

Lack, D. (1947). *Darwin's Finches*. Cambridge University Press, Cambridge.

Lannoo, M. J. (2010). *Leopold's Shack and Rickett's Lab: The Emergence of Environmentalism*. University of California Press, Berkeley, CA.

Leroi, A. M. (2014). *The Lagoon: How Aristotle Invented Science*. Bloomsbury, London. See also www.youtube.com/watch?v=e12pbSHrzAs.

Lew, J. (2017). 10 women who changed the way we see nature. See www.mnn.com/earth-matters/wilderness-resources/stories/women-who-changed-way-we-see-nature.

Matless, D. (2014). *In the Nature of Landscape. Cultural Geography on the Norfolk Broads*. Wiley Blackwell, London.

Merian, S. (1705). *Metamorphosis Insectorum Surinamensis (Veranderingen der Surinaamsche Insecten)*. G. Valck, Amsterdam.

Oliver, F. W. (1913). *Makers of British Botany. A Collection of Biographies by Living Botanists*. Cambridge University Press, Cambridge.

Oliver, F. W. and Tansley, A. G. (1904). Methods of surveying vegetation on a large scale. *New Phytologist* 3, 228–237.

Packham, J. R., Randall, R. E., Barnes, R. S. K. and Neal, A., eds. (2001). *Ecology and Geomorphology of Coastal Shingle*. Westbury Academic and Scientific Publishing, Otley.

Paine, R. T. (1966). Food web complexity and species diversity. *The American Naturalist* 100, 65–75.

Paine, R. T. (1969). A note on trophic complexity and community stability. *The American Naturalist* 103, 91–93.

Patterson, A. (1930). *A Norfolk Naturalist. Observations on Birds, Mammals and Fishes*. Methuen, London.

Pearsall, W. H. (1950). *Mountains and Moorlands*. Collins, London.

Pignatti, S. and Ubrizsy Savoia, A. (1989). Early use of the succession concept by G. M. Lancisi in 1714. *Vegetatio* 84, 113–115.

Polunin, N. (1960). *Introduction to Plant Geography and Some Related Sciences*. McGraw-Hill, New York, NY.

Raven, C. E. (1942). *John Ray, Naturalist*. Cambridge University Press, Cambridge.

Raven, C. E. (1947). *English Naturalists from Neckam to Ray*. Cambridge University Press, Cambridge.

Ray, J. (1660). *Catalogus Plantarum circa Cantabrigiam nascentium*. Cambridge University Press, Cambridge.

Richards, P. W. (1996). *The Tropical Rain Forest*, 2nd edn. Cambridge University Press, Cambridge.

Ricketts, E. F. and Calvin, J. (1939). *Between Pacific Tides*. Stanford University Press, Palo Alto, CA.

Schimper, A. F. W. (1888). *Botanische Mitteilungen aus den Tropen*. G. Fischer, Jena.

Schimper, A. F. W. (1898). *Pflanzengeographie auf physiologischer Grundlage.* G. Fischer, Jena.

Schmarda, L. K. (1853). *Die Geographische Verbreitung der Thiere.* C. Gerold und Sohn, Vienna.

Shelford, V. E. (1925). Animal communities of the San Juan Channel and adjacent areas. *Publications of the Puget Sound Biological Station* 5, 3773.

Shelford, V. E. (1929). Geographical extent and succession in Pacific North American intertidal barnacles (*Balanus*) communities. *Publications of the Puget Sound Biological Station* 7, 217–223.

Shelford, V. E. (1935). Some marine biotic communities of the Pacific coast of North America. *Ecological Monographs* 5, 251–354.

Smith, W. (2010). *Stratigraphical System of Organized Fossils.* Cambridge University Press, New York, NY. First published in 1817. https://doi.org/10.1017/CBO9780511702167.

Sprugel, A. (1980). 'Pedagogical genealogy' of American plant ecologists. *Bulletin of the Ecological Society of America* 61, 197–200.

Steenstrup, J. J. S. (1842). Geognostisk-geologisk Undersøgelse af Skovmoserne Vidnesdam og Lillemose i det nordlige Sjælland, ledsaget af sammenlignende Bemærkninger hentede fra Danmarks Skov-, Kjær og Lyngmoser i Almindelighed. *Kongelige Danske Videnskabernes Selskabs Afhandlinger* 9, 17–120.

Steinbeck, J. and Ricketts, E. F. (1941). *The Log from the Sea of Cortez.* Penguin Publishing, New York, NY.

Stone, E. (1968). *Ted Ellis: The People's Naturalist.* Jarrold Colour Publications, Norwich.

Tamm, E. E. (2004). *Beyond the Outer Shores: The Untold Odyssey of Ed Ricketts, the Pioneering Ecologist Who Inspired John Steinbeck and Joseph Campbell.* Thunder's Mouth Press, New York, NY.

Tansley, A. G. (as AGT) (1909). Review: [untitled]; Reviewed work: The Oecology of Plants and Introduction to the Study of Plant Communities by Eug. Warming. *New Phytologist* 8, 218–227.

Tansley, A. G. (1947). The early history of modern plant ecology in Britain. *Journal of Ecology* 35, 130–137.

Tansley, A. G. (1987). What is ecology? In *Nature, Natural History and Ecology*, eds. R. J. Berry and J. Crothers. Linnean Society/Academic Press for the Field Studies Council, London, pp. 5–16. Reprinted from *Biological Journal of the Linnean Society* 32. Originally published in 1951 as a pamphlet produced for the Council for the Promotion of Field Studies, which became the Field Studies Council.

ten Cate, C. J. (2009). Essay. Niko Tinbergen and the red patch on the herring gull's beak. *Animal Behaviour* 77, 785–794. https://doi.org/10.1016/j.anbehav.2008.12.021.

Tinbergen, N. (1953). *The Herring Gull's World.* Collins, London.

Treub, M. (1888). Notice sur la nouvelle flore de Krakatau. *Annales du Jardin Botanique de Buitenzorg* 7, 213–223.

Trudgill, S. T. (2003). Meaning, knowledge and constructs and fieldwork in physical geography. In *Contemporary Meanings in Physical Geography: From What to Why?*, eds. S. T. Trudgill and A. Roy. Arnold, London, ch. 2.

Trudgill, S. T. (2007). Classics in physical geography revisited. In Tansley, A. G. (1935) The Use and Abuse of Vegetational Concepts and Terms. Ecology 16(3): 284–307. *Progress in Physical Geography* 31, 501–507.

Trudgill, S. T. (2012a). Do theories tell us what to see? The 19th-century observations of Darwin, Ramsay and Bonney on glacial features. *Progress in Physical Geography* 36, 558–566.

Trudgill, S. T. (2012b). The interactions between all neighbouring organisms: the roles of Charles Darwin, Ernst Haeckel and Eugenius Warming in the evolution of ideas on plant dynamics. *Progress in Physical Geography* 36, 853–861.

Trudgill, S. T. (2012c) Nature's clothing and spontaneous generation? The observations of Thoreau and Dureau de la Malle on plant succession. *Progress in Physical Geography* 36, 707–714.

von Frisch, K. R. (1927). *Aus dem Leben der Bienen*. Springer, Berlin.

von Humboldt, A. (1807). *Ideen zu einer Geographie der Pflanzen nebst einem Naturgemälde der Tropenländer. Mit einer Kupfertafel.* Cotta/F. Schoell, Tübingen/Paris.

Wallace, A. R. (1876). *The Geographical Distribution of Animals*. Macmillan, London.

Walters, S. M. and Stow, E. A. (2001). *Darwin's Mentor: John Stevens Henslow. 1796–1861.* Cambridge University Press, Cambridge.

Warming, E. (1891). Botaniske Exkursioner 2. De psammophile Formationer I Danmark. *Videnskabelige Meddelelser fra den Naturhistoriske Forening I Kjøbenhavn* 1891, 153–202.

Warming, E. (1895). *Plantesamfund: Grundtraek af den ökologiske Plantegeografi.* P.G. Philipsen. Copenhagen. Translated into German by Emil Knoblauch (1896): *Lehrbuch der ökologischen Pflanzengeographie: Eine Einführung in die Kenntniss der Pflanzenvereine.* Ciebruder Borntraeger, Berlin.

Warming, E. (1906). Cited in Cowles (1911): Botaniske Exkursioner 2. De psammophile Formationer I Danmark (Botanical Excursions 2. Psammophilous formations in Denmark). *Videnskabelige Meddelelser fra den Naturhistoriske forening i Kjøbenhavn (Scientific Communications from the Natural History Society of Copenhagen)* 1891, 153–202. More widely cited as the republished book: *Dansk Plantevækst. 1. Strandvegetationen (Danish Plant-Life: Beach Vegetation)*. Gyldendal Bookshop Nordic Publishers, Copenhagen and Kristiania (Oslo).

Weaver, J. E. (1918). The quadrat method in teaching ecology. *The Plant World* 21, 267–283.

White, G. (1789). *The Natural History and Antiquities of Selborne.* Cassell & Company, London.

Worster, D. (1994). *Nature's Economy: A History of Ecological Ideas*, 2nd edn. Cambridge University Press, Cambridge.

Wulf, A. (2015). *The Invention of Nature: The Adventures of Alexander von Humboldt, the Lost Hero of Science.* John Murray, London.

Yapp, W. B. (1962). *Birds and Woods*. Oxford University Press, London.

Yonge, C. M. (1949). *The Sea Shore*. Collins, London.

5 · *The Educational Benefits of Out-of-Classroom Learning*

MICHAEL J. REISS

5.1 Locating Fieldwork

Fieldwork can be defined as any activity that involves leaving the school classroom, university lecture hall, laboratory or the confines of one's own home and getting outside, for some purpose connected to learning. As with any definition, there are uncertainties at the edges. For example, what precisely is meant by 'outside'? Can going round a large aquarium (when one is under a roof) count as fieldwork? I am not too bothered about providing a watertight definition; this is not a situation where the uncertainties are that important. Watching a computer simulation of river pollution, however well designed and educationally valuable, is not an example of fieldwork (though used before, after or even concurrently with fieldwork it may well help with learning). Determining the distribution of plants on a moorland (biology) is, as are mapping rock types in a location (physical geography or earth science), undertaking a survey of traffic in a city centre (human geography, sociology or social studies), asking people on a beach about their use of sunscreen (psychology) or surveying grave stones in a churchyard (history). To all intents and purposes, therefore, fieldwork can be equated with out-of-classroom learning, though the term is most often used in a way that implies being in the natural environment – without wanting to get into a long discussion about what is meant by 'natural', which I am not equating with 'wilderness'.

The first-hand experience of phenomena (Boyle *et al.*, 2007) has a long and firmly established place in British geography education (Cook, 2011; Lambert and Reiss, 2014). As Alastair Bonnett puts it, 'Geography wants to take children outside the schools and into the streets and fields . . . and into the rain or the sunshine' (Bonnett, 2008, p. 80). In the sciences, too, fieldwork plays a major role, though this has always been more the case for biology and geology than for chemistry and physics (Braund and Reiss, 2004). Fieldwork also plays an important role in astronomy. It is difficult, for example, to imagine a worthwhile astronomy qualification for 14–16- and 16–19-year-olds that did not give a central place to observation of the night sky, whether with the naked eye, through hands-on telescopes or by the use of robotic telescopes.

5.2 The Relationship between Fieldwork and Laboratory Work

Fieldwork is sometimes likened to laboratory work such that 'the field' for geographers is the equivalent of 'the lab' for scientists. However, this is completely to misunderstand the purpose of both laboratory work and fieldwork. As I will discuss in more detail below, one of the most important educational reasons for undertaking fieldwork in geography or the sciences is that such work provides authentic case studies of the world in action (e.g. Figure 5.1). A laboratory, however, provides a very interesting, and often crucial, space that differs fundamentally from that of 'the real world'.

The point of a laboratory is that it allows us to do two things: first, to control the circumstances under which events take place; and second, to control certain variables. Consider, for instance, something as humdrum as the temperature at which liquid water freezes. In the field, one rarely, perhaps never, finds 'pure water'. In a laboratory, water of very high purity is relatively straightforward to procure and research. In the field, one faces such everyday problems as a suitable vantage point from which to insert one's thermometer into the water, and then there are problems with the wind, with the temperature (which one cannot really control) failing to move from above 0 °C to below it, etc. In a laboratory, of course, even one no more sophisticated than that found in every school for 11–16-year-olds, one can easily control and simplify matters so that the freezing point of water can be measured.

Without wishing to labour the point too much, one then finds in a laboratory some interesting phenomena – such as the ability of liquid water to exist in a supercooled state at temperatures as low as –48 °C. The question then arises as to whether such phenomena are merely there to satisfy the curiosities of laboratory scientists or whether they have relevance in day-to-day life. This is where fieldwork comes in. In the

Figure 5.1 Students learning about human geography at the Olympic Stadium in East London. (Photograph provided by the Field Studies Council.) A black and white version of this figure will appear in some formats. For the colour version, please refer to the plate section.

case of supercooling, it turns out that the phenomenon is important; it not infrequently occurs, for example, in certain types of clouds. This is one of the reasons why aircraft need effective de-icing systems, as an aircraft flying through a cloud with supercooled water provides the nucleation sites that cause the water rapidly to freeze – potentially disastrous for an aircraft.

So, there can be a valuable relationship between fieldwork and laboratory work in which they both combine, in turn, to provide a better understanding of the phenomenon that one is studying.

5.3 The Reasons Why Fieldwork Is a Key Part of Learning for Subjects Like Geography and Biology

There are a number of reasons why we might expect that fieldwork would be a key part of learning for subjects like geography and biology. For a start, there is the fact that, in the field, reality does not present itself as clearly as it does in the laboratory (as discussed in the previous section), in one's textbook or via one's teacher. This is not to criticise textbooks and teachers any more than I have criticised laboratories. A good textbook and a good teacher present concepts in an order that helps the learner to make sense of them, to learn effectively and reasonably rapidly, without going up too many blind alleys. In the field, of course, while these concepts generally still apply, they are less evident – so that one has to work harder to see them.

To give an everyday example, a textbook may give the impression that it is relatively easy to undertake a geological mapping exercise once one has learnt the fundamentals. However, once in the field, almost everyone finds that the reality is far harder than they supposed! Vegetation gets in the way, bedrocks are hard to discern, suddenly the differences between textures and colours do not seem so clear-cut – and then it starts to rain. The same is true in ecology. Patterns of succession that look crystal clear when described in a textbook, a PowerPoint or on a whiteboard become far less apparent when one is out among sand dunes with the wind blowing and plants apparently appearing in all the wrong places.

Related to this is the fact that fieldwork requires the learner to move over space and over time. The importance of this depends on what one is studying. For example, if one is looking in geography at the concepts of place and space, the reality of walking from a town centre out towards the suburbs, and appreciating the reasons for the changes in the use of space, can substantially help a learner to appreciate these concepts. Similarly, the experience of mapping the distribution of seaweeds and molluscs on a rocky shore as the tide retreats requires several hours of work and is the more memorable for that. Of course, simulations and more convincing instances of virtual reality can mimic such experiences but most advocates of fieldwork will always hold that there is something significant about an actual town centre or an actual sand dune that software packages cannot replicate.

This leads on to a related point, namely that the place-based education that fieldwork entails can lead to shifts in how learners value their environments. Fieldwork, therefore, can promote an ethic of care and lead to changes in how learners not only understand but appreciate the world.

5.4 Types of Fieldwork

Sadly, many students studying the sciences in the 14–19 age range never get out of their classrooms; thankfully, this is less the case for students studying geography (Tilling, 2018). However, within biology, for example, while it is possible to teach ecology very well in a school laboratory, the subject is hugely enhanced by complementing such activity outside of the classroom, in the field. At its most modest, this may consist simply of mapping and identifying a few species of plants using quadrats on a school lawn. However, even in a city, ecology undertaken on school grounds, or within a very short distance of them, can be made more exciting and intellectually stimulating. Travelling further afield, while it carries the risk of giving the message to students that ecology can only really be undertaken in distant, exotic locations, has a number of advantages. For one thing, it tends to mean that learning is not restricted to a single or double period, thus opening up a much greater range of learning possibilities. Then there is the fact that this allows students to develop such character traits as persistence and patience, and such 'soft skills' as working in groups over more sustained periods than typically required in schools.

Given the examination-oriented nature of much fieldwork, one could argue that the focus on 'what has been learnt' dominates other possible questions about the benefits of fieldwork and out-of-classroom experiences (Amos and Reiss, 2012). Falk and Dierking (2000) challenged the 'what has been learnt?' mindset in a museum context. They asserted that such a question is fundamentally flawed and that a more useful approach would involve discovering how an out-of-classroom experience *contributes* to what a person 'knows, feels, believes or is capable of doing' (p. 13). As a result of their research, Falk and Dierking developed a contextual model, which puts forward the view that all learning takes place within three interlinked contexts: the personal, the socio-cultural and the physical. After 10 years of working with this model, they concluded that a fourth dimension – time – played an essential part in the learning process, arguing that to look at any learning experience as a 'snapshot' (p. 10) in time was not adequate for a full understanding of how learners make meaning.

These benefits, of course, are magnified by residential fieldwork. In a series of studies that a colleague, Ruth Amos, and I conducted on the possible benefits of residential fieldwork, we built on an influential review of international research into the impacts and benefits of fieldwork undertaken by Rickinson *et al.* (2004). Not infrequently, fieldwork researchers have taken a rather narrow view of the definition of learning, with the acquisition of facts and skills related directly to school curricula being the main research focus (e.g. Lai, 1999). Other studies examining outcomes during adventure-based outdoor learning have centred

Figure 5.2 Fieldwork can have benefits for cognitive, affective, social/ interpersonal and physical/behavioural development (Tom Stamp).

primarily on students' abilities to solve problems and to develop teamwork skills. Rickinson *et al.*'s (2004) meta-analysis, based upon 150 studies published in English from 1993 to 2003, provides a valuable overview. Four key developmental domains were defined to categorise multiple impacts from different kinds of outdoor learning: those pertaining to cognitive, affective, social/interpersonal and physical/behavioural development (Figure 5.2).

The London Challenge Fieldwork Programme

In 2003, the London Challenge initiative (a government-funded inner-city project) pledged to provide a fully-funded residential experience for 11–14-year-old state secondary school students from socially deprived areas in London. The courses were of three types:

- curriculum (traditional fieldwork, in ecology or geology, for example);
- adventure;
- combined curriculum–adventure.

Figure 5.3 Students studying ecology on a rocky shore. (Photograph provided by the Field Studies Council.)

All courses offered a range of structured or semi-structured activities. Curriculum activities began and ended with indoor classroom sessions, whereas students were briefed outside during adventure activities. Three types of science curriculum course were available: seashore ecology (e.g. Figure 5.3); seashore ecology with geology; and animal and plant ecology. Adventure courses included climbing, abseiling, canoeing or hill walking. On combined courses, students most commonly climbed, abseiled, sea-level traversed and built rafts or shelters; these activities were then linked to, or interspersed with, ecology and geology fieldwork. Typically, a school took 25–30 students on a field course.

Initially, the programme addressed the perceived needs of students aged 11–14 in London boroughs with high levels of socio-economic deprivation. After three years, the programme was opened to all state secondary schools in the 33 London boroughs. In the programme's midst, the UK government launched its 'Learning outside the Classroom Manifesto' (Department for Children, Schools and Families, 2006), which constituted the first commitment of its kind to put learning outside the classroom at the heart of the school curriculum. The evaluation of London Challenge that Ruth Amos and I undertook therefore acquired a greater significance.

The study involved a mixed-methods approach in which data from questionnaire surveys, interviews, samples of work, student journals and observations of field courses were used to capture evidence of impact. Approximately 33,000 11–14-year-olds from 849 schools took part in the London Challenge initiative between 2004 and 2008. Our sample included 2,706 participating students from 46 state secondary schools, involved in 77 field courses, their teachers ($n = 70$) and parents ($n = 869$). Courses took place at 20 field centres across the UK. We evaluated 13 science, 10 'other' curriculum (for example, geography or literacy), 27 adventure and 27 combined curriculum–adventure courses.

Analysis of data from surveys, interviews and field notes confirmed that students, teachers and parents were all most highly focused on expectations for, and impacts in, the affective and social/interpersonal domains. Historically, this has been a common finding even for out-of-classroom day courses (Eshach, 2007; Stocklmayer *et al.*, 2010). The small number of research studies reporting specific benefits associated with residential curriculum fieldwork, all situated in geographical contexts (for example, Nundy, 1999), suggest that a residential setting affords students and teachers the opportunity to establish productive living and learning environments. Ruth Amos and I observed similar outcomes during curriculum courses; the establishment of 'working' trust between students who would not normally spend prolonged time with one another in school (where social tensions ran high):

> At the start of the week, some people weren't listening; they were taking over and doing their own thing. But then, we realised the first thing you need to do is choose a leader and everyone needs to take turns. [Students]

Planned collaborative activities prior to curriculum tasks paved the way for effective teamwork:

> The afternoon before, students played team games such as the spider's web [each student in the team had to go through a segment in the web, without touching it, with the help of team members]; then they did a scavenger hunt round the centre where they had to decide who collected what in a certain time – both activities were great for getting them to see that relying on each other made them more effective as a team. Next day, I was amazed at the levels of cooperation during the investigation; much better than at school. [Teacher]

Activities such as climbing and abseiling, during which students readily recognised the implicit need for trust and teamwork, also fostered cooperative approaches to subsequent curriculum tasks. Positive physical impacts resulted from challenging activities such as climbing and hiking.

Emergent expectations and impacts in the four Rickinson *et al.* developmental domains are shown in Table 5.1, in rank order of importance, from the perspectives of students, teachers and parents, respectively. The findings are generated from pre- and post-questionnaire data for students and parents, by frequency analysis; and from pre- and post-interviews for students and teachers, by coding transcripts.

One of the conclusions that can be drawn from Table 5.1 is that while parents and, to a lesser extent, teachers both expected before the courses that benefits would include cognitive ones (though less than social/interpersonal ones), after the course both parents and teachers felt that the benefits had been greater in other domains than they had expected.

Fieldwork Can Allow for Independent Research Projects

One of the educational benefits that fieldwork can provide results when students undertake independent research projects. Such projects, whether undertaken in the classroom or in the field, are a feature of school geography and science in a number of countries. They may be taken individually (as they often are when assessment criteria so dictate) or in pairs or small groups. Their central feature is that the outcomes of the projects are open in that neither the students nor their teacher know in advance exactly what the investigations will yield (Bennett *et al.*, 2018). One of the reasons why one might want students to undertake such projects is that they can help them to appreciate the complexities of undertaking an investigation from start to finish; furthermore, an investigation where they have the prime responsibility for identifying the problem to be tackled and the methods to be used.

Fieldwork provides an excellent opportunity for students to undertake independent research projects and feed their findings back to their peers. A systematic review of the literature on independent research projects in school science has recently been undertaken (Bennett *et al.*, 2018). Thirty-nine papers met the review inclusion criteria, reporting on work from 12 countries. The review indicated that such projects are often associated with wider initiatives, such as authentic science, problem-based learning and project-based learning. The majority of the research reviewed explored areas such as conceptual understanding, motivation to study science once it is no longer compulsory, attitudes to science and the development of practical skills. Benefits were identified in relation to the learning of science ideas, affective responses to science, views of pursuing careers involving science and the development of a range

Table 5.1 *Ranked importance of pre-course expectations for, and post-course reported gains from, fieldwork experiences within the four Rickinson* et al. *(2004) developmental domains (1 = highest; 8 = lowest)*

Developmental domain	Expectation/ reported gain	Student pre-	Student post-	Teacher pre-	Teacher post-	Parent pre-	Parent post-
Cognitive	Learning new ideas	–	–	5	8	2	3
Affective	Having fun	4	4	–	–	6	–
	Building self-esteem	–	6	3	5	–	–
	Building self-confidence	–	5	–	4	4	5
	Developing independence	–	7		6	5	4
	Improving motivation to learn	–	–	4	7	–	–
	Seeing new places	5	8	6	–	–	–
Social/interpersonal	Improving cooperation	6	–	7	2	–	–
	Improving teamwork	–	3	1	1	3	1
	Building relationships	1	1	2	3	1	2
	Being away with friends	3	2	–	–	–	–
Physical/behavioural	Trying (a variety of) new activities	2	1	–	–	–	6

of skills. Studies focusing on traditionally under-represented groups indicated that such students felt more positive about science as a result of undertaking independent research projects. These findings are encouraging. One imagines that they are likely to be replicated in geography.

5.5 Ensuring the Right Circumstances Are in Place to Encourage Fieldwork

The educational benefits of fieldwork are most likely to be realised if there are curricula that stipulate that such fieldwork be undertaken, if teachers/lecturers are confident in undertaking fieldwork and are supported by their institutions in taking their students on fieldwork and if terminal assessment regimes in some way reward students for undertaking fieldwork.

In England, a number of authors have examined why fieldwork is not undertaken more than it is. The most recent analysis is provided by Tilling (2018). Tilling shows that advanced-level biology specifications (for 16–19-year-olds) contained slightly less ecology in 2017 than they did in 1987 and slightly less ecology in 1987 than they did in 1957. In addition, the amount of time required or recommended for fieldwork was *much* less in 2017 than in 1987 which, in turn, was slightly less than in 1957.

Of course, such changes to specifications are the result of wider cultural shifts. In biology, there has been a move towards more molecular and cell biology in recent decades; today's biology graduates are more likely to have little knowledge of ecology and no experience of fieldwork compared to the case a generation ago. When I look back on my undergraduate education (which included some physics and mathematics as well as biology), we spent many days out in the field and were expected to have a solid knowledge not only of many standard UK ecosystems (terrestrial, freshwater and marine) but also of the role of humans in history, in affecting the British landscape.

Indeed, 20 years ago, I wrote the following:

> I bought recently at a church fête a second-hand edition of *The Ladybird Book of the Seashore and Seashore Life* authored by Nancy Scott. It is an excellent book, stuffed full of the best type of natural history. For example, it has an entire double page spread on the periwinkles. This not only tells the reader how to distinguish small, rough, common and flat periwinkles, but gives a certain amount of detail about their feeding ecology and reproductive biology and

provides an excellent colour drawing which indicates the intraspecific variation in colour and shape one finds in most periwinkle species. Nancy Scott's book was published in 1964. Nowadays I suspect most teachers would expect only 'A' level students and undergraduates to be tackling this level of detail. [Reiss, 1998, p. 21]

My point was not to complain about a decline in standards – as I noted in the same article, *Jurassic Park* had more molecular biology in it than most undergraduate biology texts did in 1964. Rather, it was simply to illustrate the shift in our expectations of the biology that school children should know. School students increasingly have little or no knowledge about their immediate biological environment even though they may have a better understanding than their predecessors did of global biology. Many 16-year-olds leave school unable to identify a red admiral – or even to know that it is a butterfly – yet can tell one quite a bit about global climate change.

More mundanely, the assessment regimes that operate can constrain fieldwork. Students' work can be assessed for a number of reasons. For a start, there is the ongoing, informal assessment by a teacher (assessment for learning, also known as formative assessment), which enables the teacher to make fine-tuned adjustments to their teaching depending on the feedback they receive from their students. The role of formative assessment in enhancing the learning process is of particular and special significance in fieldwork, for when done well, a close, genuinely iterative relationship can build between teachers and students – and between students – who together see an enquiry through from start to finish; typically, this will involve much discussion, drafting and redrafting (Lambert and Reiss, 2014).

However, the problems come with a summative assessment, where a student's learning (whether for knowledge, understanding, skills or whatever) is determined at the end of some unit of work or course of study. Such an assessment may be used for a number of purposes, including accountability measures of teachers or schools, but principally to grade students, as in formal examinations in the school system.

There are two main views with regards to the place of fieldwork within formal, summative assessment. One view is that it is not easy to assess fieldwork well, in terms of a contribution to the final grade, and that, rather than getting bogged down in such assessment, better learning results simply from ensuring that students engage in high-quality fieldwork as a core part of their course, without this contributing directly to their final grade. The other view is that teachers and students in many countries, including England, now operate within a system that is so heavily influenced by accountability considerations that if fieldwork is not a required part of the assessment system, it risks being devalued, with substantially less time and attention being paid to it.

However, there are problems when fieldwork is a required part of the assessment system. In both school geography and school biology, the desire by teachers to ensure that their students do as well as they can means that too much fieldwork may become formulaic, where students tread the same safe paths as their predecessors have in previous years. The result is that students may obtain good grades but never realise the full benefits of fieldwork discussed above.

5.6 Conclusions

In an article I once wrote with Martin Braund on the contribution of out-of-school learning for a more authentic science curriculum (Braund and Reiss, 2006), we identified five ways in which out-of-classroom contexts can add to and improve the learning of science:

- improved development and integration of concepts;
- extended and authentic practical work;
- access to rare material and to 'big' science;
- changed attitudes to school science, stimulating further learning;
- increased collaborative work and student responsibility for their learning.

Much the same list applies to geography. Fieldwork, along with what some would consider to be other instances of out-of-school learning, such as museums, can do a great deal to complement the education that students receive at school. I stress the word 'complement' – just as fieldwork has an especial place in learning, so do other sites of learning, including schools, and such tools as textbooks. The skill of those responsible for the education of learners over a number of years is to provide a wide range of learning opportunities, at opportune moments. Recent years have seen a great enthusiasm in some quarters for Forest Schools, (e.g. Knight, 2009; Passy and Waite, 2011) in which learning takes places over extended periods of time in an outdoor setting (canonically, a wood). There is much to be said for this but I suspect, once we eventually see high-quality evaluations, that high-quality fieldwork can reap comparable benefits.

For all that the learning concepts are important – and as an ecologist by background, there is nothing I enjoy more than taking a group of learners, of whatever age, on a field trip to help them learn about the landscape and the organisms that live there – perhaps the most important contribution that out-of-classroom learning can make is to learners' understanding of themselves and their appreciation of the worth of both the natural and many constructed environments. One only has to read the latest in a long line of depressing reports on the state of wildlife in the world (e.g. World Wildlife Fund, 2018) to be reminded of both the scale and the urgency of the task.

Finally, let me say that, given the nature of this book, I have concentrated on fieldwork but much valuable out-of-classroom learning can take place in other ways. To give just two examples, school gardens and school farms, though sadly all too rare, can be wonderfully powerful ways of facilitating learning, helping students not only to appreciate the life cycles of organisms but to derive satisfaction from helping to look after plants and animals, enabling them to thrive and grow (Sanders, 2007; Corbett *et al.*, 2017; Dunne, 2018).

References

Amos, R. and Reiss, M. J. (2012). The benefits of residential fieldwork for school science: insights from a five-year initiative for inner-city students in the UK. *International Journal of Science Education* 34, 485–511.

Bennett, J., Dunlop, L., Knox, K. J., Reiss, M. J. and Torrance-Jenkins, R. (2018). Practical independent research projects in science: a synthesis and evaluation of the evidence of impact on high school students. *International Journal of Science Education* 40, 1755–1773.

Bonnett, A. (2008). *What is Geography?* Sage, London.

Boyle, A., Maguire, S., Martin, A., *et al.* (2007). Fieldwork is good: the student perception and the affective domain. *Journal of Geography in Higher Education* 31, 299–317.

Braund, M. and Reiss, M. J., eds. (2004). *Learning Science Outside the Classroom.* RoutledgeFalmer, London.

Braund, M. and Reiss, M. (2006). Towards a more authentic science curriculum: the contribution of out-of-school learning. *International Journal of Science Education*, 28, 1373–1388.

Cook, V. (2011). The origins and development of geography fieldwork in British Schools. *Geography* 96, 69–74.

Corbett, M., Brett, P. and Hawkins, C. (2017). What we're about out here: the resilience and relevance of school farms in rural Tasmania. *Journal of Research in Rural Education* 32, 1–12.

Department for Children, Schools and Families (2006). The Learning outside the Classroom Manifesto. See www.lotc.org.uk/wp-content/uploads/2011/03/G1.-LOtC-Manifesto.pdf.

Dunne, R. (2018). Why we need an education revolution. *Learning for Well-being* 5, 1–8.

Eshach, H. (2007). Bridging in-school and out-of-school learning: formal, non-formal and informal learning. *Journal of Science Education and Technology* 16, 171–190.

Falk, J. and Dierking, L. (2000). *Learning from Museums*. Alta Mira Press, Rowman and Littlefield, Walnut Creek, CA.

Knight, S. (2009). *Forest Schools and Outdoor Learning in the Early Years*. Sage, Los Angeles, CA.

Lai, K. (1999). Freedom to learn: a study of the experiences of secondary school teachers and students in a geography field trip. *International Research in Geographical and Environmental Education* 8, 239–255.

Lambert, D. and Reiss, M. J. (2014). *The Place of Fieldwork in Geography and Science Qualifications*. Institute of Education, University of London, London. See www.field-studies-council.org/media/1252064/lambert-reiss-2014-fieldwork-report.pdf.

Nundy, S. (1999). The fieldwork effect: the role and impact of fieldwork in the upper primary school. *International Research in Geographical and Environmental Education* 8, 190–198.

Passy, R. and Waite, S. (2011). School gardens and Forest Schools. In *Children Learning Outside the Classroom*, ed. S. Waite. Sage, Los Angeles, CA, pp. 162–175.

Reiss, M. J. (1998). The future of lifescience education. *School Science Review* 79, 19–24.

Rickinson, M., Dillon, J., Tearney, K., *et al.* (2004). *A Review of Research on Outdoor Learning*. National Foundation for Educational Research, Slough.

Sanders, D. L. (2007). Making public the private life of plants: the contribution of informal learning environments. *International Journal of Science Education* 29, 1209–1228.

Stocklmayer, S., Rennie, L. and Gilbert, J. (2010). The roles of the formal and informal sectors in the provision of effective science education. *Studies in Science Education* 46, 1–44.

Tilling, S. (2018). Ecological science fieldwork and secondary school biology in England: does a more secure future lie in geography? *The Curriculum Journal* 29, 538–556. https://doi.org/10.1080/09585176.2018.1504315.

World Wildlife Fund (2018). *Living Planet Report – 2018: Aiming Higher*, eds. M. Grooten and R. E. A. Almond. World Wildlife Fund, Gland.

Part II
Essays: Inspiring Fieldwork

Forty-five authors write enthusiastically about their fieldwork experiences. Many write about influential people who significantly influenced their interest and love of fieldwork. Several share a defining moment that made them realise that this was the sort of professional work they wished to undertake. Some describe projects they have been involved with, an important breakthrough in their subject area, or a new contribution in terms of a technique or approach.

The essays range over very broad areas of the geophysical, life and human sciences: from the Cambrian explosion to today's Great Barrier Reef; from Roman forts in Scotland to graffiti in an Indian city; from species new to science in the deep ocean to the reclamation of derelict land; from field sketching to the use of bat detectors. What links the authors and their amazing array of stories is careful, close, sustained observation of the world around them. Their enthusiasm for observation is undiminished, even if the supporting technology has changed out of all recognition since the days of Gilbert White and Charles Darwin.

6 · *Understanding the Decline of Hen Harriers on Orkney*

ARJUN AMAR

The Hen Harrier – Harried

The hen harrier is well known in the UK for the wrong reasons – it has been dubbed 'the most persecuted bird of prey in Britain' due to its role in the conflict with driven red grouse shooting (Redpath *et al.*, 2010). The numbers and the distribution of this species on mainland Britain are hugely restricted due to the illegal killing of these magnificent birds (Murgatroyd *et al.*, 2019). Thankfully, most of my research on this species has not been on the intensively managed grouse moors of mainland Britain, where these birds are frequently killed by gamekeepers, but on the wild unmanaged moors of the confusingly named Orkney Mainland, the largest of the Orkney Islands in northernmost Scotland.

Orkney's Hen Harriers

The population of hen harriers on Orkney is known to many in the ornithological world. It was widely believed to have prevented the species from becoming extinct in the UK. In the early part of the twentieth century the species was eliminated as a breeding species on mainland Britain due to persecution by gamekeepers; however, a few birds clung on in the far flung western and northern isles of Scotland, and from there recolonised mainland Britain when persecution relaxed during World War II. The hen harrier population in Orkney is also one of the longest studied raptor species in the world. Research on this population was initiated by the late Eddie Balfour, in the 1940s (Balfour and MacDonald, 1970), who was one of the first employees of the

RSPB in Scotland, and after that by Nick Picozzi from the Institute of Terrestrial Ecology (Picozzi, 1984), followed by the late Eric Meek who for over 30 years was the RSPB's Orkney Area Manager (Meek *et al.*, 1998).

By the late 1990s, this population was in trouble. Long-running data showed that the numbers of young produced had fallen by two-thirds, from around 60 young per year in the 1970s to less than 20 per year in the 1990s (Meek *et al.*, 1998; Amar *et al.*, 2005). Eric Meek was so worried by this trend that he urged his RSPB bosses to start an investigation to determine the cause of this decline. Ian Bainbridge, who was then the Head of Research for RSPB Scotland, contacted a raptor researcher, Steve Redpath, and together with another young researcher, Xavier Lambin, and Ron Summers, a research scientist working with Ian in the RSPB, they devised a PhD study of the harrier population on Orkney to diagnose the cause of the decline. In 1998, I started my PhD and arrived on Orkney to begin a study of this magical bird.

Figure 6.1 The author with a martial eagle (from the author's photo collection)

From Africa to Orkney

I arrived with relatively little experience of research on British moorlands. My field research experience derived from the savannas of East Africa, working with Augur buzzards, and the agricultural fields of western France studying Montagu's harriers. Fieldwork on Orkney was wholly different to what I had experienced previously, and far more challenging.

The first thing that struck me was the weather on Orkney, especially the wind! Orkney is famous for its treeless landscape, which provides dramatic, open skies and crystal-clear vistas. Unlike the grouse moors of mainland Britain further south, this treeless landscape is not due to intensive burning, but the power of the wind – nothing taller than a heather bush has a chance of sustained growth.

My fieldwork season started each year in March, which is still full-on winter on Orkney. The first challenge was simply to observe birds in such cold and windy conditions. When watching harriers, the constant use of binoculars means that one's hands feel the full brunt of the wind chill. My first pair of fleece gloves were wholly inadequate – and my hands were chilled to ice blocks within 15 minutes. I then purchased the most expensive mountaineering gloves I could find. Now I had slightly warmer hands, but could not hold a pencil to take notes, or press the buttons on the Dictaphone to record observations!

Learning through Fieldwork

Fieldwork on Orkney consisted of finding all the territory-holding birds and monitoring their breeding output. Orkney harriers were renowned for being extremely polygynous, with males in the past provisioning up to five breeding females at the same time (Picozzi, 1984). However, I quickly found that this behaviour had ceased on Orkney. Although males would still often have harems of several females, the extra females rarely laid eggs and never fledged young. We suspected that this might be due to a shortage of food. To test this, over a two-year period, we provided food every day during the breeding season to different males. We put out dead, day-old cockerel chicks on feeding posts in the centre of the male's territory to directly manipulate the amount of food available to the males. This large-scale field experiment was the highlight of my fieldwork. Fed males started to breed with more females, and we began to observe polygyny again in the population, showing that it was now limited by food. This provided an explanation for the low number of chicks being produced (Amar *et al.*, 2002).

Being constantly in the field, repeating similar tasks each day, made me think about the best and most efficient ways of accomplishing the research chores. However, the motivation for this was not always straightforward – although fieldwork is often physically and mentally challenging, it is not necessarily mentally simulating. Thus, during my supplementary feeding experiment, I became obsessed with exploring the most efficient routes for walking and driving to get to each territory in which I was putting out supplementary food. I kept detailed notes on times of different routes and would try and achieve a new 'personal best' each day.

Apart from a few peat-cutting tracks, there are not many tracks that traverse the moorland of Orkney, which means to get across the sites you have to walk though knee-high heather and bog. Walking through this vegetation is like walking through mashed potato, with each step being dogged by the resistance of the plants that seem to sap your energy.

Over the four years, I came to cue into the harriers' subtle behaviours which helped reveal their breeding status and nest locations. Finding a new nest of a harrier is exhilarating, particularly when the location has evaded you through several days of watching. I experienced some of the other wildlife sharing this moorland habitat: the huge and menacing great skuas (aka bonxies) would dive-bomb you if you got too close to their nests; the dunlin and golden plover, with their beautiful calls, breed on the shoulders and tops of the hills; and the ethereal red-throated divers bred in the lochans at the highest points of the islands. I became familiar with one of the harrier's main prey – the endemic Orkney vole. I spent a great deal of time looking for their tunnels (and their poo), which they build in the mats of dead grass that form in areas with less grazing.

I am now an academic at the FitzPatrick Institute of African Ornithology at the University of Cape Town, South Africa, where my research is mainly focused on the conservation and ecology of African raptors. Within my current role, I supervise many students engaged in fieldwork in some fantastic locations throughout Africa, from the tops of the Drakensberg Mountains to the deserts of the Kalahari. Unfortunately, I no longer get to spend much time in the field. However, I feel my time in search of harriers in Orkney gives me both empathy and an understanding of the challenges that students face, but also a deep-rooted appreciation of wild predators in an environment where they face complex challenges.

References

Amar, A. and Redpath, S. M. (2002). Determining the cause of the hen harrier decline on the Orkney Islands: an experimental test of two hypotheses. *Animal Conservation* 5, 21–28.

Amar, A., Picozzi, N., Meek, E. R., Redpath, S. M. and Lambin, X. (2005). Decline of the Orkney hen harrier *Circus cyaneus* population: do changes to demographic parameters and mating system fit a declining food hypothesis? *Bird Study* 52, 18–24.

Balfour, E. and Macdonald, M. A. (1970). Food and feeding behaviour of the hen harrier in Orkney. *Scottish Birds* 6, 57–66.

Meek, E. R., Rebecca, G. W., Ribbands, B. and Fairclough, K. (1998). Orkney hen harriers: a major population decline in the absence of persecution. *Scottish Birds* 19, 290–299.

Murgatroyd, M., Redpath, S. M., Murphy, S. G., *et al.* (2019). Patterns of satellite tagged hen harrier disappearances suggest widespread illegal killing on British grouse moors. *Nature Communications* 10, 1094.

Picozzi, N. (1984). Breeding biology of polygynous hen harriers *Circus c. cyaneus* in Orkney. *Ornis Scandinavica* 15, 1–10.

Redpath, S. M., Amar, A., Smith, A., Thompson, D. B. and Thirgood, S. (2010). People and nature in conflict: can we reconcile hen harrier conservation and game management. In *Species Management: Challenges and Solutions for the 21st Century,* eds. J. Baxter and C. A. Galbraith. Scottish Natural Heritage, Perth, pp. 335–350.

7 · *Rocky Shores Are Not Just for the Able-Bodied*

JOHN ARCHER-THOMSON

I have had the privilege to be a tutor, and latterly Associate Tutor, for the Field Studies Council for 37 years, and counting (Archer-Thomson and Cremona, 2019). In this time, I have introduced groups to the wonders of the coastal zone, namely, offshore islands, saltmarshes, sand dunes, mangrove swamps, coral reefs, cliff tops and the sublittoral and, in particular, rocky shores. The latter, in my opinion, is one of the most fascinating, biodiverse and downright weird habitats available for study anywhere and it has the additional merit, especially in a British context, of being natural, at least in the main.

Students' reactions to the rocky shore vary of course. Many are enthusiastic from the start, some need encouragement to appreciate the finer points of the intertidal zone (especially if it is cold, windy and raining) but even the most reticent participants get hooked after a while. One group of inner-city lads comes to mind. They had been sent to Dale Fort field centre unaccompanied by a member of staff (this was many years ago) and were rather reluctant attendees: polite but definitely looking forward to the train journey home. Their aversion to getting involved began to waver, on their first visit to the rocky shore, at the discovery of various, rather large, shore crabs: this, they decided, was at the acceptable end of what constituted 'interesting'. What really got them, however, was the encounter with a small-spotted catshark (*Scyliorhinus canicula*) egg case, replete with yoke sac and viable embryo. At their insistence we took this stranded life form back to the lab and set it up in an aquarium tank, with aeration and habitat embellishments such as seaweed, small stones and various molluscs, just so that our perinatal guest felt at home. The students continued to monitor the progress of

Figure 7.1 Dale Fort staff, including Julian Cremona (Director of Studies) at the helm and the author at the back of the boat, with students and helpers freshly returned from their plankton trip

'their' embryo throughout the rest of their time at Dale Fort and seemed genuinely sad to leave it at the end of their stay, especially as it seemed that escape from the egg case was imminent.

I have lost count of the number of times that students have said their field course made the subject come alive. I have enjoyed their astonishment when they discover that barnacles and limpets are not, in fact, rocks; that there are wriggly things under boulders on the shore (porcelain crabs (*Porcellana platycheles*), worms, brittlestars and sand hoppers of various species) and that the slimy stuff is an animal (sponges, sea mats, sea squirts). Maybe most importantly of all, and especially in long teaching runs when fatigue threatened, it was vital to be aware that a field course could, and frequently did, alter the student's choice of further education, career path and, effectively, their lives.

Most of the students I have taught have been able-bodied; even so, great care is needed on the shore as seaweeds and wet rocks are inherently slippery and worthy of caution. But with the right shore and plenty

Figure 7.2 Student (left) feeling the various life forms in a rock pool with helper (right)

of help we (Field Studies Council staff) have had wheelchair-bound pupils in the intertidal zone. Rocky shore animals are, for the most part, of a size where they can be handled with ease; it was a simple thing to enhance our seated guests' enjoyment of the shore by giving them periwinkles, topshells, dogwhelks, small crabs and even blennies to handle. Small stones with attached limpets or anemones, similarly so. Making the effort to get these children to a wild place, and watching their delight and excitement, was reward enough but then to top it with live biology took the experience to another level.

There is one group of youngsters that we took on to the shore (and, additionally, up a shallow, sheltered estuary for population studies; out to sea on the Centre's rigid-hulled inflatable boat (RIB) for plankton sampling, and through coastal grassland and woodland) that may prove to be the most memorable of all in terms of what fieldwork can mean for all those involved. This group (we took a number of different ones over the course of a few years, as funding and other constraints allowed) were either partially sighted or completely blind. One young girl who

I remember in particular experienced a beach for the first time in her life through the medium of touch, sound and smell. She felt pebbles, seaweeds, snails, rocks, barnacles and limpets, anemone tentacles and a (small!) crab's legs and pincers; she heard water movement and waves break. She sensed the ocean as it moved beneath her while she was on the Centre's RIB, felt the sensation of speed as the boat gathered momentum, felt the cold seawater halfway down the outer tubes of the boat. This was an experience so profound that it not only enriched this young person's life but also that of her mother, who accompanied her, and all the teaching staff who helped with these demanding, tiring but incredibly rewarding courses. The teacher/helper student ratio for these events was one to one; even so, it was necessary to be fully alert at all times as these brave little souls, with extremely limited visual capability, were off over the rocks like mountain goats. Their enlightened group leader wanted the children to push the envelope of their capabilities and manage risk accordingly. With attendant, committed staff 'en garde' at all times these youngsters were able to explore this new, exciting environment in safety and we accomplished all of this without recourse to so much as a sticking plaster.

Reference

Archer-Thomson, J. and Cremona, J. (2019) *Rocky Shores*. Bloomsbury Publishing, London.

8 · *Life, Love and Longing to Survive*

ALISON AVERIS

> Thank you for the days. Those endless days, those sacred days, you gave me.
>
> Ray Davies 'Days'. The Kinks, 1968

Islay in May: the brown hills shimmering under a white spring sun. There are willow warblers and cuckoos among the new fragrant leaves of the birch woods; the sea scintillates with light. On the steep slopes of tall shaggy dark heather and sweeps of white quartzite scree, the bright flowers of milkwort and tormentil are beginning to spread out their frail petals for the first insects of spring.

I climb the steep, stony north face of Beinn Bheigier, looking for a special group of liverworts. They are plants that grow only in places like this – shaded rocky hillsides in the parts of the country where the climate is wet and equable. I start to see wine-dark patches of *Sphagnum capillifolium*, showing me I'm on the right track, and then there they are: cornelian cushions of *Herbertus aduncus*, spangled mounds of *Mylia taylorii*; patches of *Scapania gracilis, Anastrepta orcadensis* and *Bazzania tricrenata* in subtle shades of gold. But I am looking for something else: a plant I saw a year ago in the quartzite hills of the west of Ireland. It's the right habitat, the right climate, the right altitude . . . and then suddenly, just as the slope of the hill curves over to the summit ridge, there is a clump of something that looks a little bit different: warm reddish-brown, shining. I hold a wee tuft to my hand lens and there it is: the fold of the leaf, the curled marginal teeth like eyelashes on the lid of a tiny eye. It's *Adelanthus lindenbergianus*, never seen before in Scotland.

Figure 8.1 Late snowbed vegetation on the col between Aonach Mhòr and Aonach Beag in Lochaber (Ben Averis)

It's August. I am standing on a high, steep slope, 1,000 metres up and 15 kilometres from the road end. The north-facing cliff of Aonach Beag leans over me like a thundercloud; 400 black implacable metres of steep shelving ledges, streaming with water. Vegetated, unstable, unclimbable cliffs where no sheep or deer has ever ventured. The silence beyond the ripple of rills and streamlets is the silence of 10,000 years ago; the vegetation of this corrie – rock-jagged, scree-scattered, snow-patched even now – is still late-glacial.

Since Steven and Carlisle wrote *The Native Pine Woods of Scotland* in 1959, people have been repeating their mantra: 'To stand in the pine-woods is to feel the past'. But the pine woods are young in comparison to this icy antiquity. Here, in this chill air, on these sunless slopes, is what remains of Scotland's first fragile skin of vegetation, here since the ice released its grip, 10,000 years ago.

On the high inaccessible ledges there are silver spreads of downy willow *Salix lapponum*, and at my feet a shimmering sward of the montane grass *Phleum alpinum*. Astonishingly, in this thoroughly explored country of ours, it seems that nobody before me has ever been here and found these plants here, in this place.

Fieldwork is not all about these extraordinary privileged moments. Wherever you go, however many times you go, there is always something new to find; some new experience; some addition to your knowledge and understanding. It's about the landscape changing from beautiful background scenery to a legible palimpsest: layer upon layer of geology, soil, climate, vegetation, plants and human settlement and management. It's the knowledge, understanding and experience that add immeasurably to that first deep emotional response: I love it here. This is where my spirit belongs.

So the true value is the personal one: what this immersion in landscape and vegetation gives back in return for the years of patient exploration. Beyond the postcards and the calendars and the tourist experiences, the hills are the real world, where human survival can't be taken for granted. It depends on knowing when to put your coat on; when to find shelter and have something to eat and drink; when to turn back; how to cope without panicking when the mist comes down on the remote plateau; how to find your way down through a staircase of crags in slashing rain; how to recognise the first signs of hypothermia or heat stress and what to do about it. It's about reaching the absolute limits of mind and body, day after day after day. And then the day comes when you're standing in some deep-shadowed corrie under a cliff wall like doomsday, miles from anywhere, from anything, from anyone in the human world, with the knowledge and the confidence that you are as safe here as you are anywhere.

There's the quietness inside: the memory of being in that state of grace. The exultation of the high summits and plateaux; the walls and ridges strung across the skies. The feeling of hard rock under boot soles; *Racomitrium* heath, crisp and warm under bare feet. The taste of spring water after hours of thirst. The feel of loch water on naked sweating skin. The calm regard of eagles. The scents of birch trees and bog myrtle, heather and bracken and peat.

It's a metaphor for life. For every glorious hour on the summit there are five hours of slogging through blanket bog in a cloud of flies: the times when you just have to take hold of your rucksack straps in your two hands and keep shoving one foot in front of the other for as long as it takes.

There are days of mist and rain falling, when I'm walking for what seems like forever in a white world of water, on soaked hillsides of grey

Figure 8.2 The summit ridge of Glas Bheinn, looking towards Ben More Assynt in West Sutherland (Elen Averis)

grasses, the faces of the flowers turned down. The times that I walk in air like wine and honey, in green pastures beside still waters, beside brown and amber rivers, running sleek under a fringe of alders, willows, birches. And there is coming down from the hill in the flaring gold of sunset into woods of singing birds, bringing the memories of stony, moss-clad summits and immense views back with me; part of me forever.

The title of this essay comes from the song 'Visionary Mountains' by Joan Armatrading.

9 · *Bringing Palaeoecology Alive*

HILARY H. BIRKS

For many years, I ran a course on 'palaeoecology' at the University of Bergen, in western Norway. Palaeoecology has an unbroken tradition here, since Knut Fægri in the 1940s. Palaeoecology is the study of past ecological events and processes behind current ecosystems. We cannot see the past directly, but must reconstruct it from evidence stored in sedimentary and archaeological archives, rather like a cold-case detective investigation. Of course, this evidence is incomplete, and we use 'proxies' as clues about past ecological conditions. Plant proxies are usually pollen grains and larger remains such as seeds, fruits and leaves (macrofossils). Animal proxies are often remains of invertebrates, but fish scales and vertebrate bones are also studied. Palaeoecological archives are usually lake or bog sediments that have continuously deposited a timeline of events and ecological developments through the past. The students learn how we collect sediments with a coring device that samples this time sequence (Figure 9.1).

How can we obtain lake sediments from under several metres of water? In Bergen, where thick ice is rare, we must core from open water using a raft. During the course, we transport the student class and coring equipment to a small lake, 20 km from Bergen, at Fjell. We use two inflatable boats linked by a plank raft as our coring platform. A third tender boat holds a board for extruding the cores, wrapping material and a core box.

Two more boats take the students. The first excitement is inflating the boats with foot pumps – which crew will win? An instructor and two students launch a boat and row around the lake with an echo sounder to locate the deepest spot where we will take our core. They also take a

Figure 9.1 Fægri and Iversen support the coring raft, the corer and the extension rods. The tender boat rows away with the core-storage box, ready to tow the raft to the cross-lake rope (Hilary Birks).

short core of the topmost sediments with a specialised sampler and bring it ashore. The students take it in turns to extrude it on the spot into centimetre-deep cross sections, collecting the soft sediment in plastic bags. They are intrigued by the smell and dark colour at the top. This is due to abundant organic matter preserved in the anaerobic sediments, probably originating from the small farm in the catchment. Fjell Lake is relatively unpolluted; polluted sediments can smell like sewage! The uppermost sediment records the ecological responses of the lake to a known recent history of tree felling or planting, cultivation, grazing and other human activities like road- and house-building, etc. These reconstructions can help to form predictive models for interesting biodiverse lakes that are about to be impacted ecologically. The students love to see life in the top sediments, wriggling chironomid larvae and water fleas. Animals always have a high impact!

Now we start the deep-sediment coring. First, a rope is suspended across the lake above the deepest point, an activity that involves more students rowing boats. The coring raft is launched on its inflatable boats, called Fægri and Iversen after the pioneers of palaeoecology, who wrote a famous textbook (Fægri and Iversen, 2000). The raft is towed hand-over-hand along the rope to the spot and anchored to the rope. The tender boat is launched and tethered alongside the raft. The remaining students row out the other boats, to be spectators to the action (Figure 9.2).

We core with a 10-cm-diameter, 1-m-long Livingstone–Wright piston corer with an internal square rod (Wright, 1967). The piston, attached to a long wire, is held at the bottom of the core tube by the square rod inside. Extension rods are added as the corer goes down, similar to a chimney sweep's brush. Once the desired depth is reached, the square rod is lifted up the tube and twisted. The piston is held firm on the wire and the core tube is pushed past it, into the sediment. The full corer is lifted out of the water, the sediment being held within it by the vacuum created by the tight piston. The extension rods are detached and it is passed to the tender boat. By reversing the coring process, the sediment is extruded onto a board to be wrapped and labelled.

Figure 9.2 Coring the sediments of Fjell Lake, near Bergen, Norway. The student palaeoecology class watches keenly (Hilary Birks). A black and white version of this figure will appear in some formats. For the colour version, please refer to the plate section.

The students note the sediment type and the depth of the core segment, being certain to label which end is the top!

Coring requires a strong person to push the corer down and pull it up again. Another person holds the piston wire, and a third passes rods. Students have undertaken all these tasks under supervision, and once a super-keen Bangladeshi student even had to take his own personal core!

All the students see the sediment in the field. The Holocene sediment is organic brown mud. The students get excited when they see fresh-looking pine needles and birch leaves sticking out of it and are astonished when they are told that they are about 5,000 years old (as old as the pyramids). Below about 4 m of sediment, it becomes inorganic, with a section containing small stones and sand. In this section is a half-centimetre layer of dark sand. Below the stony layer is a slightly browner, smoother layer before the core bottoms on gravel. How did these stones get into the lake? What are these different layers? The students have learned that at the end of the last glaciation, about 15,000 years ago, the climate ameliorated. Lake productivity and organic matter increased. Back in the lab, students discover marine foraminifera shells near the bottom. How did they get into the lake, which is about 50 m above sea level? They discover that the huge weight of the ice sheet depressed the land. As it melted, the lake basin was a marine bay. As the land rebounded, the lake was isolated and raised to its present elevation above sea level. In a subsequent cold-climatic reversal, the Younger Dryas stadial, the lake froze and stones skidded onto the ice from the unstable slopes. Microscopic examination reveals that the dark layer within the Younger Dryas is volcanic glass; it is a tephra. It originated from an eruption on Iceland about 12,000 years ago. About 11,700 years ago, the climate warmed permanently at the start of the Holocene. Biological succession started. In the lab, the students learn to identify and count the different types of pollen preserved in the sediment. They discover the plant succession that occurred, both on land and in the lake. Open tundra was colonised by dwarf shrubs and then trees, birch followed by pine with some hazel and then some oak and deciduous trees. About 5,000 years ago, they detect signs of Neolithic occupation round the lake as forest clearance started for agriculture. This intensified through the Bronze Age, the Iron Age and Medieval times, and up to the present. Recently, seedlings from planted spruce are invading the formerly open heathlands of the west Norwegian cultural landscape.

When we go coring in September we usually enjoy fine weather. Even in cold, wet conditions, the students enjoy themselves. They come from many countries, not only Scandinavia. It is always gratifying to see their amazement at the lake sediments, their age, their depth and what they contain. They have a great day messing about in boats and it brings palaeoecology alive for them.

References

Fægri, K. and Iversen, J. (2000). *Textbook of Pollen Analysis*, 4th edn. John Wiley & Sons, London.

Wright, H. E., Jr (1967). A square-rod piston sampler for lake sediments. *Journal of Sedimentary Petrology* 37, 975–976.

10 · *Expedition Botany/Hobby Botany*

JOHN BIRKS

My botanical activities in the last 55 years divide into three main parts – palaeoecological research involving pollen analysis, vegetation history and statistics; lecturing and field courses; and 'hobby botany'. This hobby has encompassed over 90 botanical expeditions to alpine and arctic areas on all continents except Antarctica. Although retired, I remain active in research and have more time for my hobby of expedition botany. I define an expedition as an adventure with a purpose.

How did I conceive and put together my first botanical expedition in 1965? There were three main contributing factors. (1) I was very fortunate to have an inspiring sixth-form biology teacher who had all the *New Naturalist* volumes then published. I avidly read these, especially the botanical volumes by Turrill, Gilmore and Walters, Summerhayes, Lousley, Hepburn, Pearsall, and, most importantly, John Raven and Max Walters's *Mountain Flowers*. (2) My school had just hired a new physical education teacher who was also an ace climber. He started a school climbing club and we learnt the basics of rock-climbing on the Gritstone Edges in Derbyshire. They were not very interesting botanically but I was inspired to go to public lectures by great climbers such as Chris Bonnington, Ian Clough, Doug Scott, Joe Brown, Don Whillans, Dougal Haston and others, about climbing in the Alps, Scotland, Wales and, most memorably, Torres del Paine in Chile (it was a great thrill to visit the Torres in 2000). (3) In August 1963, I attended a Field Studies Council (FSC) week-long course on bryophytes, run by Michael Proctor at Malham. During the week, Michael's endless enthusiasm and knowledge spilled over from bryophytes to plant photography, botanising in the Alps and Lapland, pollination, climbing and caving, amongst others. These three factors led me to plan a student expedition to Abisko in Swedish Lapland, in the summer of 1965, comprising myself and four other very keen botany students, including Hilary Lees (we married in September 1966, and she is also infected with the botany expedition bug).

The first hurdle for our 1965 expedition was to get official approval from the Cambridge University Expedition Society (CUEX) and the Explorer's and Traveller's Club. I prepared a proposal and was interviewed by five elderly male professors. To get approval, an expedition should be imaginative, unusual, student-organised, adventurous, fun and purposeful. The panel asked why there a woman in the party. I answered that Hilary was an excellent field botanist. I could hear murmurings amidst the panel of 'an orgy under the midnight sun'! Our expedition was approved by CUEX and by the Royal Geographical Society and we obtained large amounts of Ryvita, Marmite, Bovril, chicken supreme ready meals, etc. We travelled by boat to Göteborg and then had a 30-hour train journey through the endless boreal conifer forests of Sweden. We finally reached Abisko at 04:00 but started botanising on the way from the railway station to the research station thanks to the midnight sun. The research station consisted then of the Director's

Figure 10.1 Bartsia alpina (John Birks)

Figure 10.2 Buxbaumia aphylla (John Birks). A black and white version of this figure will appear in some formats. For the colour version, please refer to the plate section.

house and three small cabins – *Salix* (willow), *Carex* (sedge) and *Culex* (mosquito) – and we rapidly became acquainted with all three genera! There were so many exciting plants to see on that first walk – *Bartsia alpina*, *Rubus arcticus*, *Linnaea borealis* and the unusual moss *Buxbaumia aphylla*. I have never seen that moss again!

We spent six weeks doing simple ecological projects – mire vegetation and chemistry, tree rings of pine in the Abisko Valley, vegetation patterns on frost-heaved areas and nutrient deficiency in highly calcareous soils. We did not find out much but we learnt a huge amount, particularly about how not to do ecological projects! We visited many habitats, collected vascular plants and bryophytes (now all in the Bergen and Edinburgh herbaria) and vowed never to eat Ryvita or chicken supreme ever again! Having had advice in 1963 from Michael Proctor in plant photography, I photographed much of the flora. I was delighted when Oleg Polunin selected several of my Abisko pictures for his *Flowers of Europe* book (e.g. *Linnaea borealis*, *Cornus suecica*, *Pedicularis hirsuta*). This greatly boosted my photographic morale, as did Michael describing some of my efforts as 'very acceptable' or even 'quite pleasing'. The expedition received funds from various organisations and our colleges and the whole expedition cost £361, including £62 personal contributions – pretty good for five people and six weeks in Sweden.

I have returned to Abisko six times (once with Michael in 1988) to teach and to core lakes for pollen analysis. It is now a major research station next to a tarmacked road from Kiruna to Narvik (although the train line is still operational). Although I always enjoy being at Abisko, the 1965 trip remains a major personal highlight.

After Abisko 1965, Hilary and I started to explore other parts of the world: the Mediterranean and the Alps in 1969; North America in 1970 and 1971; and Montenegro in 1971. Our first expedition to South America was in 2000, Asia in 2001, Africa in 2002 and Oceania in 2003. Even in 1969, I was well and truly infected with the expedition bug and this infection continues to this day. Every expedition is as exciting and challenging as Abisko 1965 was. We gain great enjoyment and uplift by exploring new areas and finding plants new to us and even new to science. Photographing attractive plants in wonderful wild settings and identifying plants using floras that are not always written in English remain challenging and rewarding. I was very fortunate that my student botanical studies included taxonomy, morphology and ecology, and involved many field courses led by outstanding scholars such as Max Walters, Donald Pigott, David Coombe and Oliver Rackham, as well as Michael at the FSC at Malham. I quickly learnt how to use floras. Back then, there were no apps, Google or internet! In my final-year practical exam in taxonomy, I had to identify four species of *Potentilla* using a German flora – an invaluable experience for identifying plants from different parts of the world in future years.

Hobby or expedition botany has been and continues to be a major activity for us. We have gigantic libraries of photographs, floras, maps and notebooks. We enjoy reliving some expeditions through lectures to botanical societies and we prepare every year a calendar of alpine plants from a particular area. Our calendars and some expedition lectures, reports and photo galleries can be found at www.eecrg.uib.no/presentations.htm and www.eecrg.uib.no/projects/AGS_BotanyExp/Introduction.htm.

I consider myself fortunate to have caught the expedition bug early and to have received a wonderful training in what, sadly, some people call 'old-fashioned botany', namely basic taxonomy and ecology. Even if they are old-fashioned, they are as important in environmental science and field studies today as they were when I was a student in 1963–1966.

It is very gratifying to see that groups like CUEX still support student expeditions to remote places, as CUEX has done since 1904. Given the current accelerating rate of global change in arctic and alpine areas, there is an ever-increasing urgency for such explorations and careful recording of species occurrences and their elevations. There is still so much to explore on our planet.

11 · *The Illisarvik Drained-Lake Field Experiment: a Legacy of J. Ross Mackay*

CHRIS BURN

The Illisarvik drained-lake field experiment (Figure 11.1) was the brainchild of J. Ross Mackay (1915–2014), Canada's pre-eminent authority on permafrost (Figure 11.2) (Mackay, 1997; Burn, 2015). Ross drained the oval-shaped, 300-m × 600-m, tundra lake in eight hours on 13 August 1978, to study the effects of permafrost development *ab initio*. The site, on Richards Island in the Mackenzie delta area, lies at the western Arctic coast. Lake Illisarvik drained out to the Beaufort Sea by thermal erosion of an ice-wedge system, facilitated by thaw beneath a trench, dug earlier in the summer above the ice wedges. Prior to drainage, the bowl-shaped body of unfrozen, lake-bottom sediments, surrounded by permafrost, was 32 m deep, beneath the middle of the lake. The site has been monitored continuously for 40 years, with observations in summer and winter. In Inuvialuktun, the language of the indigenous Inuvialuit, *Illisarvik* means 'a place of learning'.

Mackay's fieldwork in the western Arctic spanned 60 years (1951–2011). He began his research before field radios were used, with few maps, and when transportation was almost entirely by boat or dog team. Under these conditions, field seasons require detailed planning and careful attention to logistics. He led the transition in periglacial studies from historical analysis of landform development to application of physical principles, especially concerning the effects of temperature on rheological behaviour of frozen soil and the effects of phase transition on ground behaviour and deformation (Church, 2017). He focused on field measurement, and the explanation of observed ground behaviour in analytical terms, rather than numerical simulation. He studied the inherited characteristics of the landscape and was always looking for clues

Figure 11.1 The Illisarvik experimentally drained lake, Richards Island, Northwest Territories, at the western Arctic coast of Canada (69° 28′ 51″ N, 134° 35′ 04″ W). (Photograph by Sam Cornish, August 2015.)

Figure 11.2 Professor J. Ross Mackay on the tundra, near Illisarvik, August 2001. The dowel rods have been pushed downwards to the frost table in the ground and reveal the undulating surface of permafrost in this hummocky microtopography (Mackay, 1980). The frost table forms a mirror image of the surface due to concentration of organic matter at the sides of hummocks (Chris Burn).

to long-term behaviour of landforms, as in his paper on the rate of growth of pingos, the ice-cored hills that are abundant in the outer Mackenzie delta area, which he inferred from analysis of seasonal growth bands in the ice (Mackay, 1990).

Ross Mackay was an imaginative field scientist. Perhaps his most remarkable field experiments were, indeed, conducted on pingos, which grow in the sandy sediments of drained lakes. He tested the theory of pingo growth by pore-water expulsion from closed-system freezing of the sands by recognising that the ice core of these hills likely forms from a water lens. The water lens originates from the increase in volume of pore water during freezing; it is under pressure, and accumulates where the ground deforms to accommodate the rising volume. By drilling into a pingo and penetrating the ice core, he released pressurised water and the pingo settled. After the drill hole froze back, the pingo began to grow upwards again (Mackay, 1977). Two small pingos have grown at Illisarvik since drainage and water has flowed out of each when they were punctured by drilling.

Illisarvik has also provided a unique opportunity to study the growth of ice wedges, the veins of ice which grow in cracks that open as the ground contracts in winter and which are subsequently filled by snow meltwater in spring. The wedges form the polygonal patterns that cover tundra lowlands like a quilt. The ice wedges at Illisarvik grew more rapidly at first than in any previous measurement of growth rates due to the frigid ground conditions in the unvegetated drained lake bottom. But as vegetation colonised the surface and trapped a 40-cm snow cover, the minimum ground temperatures rose and the ice-wedge cracking unexpectedly ceased (Mackay, 1986). The study demonstrated the value of field investigation in revealing ground behaviour that we did not anticipate beforehand.

Illisarvik has presented the opportunity to study ice wedges and pingos from their initiation in the lake bottom. Timescale is one of the elusive properties of landscape science. Most of our investigations are on human timescales, but the landscape and the development of landforms require longer consideration. There are few long-term records of landform development, as we have for weather or streamflow at some locations, but we may study features of different ages and measure their characteristics at different points in their development. It is rare to be able to watch a landform grow from its beginning.

Arctic fieldwork is inhibited by logistical constraints, the arduous conditions encountered, especially in winter, and minimal research infrastructure. A field cabin at Illisarvik provides shelter in summer and accommodation in winter. It is a powerless and rudimentary structure, but essential and sufficient. Much of the current motivation for Arctic research stems from the increasingly recognised progress of climate change and its amplification in the polar regions. For the permafrost environment, key factors that complicate research plans stem from the dominant occurrence of both the warming and the changes to precipitation being in autumn and winter. Illisarvik provides a unique location to assess changes in ground temperature, in snow cover and in seasonal thawing at the ground surface, because continuous records extend from the late 1970s (e.g. Mackay and Burn, 2002). These records have been shown to be consistent with more intermittent regional information, so Illisarvik has become a sentinel site with respect to permafrost change in the Canadian western Arctic (Burn and Kokelj, 2009).

The fundamental studies of permafrost processes, the motivations that led Ross Mackay to drain Lake Illisarvik, may have been eclipsed in terms of scientific interest by the response of the undisturbed tundra landscape to the unprecedented and continuing rate of climate change. But his foresight in the necessity of field investigation, and the wide-ranging nature of his data collection, have provided us with a place where the effects of current interest can be evaluated in a long-term context. The capacity of remote sensing to detect changes at the ground surface both in terms of cover and elevation seems to be increasing significantly, so that the widespread current effects of climate change on permafrost landscapes can be observed in a way that was unimaginable 15 years ago (e.g. Kokelj *et al.*, 2017). However, changes to permafrost itself and the depth of thawing above permafrost are not yet detectable by remote sensing. This means that the legacy Ross bequeathed us, of long-term field sites and an investigative approach that recognised the timescales applicable to landscape processes, should have a secure future.

Acknowledgements

Field studies at Illisarvik have been supported by the Natural Sciences and Engineering Research Council of Canada; the Polar Continental Shelf Project, Natural Resources Canada; and the Aurora Research Institute, Aurora College, Inuvik. The project was initiated with considerable assistance from J. A. Hunter and A. S. Judge, Geological Survey of Canada. Many assistants, especially Andrew Burn, Douglas Esagok, Graham Gilbert and Alice Wilson have contributed to the research.

References

Burn, C. R. (2015). J. Ross Mackay (1915–2014). *Arctic* 68, 129–131.

Burn, C. R. and Kokelj, S. V. (2009). The environment and permafrost of the Mackenzie delta area. *Permafrost and Periglacial Processes* 20, 83–105.

Church, M. (2017). Reconstructing periglacial geomorphology: the contribution of J. Ross Mackay. *Permafrost and Periglacial Processes* 28, 517–522.

Kokelj, S. V., Lantz, T. C., Tunnicliffe, J., Segal, R. and Lacelle, D. (2017). Climate-driven thaw of permafrost preserved glacial landscapes, northwestern Canada. *Geology* 45, 371–374.

Mackay, J. R. (1977). Pulsating pingos, Tuktoyaktuk Peninsula, N.W.T. *Canadian Journal of Earth Sciences* 14, 209–222.

Mackay, J. R. (1980). The origin of hummocks, western Arctic coast, Canada. *Canadian Journal of Earth Sciences* 17, 996–1006.

Mackay, J. R. (1986). The first 7 years (1978-1985) of ice wedge growth, Illisarvik experimental drained lake site, western Arctic coast. *Canadian Journal of Earth Sciences* 23, 1782–1795.

Mackay, J. R. (1990). Seasonal growth bands in pingo ice. *Canadian Journal of Earth Sciences* 27, 1115–1125.

Mackay, J. R. (1997). A full-scale field experiment (1978–1995) on the growth of permafrost by means of lake drainage, western Arctic coast: a discussion of the method and some results. *Canadian Journal of Earth Sciences* 34, 17–33.

Mackay, J. R. and Burn, C. R. (2002). The first 20 years (1978/79 to 1998/99) of active-layer development, Illisarvik experimental drained lake site, western Arctic coast, Canada. *Canadian Journal of Earth Sciences* 39, 1657–1674.

12 · *In Praise of Meteorology Field Courses*

STEPHEN BURT

It's early September, just before the start of the autumn university term. I'm in western Scotland with a group of undergraduate and Masters students, just about to launch a radiosonde (a device to measure vertical temperature, humidity and wind profiles through the bottom 20 km or so of the atmosphere). A cold front is imminent, and the rain is both horizontal and cold. No doubt about it – it must be a meteorology field course.

Field courses are an essential component of teaching meteorology, whether at undergraduate or postgraduate level. There will no doubt be many students who go on to research or modelling careers, but it's a poor show if a meteorology student graduates without knowing one end of a thermometer from the other or fails to grasp the many sources of error in making surface wind measurements. Real, sharp-end experience of making weather observations, drawing conclusions and fitting local observations into the bigger picture is very important – after all, weather doesn't just happen in computer models. There's no rule that says weather on field courses has to be dire, if not life-threatening, for it to be valuable, but in reality there's only so much you can say during a week dominated by 5/8 stratocumulus and a light south-westerly breeze. Hence my current sojourn to the Lochranza Field Centre on the beautiful Isle of Arran, a week-long field course run jointly between the universities of Leeds, Edinburgh and Reading.

Meteorology field courses differ fundamentally from, say, geology. A geology field course on Arran, for example, would know where to find which rocks; it would be reasonable to expect that this year's field course would find particular rock types in much the same locations as last

Figure 12.1 Malham Tarn House, home to the Malham Tarn field centre in North Yorkshire photographed on 27 May 2017 (Stephen Burt).

year. Not so with weather: any one week's weather in one year is very unlikely to be identical to the weather in that week the previous, or following, year – particularly in western Scotland. So, the challenge for field course leaders in meteorology is to be flexible enough to adapt the programme, not only to the vagaries of the day-to-day weather, but to be equally prepared to adapt topics to remain relevant to the current weather situation.

One particular field course ignited my decades-long professional and personal interest in meteorology and meteorological science. That course was one run jointly at that time by the Field Studies Council (FSC) and the Royal Meteorological Society (RMetS), and held in the beautiful surroundings of Malham Tarn in North Yorkshire, in September 1972. My school had a long-established weather society, the Rugby Weather Bureau (sadly, now defunct), and myself and a small group of us managed to persuade our parents to cough up the fees. I was 14.

Malham Tarn is an impressive, solid stone-built building with miscellaneous outbuildings, set just back from the tarn itself (Figure 12.1). A small climatological station (Figure 12.2), established here in November 1948, continues today, having accumulated several decades of extremely valuable high-quality upland weather records (Burt and

Figure 12.2 The climatological station at Malham Tarn (54.10° N, 2.16° W, 395 m above mean sea level), photographed on 27 May 2017 (Stephen Burt). The louvred white box is a Stevenson Screen. A black and white version of this figure will appear in some formats. For the colour version, please refer to the plate section.

Horton, 2003). With a similar climatological station set-up at our school site, those of us from Rugby were on familiar ground – at least on the observational side. Even today, Malham is pretty remote from the trappings of modern life. Back in 1972, long before Wi-Fi, mobile phones and the internet, Malham really was 'the back of beyond' and pretty much cut off. (It was the week of the Munich Olympics and the terrorist attack which killed 11 Israeli athletes, but I knew nothing of this until we were on the way home and heard the news bulletins on the car radio.) For current weather information, we had our own observations and the Shipping Forecast four times daily on long-wave radio – and that was it; no real-time feed of current weather observations from across the world, no plentiful satellite images or forecast model products or wireless observations relayed from a portable mobile automatic weather station updated every few minutes, as I have at the click of a mouse this week. (This is beginning to sound a little like Monty Python's 'Four Yorkshiremen' sketch . . .) Pretty much the only thing in common was atrocious upland weather – frequent low cloud, strong winds and driving rain remain strong impressions, 45 years later, although I am sure there must have been some sunshine too.

The Malham course was led by Alastair Thom from the University of Edinburgh, together with his PhD student, Bob Riddaway. Despite the drawbacks of remoteness and (when compared with today) very limited external communication, the course leaders made both indoor and outdoor exercises stimulating and memorable. Competitive Shipping Forecast transcriptions and plotting before breakfast became a way of life; boundary-layer wind profiles on open moorland and in forest clearings in pouring rain gave us experience in instrument handling and measurement errors; synchronised temperature and wind measurements by two student groups at the top and bottom of Malham Cove provided insight into lapse rates and wind eddying around topographic obstacles. It was an inspiring and memorable (if wet and windy) week, which taught me a lot about meteorology, instruments, working in a team and 'the joy of finding things out' – not necessarily in that order.

The following year, I attended another FSC–RMetS field course, at Nettlecombe Court in Somerset, in mid-August 1973. The weather could not have been more different – a hot and sunny week, with clear days giving large diurnal variations in temperature and wind speed/direction. I can still remember my surprise at the strength and sharpness of katabatic cold air drainage into the valleys below Bird's Hill during the August twilight. Bob Harwood from Oxford led this course, with an inspiring, eclectic and engaging series of activities and exercises. For me, Malham and Nettlecombe fixed meteorology at the forefront of my mind as an accessible, interesting and endlessly varied science, which led to a clear career choice by the time I was 15. And the rest, as they say, is history!

Reference

Burt, T. P. and Horton, B. P. (2003). The climate of Malham Tarn. *Field Studies* **10**, 635–652.

13 · *Time, Place and Circumstance*

TIM BURT

What sort of scientist am I? Peter Gould (1985, p. 320) encouraged us to recognise the particular historical situation in which we find ourselves, 'to acknowledge the trap of one's own time and place and circumstance'. For me, being born on the Mendips and brought up in the south-western English county of Somerset provided the time and place; circumstance came from a charismatic schoolteacher, Jim Hanwell. Once I fell under Jim's spell, I never wanted to study anything but geography.

I recall early on being sent out to measure the width of the main road outside the school gate in order to compare with the width of the red line printed on the Ordnance Survey map. Health and Safety concerns would preclude such activity these days, and clearly Jim should have been severely reprimanded for placing me in such peril, but nevertheless, I was hooked, and Jim was destined to be stuck with me until the end of my school career, and indeed well beyond. Jim's main interest was in physical geography and geology: his teaching coincided with the quantitative revolution in fluvial geomorphology, the move towards process studies and away from qualitative denudation chronology. Attempting to understand landforms in relation to their formative processes was the new quest.

Somerset clearly played its part, imbuing me with a deep curiosity about landforms. Jim's teaching often included the Mendip periclines, contrasting with Jurassic escarpments closer to home. This all comes together at the Vallis Vale unconformity, seen in the deep valleys and quarries west of Frome in the eastern Mendips, barely a stone's throw from my birthplace in the village of Mells.

The Mendip Hills are an example of anticlinal folds, recognised on geological maps by a concentric outcrop pattern with the oldest rocks at the core. The distinctively shaped variety of anticline that is found in the Mendips is known as a pericline or 'whale back' (British Geological Survey, 2017); four periclines *en echelon* (as Jim taught me) make up the Mendips. All of the folds are more steeply dipping on their northern side with more gently dipping beds on their southern flank. The Carboniferous Limestone was deposited between 363 and 325 million years ago (MBP). The periclinal folds date from the end of the Carboniferous around 300 MBP; continental drift caused a major phase of mountain building, known as the Variscan or Hercynian Orogeny. A major time gap (the unconformity) then occurs between rocks of Carboniferous age and the younger Jurassic rocks (170 MBP) on top. The unconformity was first described by Conybeare and Phillips in 1822, but it was Sir Henry De la Beche (1846), the first Director-General of the British Geological Survey, who noted in 1846 that the unconformity surface was 'drilled into holes by lithidomous animals, which must have existed in the seas at the commencement of the inferior oolite', quoting the detailed observations of Phillips, back in 1829, who had

Figure 13.1 Jim Hanwell (white shirt) examining the surface of the Vallis Vale unconformity (Tim Burt). A black and white version of this figure will appear in some formats. For the colour version, please refer to the plate section.

Figure 13.2 Jim Hanwell and Tim Burt at the Vallis Vale unconformity (Elizabeth Burt). A black and white version of this figure will appear in some formats. For the colour version, please refer to the plate section.

recognised that the holes were of two kinds; one long and slender, the other short and squat.

One of my photos shows Jim Hanwell at a site near Great Elm (Figure 13.1). Here, the Jurassic overburden was stripped in readiness for quarrying of the Carboniferous Limestone. Happily, some wise government official intervened, having recognised how unusual it was to be able to see a wide expanse of erosion surface exposed in this way, and the feature was rescued and preserved for posterity. The other photo (Figure 13.2) shows Jim and I at the classic Vallis Vale exposure. The pale, horizontal beds of Inferior Oolite (capital letters added since De la Beche's time!) are clearly seen above the steeply dipping dark grey Carboniferous Limestone below.

Unconformities are dramatic and intriguing features of the natural landscape. Vast timescales are required to produce them. Start with the deposition of limestone. Wait some 25 million years before those rocks are uplifted and buckled as part of the mountain-building process. Then, there needs to be a long period of erosion, producing the flat erosion surface cut straight across the steeply dipping beds. Long after this (more than 100 million years), younger rocks are eventually deposited on top.

Then, there needs to be another long period when all the rocks involved are raised up or when sea level falls (or both) before the current phase of subaerial denudation can proceed. It is not hard to be curious about such features!

As it happens, my own career has been concerned with surficial processes, geomorphological and hydrological, rather than with the rocks below. But it was rocks and relief, as taught by Jim Hanwell, which got me started and made me inquisitive. From Somerset's Vallis Vale unconformity to Dorset's Jurassic coast, I was lucky enough to grow up in a fortunate time and place and circumstance.

References

De la Beche, H.T. (1846). *On the Formation of the Rocks of South Wales and South Western England. Memoirs of the Geological Survey of Great Britain, Volume 1.1.* Her Majesty's Stationery Office, London.

British Geological Survey (2017). Geological structure of the Mendips. See www.bgs.ac.uk/mendips/geology/Geological_Structure.htm (accessed 2017).

Gould, P. (1985). *The Geographer at Work.* Routledge & Kegan Paul, London.

14 · *Sampling Fish Diversity along a Submarine Mountain Chain*

INGVAR BYRKJEDAL

Creeping through a hatch above a conveyor belt into the cargo room in Bergen Airport is not the usual way to join a research cruise to the Mid-Atlantic. As the ad hoc travel leader for a chartered flight with scientists and technicians heading for the Azores, I was desperately seeking a box with scientific instruments, shipped from the UK. I had been instructed under no circumstances to go without it. And now, at the moment of departure, the box was missing, and in desperation airport staff led me into the cargo room to search for it – airport security in 2004 was somewhat relaxed then.

I had just finished organising the travel. Now, in the cargo room, I was called upon to come immediately to the departure gate, otherwise they would go without me; the time window for take-off was about to be passed. Out of breath, I just managed to slip through the door of the plane before it was locked. Once in the air I had to give a long list of last-minute instructions about the practicalities on arrival at Horta, the destination of the chartered flight. Asking my fellow travellers to observe that, while the plane was looking like a hen it would fly like an eagle, I finally wished everyone an enjoyable flight and slumped, exhausted; the damned box turned out to be on the flight as well.

The cruise had actually started a month earlier, going from the Reykjanes Ridge to the Azores to study the diversity of nekton (fish and other free-swimming organisms) along the northern part of the Mid-Atlantic Ridge (sometimes abbreviated to 'MAR'). The first leg of the cruise concentrated on pelagic life, while the second leg, which I was joining, was to sample bottom-living nekton on a northbound return route. Recognising that two-thirds of the Earth's surface is marine, and that our understanding of what lives in the ocean lags far behind our desire, leading marine scientists, in 2000, organised the Census of Marine Life (CoML), with the ambitious aim of recording life in little-studied parts of the ocean during the ensuing decade (Ausubel, 2010). A large number of expeditions set forth, involving a broad, international team of scientific experts. The two-month cruise along the Mid-Atlantic Ridge in the summer of 2004, referred to as the 'MAR–ECO' expedition, constituted one of these.

The Mid-Atlantic Ridge

The Mid-Atlantic Ridge is one of the major geological formations on our planet, stretching over northern and southern hemispheres and rising from the 4,000-m-deep seabed to the surface (McIntyre, 2010). Back in 1910, the Norwegian expedition on RV *Micael Sars* found an increasing fish diversity from the seabed to the ridge, and with its structural diversity the ridge seemed the ideal location for unravelling the underlying basis for this. Involving about 35 scientists from 13 countries, the logistics were organised mainly by the Institute of Marine Research in Bergen, not least in providing the ship and sampling equipment. The ship was the brand-new RV *G. O. Sars*. This 78-m-long vessel held cutting-edge technology for oceanographic endeavours – sampling, seabed mapping and all manner of recording. Accommodating 30 people in addition to its regular crew, there was competition for membership of the cruise party. The University Museum of Bergen was assigned to act as the main repository of collections, and being the museum's curator of vertebrates, I managed to get a place.

The plan was to have experts identifying as much of the material as possible on-board, so that analyses of communities would be undertaken rapidly. Taxonomic workshops and museum loans to taxonomists around the world were organised in parallel with, and in the aftermath of, the cruise, to maximise success.

Arriving in Horta, to embark, we soon found ourselves accommodated wonderfully. For me, used to fieldwork on arctic–alpine birds in desolate places under rather primitive conditions, this was like checking in at a top-class hotel: cabins with telephone and internet connections, splendid meals served in the mess and excellent labs. This was a habitat in which I could thrive – so long as my stomach could tolerate the rough Mid-Atlantic seas. Participants from various countries arrived. We met those who had been on the first leg, the pelagic experts. They left the ship as we came on-board, describing excitedly what they had caught and already discovered. Our first evening in Horta took us all to a formal dinner with notabilities and fine speeches, but we ended up in Pete's Bar, renowned as the world's best by yachtsmen. We could not have been more buoyed!

Leaving the Azores, we spent time organising ourselves and preparing for the work to come. Finally, we reached our first pre-determined bottom trawl station, where a haul at 3,000 m was to be done. The whole operation took many hours, as the deployment and retrieval of the trawl was massive. Rosters in six-hour shifts had been set for the cruise, and tasks allocated, but curiosity completely overwhelmed us, and nobody wanted to miss the first trawl haul.

With everybody present, the landing space became crowded. This was not to be a familiar North Sea haul for cod and haddock, but a revelation of what we talked of as the 'deep-sea mythology' in abundance: big eyes, big mouths, big teeth, light organs, black skin, red skin – biota unseen but, at least some of it, imagined. Samples were rushed to the lab and hectic investigations began. Luckily, the fridge room was large, for the temperature in the lab got perilously high for our virgin material, so only small portions could be taken from the fridges. Identifications and registrations were done, pictures taken and then quickly material was despatched to the deep freezer for storage.

Deep-Sea Discoveries

In all, 18 bottom trawls were taken as we moved north on the Mid-Atlantic Ridge. The pelagic fish trawls on the first leg amounted to 61. Lots of deep-frozen material accumulated, but the ship's 'belly' apparently could take it all.

In total, 61,000 specimens of fish were collected, representing more than 300 species of 82 families. More than 22,000 of these, almost 4,000 lots, went to the museum for permanent storage. The number of species far exceeded the ca. 175 species found in inshore waters of the British Isles. Also, a large number of invertebrates were caught and preserved, especially crustaceans and cephalopods (Boyle, 2009). We soon realised we were dealing with highly rare species. The taxonomists on-board had to expand their identification manuals to worldwide keys.

The reports summarising the fish species from the MAR–ECO (Porteiro *et al.*, 2017) often had notes like 'first record on the northern MAR', but some species that were previously considered rare turned out to be rather common. And, there were species new to science. So far, three fish (apart from a large number of invertebrates) have been described for the first time from the material collected.

Species New to Science

I had the special privilege of describing species previously unseen: a fathead sculpin, which we named *Cottunculus tubulosus*. The genus was fairly easy to tell, as I had experience with species of this group from the Arctic, but characteristics clearly differing from other species in this genus convinced us that we had a species new to science. It is a very special thing to discover a species that no one so far has known existed. One cannot help thinking about this – that there is a life form moulded by natural selection over millennia – with 'patents' for survival all of its own. Had the species been sufficiently self-conscious, it would undoubtedly consider itself the most important one on the planet, whereas for us humans it has so far remained totally unknown with all its special life-history traits. You get a strange notion that the discovery represents a gift to us all, something to wonder about and marvel over. What little we know about these new species is summed up on the label accompanying the specimen in its jar. That is humbling.

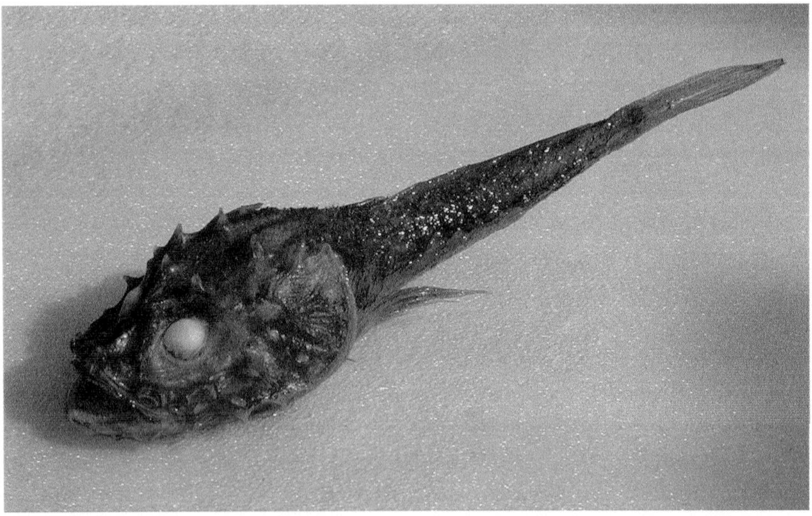

Figure 14.1 A species new to science: *Cottunculus tubulosus* (Ingvar Byrkjedal)

Figure 14.2 The weird-looking smallspine spookfish (*Harriotta haeckeli*) was described for the first time only in 1972. Known from depths below 1,500 m in parts of the North Atlantic, this rarely caught cartilaginous fish was recorded for the first time on the Mid-Atlantic Ridge by the MAR–ECO expedition in 2004 (Ingvar Byrkjedal).

The life on-board found its routines, but none of the trawls were remotely routine in nature. Each catch brought something to marvel over. It struck me and the others on-board how painfully little we actually know about the organisms in the deep. We often do not know enough to identify and classify them correctly, let alone how their life goes on below the waves. It is a forbidding environment for us. We cannot just go down there and spend time watching behaviour, although there are technical devices that can give us small glimpses into that world. How little would we have known about the life of birds, if all we had for studying them were nets lowered from aeroplanes! Just imagine a phenomenon such as the bioluminescence, so common in the ocean depths. Lantern fish, and there are a lot of species, have small circular light organs distributed in rows on their body, in patterns characteristic of the species. Light organs can individually switch the light on and off. With 20–25 such organs along your body, what a tool for detailed communication! This is just one of the many practically unstudied phenomena in the deep that triggers our curiosity.

Nonetheless, an expedition like MAR–ECO increases our insight into life in the depths. Studies of the patterns of horizontal, vertical and diurnal distributions, trophic relationships, reproductive seasons, fecundity, functional morphology and a number of other aspects give us great insights. A large number of scientific papers from the MAR–ECO expedition on such subjects have been published. All of this will be of great value as the conservation of the seas takes on added urgency. As a result of our expedition a large North Atlantic area, covering the biologically important Charlie–Gibbs Fracture Zone, has

Figure 14.3 A number of pelagic anglerfish species, such as the humpback anglerfish (*Melanocetus johnsonii*) shown here, live in the deep below the photic zone. Typically, they have black skin, only producing light in a small organ at the end of a rod pointing up from the forehead. This organ is used as a lure to attract prey. Large gapes, sharp fanglike teeth and widely distensible stomachs enable these fish to hold onto, and swallow, surprisingly large prey (Ingvar Byrkjedal). A black and white version of this figure will appear in some formats. For the colour version, please refer to the plate section.

been set aside as an international area for protection against economic exploitation.

Being on that ship, smelling, seeing and shouting about the landed catches, and then realising the importance of the samples examined and recorded with great care, is unique. Scientific endeavour fired by curiosity and comradeship – what could be better? But, one thing abides with me – daily, we saw and landed loads of rubbish, mostly plastic in all forms, and we saw every formation imaginable of styrofoam, floating everywhere in the Mid-Atlantic. We shall lose far more species in the seas than will be discovered, and that is awful.

References

Ausubel, J. H. (2010). *First Census of Marine Life 2010: Highlights of a Decade of Discovery*. Census of Marine Life, Washington, DC.

Boyle, P. (2009). *Life in the Mid Atlantic*. Bergen Museum, Bergen.

McIntyre, A. D. (2010). *Life in the World's Oceans: Diversity, Distribution, and Abundances*. Wiley-Blackwell, Oxford.

Porteiro, F., Sutton, T., Byrkjedal, I., *et al.* (2017). Fishes in the northern Mid-Atlantic Ridge collected during the MAR-ECO cruise in June–July 2004: an annotated checklist. *Arquipelago* 10 (supplement), 1–125.

15 · *Place and Placefulness*

RICHARD CARRICK

The title is a play on a geographic text that had a great impact on my early academic career. This was *Place and Placelessness* by Edward Relph (Relph, 1976).

My realisation that geography wasn't just the subject you did when you couldn't think what you wanted to do with your life came in 1974, with the publication of Douglas Pocock's *The Nature of Environmental Perception* (Pocock, 1974). A Durham geography undergraduate at the time, then as a postgraduate town planner and subsequently as a geography lecturer, this, and related texts, sustained my interest in places and their meaning, culminating in a 35-year career of sending people to places to holiday.

Edward Relph's massively influential text *Place and Placelessness* gave human geography the impetus it badly needed in the 1970s. Not possessing the scientific legacy and consequent confidence of physical geography, Relph provided much of the necessary framework to understand how places work and can best be understood. The was an essential underpinning if there was any chance of place distinguishing geography from other disciplines and giving it a unique focus.

Relph's approach was rooted in defining a 'phenomenology of places'. Phenomenology is the interpretive study of human situations, events, experiences and meanings, which mostly get taken for granted. Diving beneath the surface and getting to grips with the explanation of why and how we perceive such elements in the context of places is what Relph brought to human geography.

When I reflect on why people travel where, beyond the basics of price and opportunity, it is fundamentally to visit places. The word travel is

Table 15.1 *A classification of places people travel to*

Iconic, symbolic places	Bucket-list destinations, places you go to for celebration, or commemoration
Specific-purpose places	Places that educate and enrich; art history, archaeology, music, study
Places to do good	Environmental, aid visits, charitable causes
Places that feel good	Architectural, location, food and drink, activity, people, events
Places that do you good	Places where people live longer and more healthily than elsewhere

said to derive from the Old French 'travail' – to undergo hardship – and even, or especially, the modern experience is one to be endured in order to reach places we want to visit. Other than some rail journeys and cruises, the mode of transport is secondary to the ulterior aim of visiting places.

A current trend in the travel industry is for consumers to increasingly crave authenticity, self-actualisation, meaningfulness and wellness. In conventional parlance, in travelling to places, we seek perfect, or magical, moments. Not travel for travel's sake, but to satisfy some deeper craving, which takes us full circle and back to the meaning and the understanding of place itself.

Places carry meaning, born out of shared history, often moulded by 'planning interventions', sometimes impacted by disease, transport, food supply and population conditions.

An initial attempt to categorise places we travel to is shown in Table 15.1.

It must be stressed that this attempted classification is part of some work for a tour-operator client, looking to differentiate themselves in the market, and not one founded on any sound academic or theoretical basis. However, it does seem that it owes its existence to an early appreciation of what constitutes place and, also, it is one that lends itself to fieldwork measurement.

Each type of place can be seen to offer opportunities for various types of fieldwork calibration, as outlined in Table 15.2.

In the context of increasing levels of consumer demand for travel to places that aid or inspire well-being, desk research has been undertaken to identify those places globally where people display the highest longevity and lowest incidence of disease. This is not to suggest that a short visit to such places will immediately transmit such benefits to the

Table 15.2 *Placefulness insight: fieldwork measurement and calibration*

Attitude surveys amongst users	Users can be residents or visitors and surveys should aim to understand how a place is perceived and the components that drive satisfaction or provide appeal; filming and recording can also be employed
Built form and structure analysis	Building design, materials, components of streetscape and landscape; photography, sketching
Transport and pedestrian flow analysis	Based on collected survey data, to reveal those landscape components generating the highest and most popular flow levels
Environmental comfort indices	How habitable have places been and are today: climatic (sunlight, rainfall, temperature, etc.), pollution levels, hazard risk
Demographic and socio-economic data analysis	From available census data, supplemented by interview-based fieldwork; life expectancy indices, incidence of disease, food and diet data

Table 15.3 *Places that do you good*

Hunza Valley	North Pakistan, high altitude, low pollution levels, terrain encouraging moderate exercise, very low levels of cancer, heart disease, diabetes, obesity, local diet rich in apricot seeds, fruit, vegetables, nuts, low animal protein, yoghurt, grains
Ikaria	Greece, coastal, quality local olive oil, wine, honey, low pollution, high sunlight
Tasman	New Zealand, longevity, low incidence of disease, local organic diet
Okinawa	Japan, high female longevity, local vegetable diet high in antioxidants, local and famous healthy cuisine
Silanus	Sardinia, mountain lifestyle, low gluten, local diet, high vegetable content
Vikabama	Ecuador, 5,000 feet above sea level, perfect year-round climate, natural mineral water
Nicoya	Costa Rica, world's lowest level of middle-aged mortality, hard water area, high longevity for elderly males, lower cardiovascular mortality

traveller, but arguably it is important for geographers to attempt to understand how a local environment and its living conditions can aid well-being.

Such areas have been referred to as the world's 'Blue Zones'. Whilst not attempting to be an exhaustive list, the places listed in Table 15.3

Figure 15.1 San Vito Lo Cappa, north-western Sicily (top), and the Hunza Valley in Pakistan (bottom) both display high longevity and low instances of serious illness, apparently as a result of unique, local diets, active lifestyles and low pollution. (Photos from Unsplash.)

merit further study by geographers. The list omits those places where longevity and well-being appear to derive from a religious orientation, as opposed to an environment and associated lifestyle conferring benefits on those who live there. Loma Linda in California is one such example where membership of the Seventh Day Adventist church and strict adherence to their beliefs and practices appears to be the primary determinant of the health of the community other than any obvious place-related influences.

Geographers are well placed to contribute to a better understanding of place. The ability of the discipline to dissect the meaning and significance of place and the part it plays in the global economy is unsurpassed. Fieldwork opportunities to gain better insight into place and playfulness abound. Given current popular debate around global warming, pollution and health concerns, there has never been a stronger case for this focus. Parochially, in the travel industry, place is the commodity being promoted, sold and consumed and geographers will find a ready audience for an enhanced understanding.

References

Pocock, D. C. D. (1974). *The Nature of Environmental Perception, Occasional Publication, Number 4*. University of Durham, Department of Geography, Durham.

Relph, E. (1976). *Place and Placelessness*. Sage, London.

16 · *Ripples across the Pond*

STUART CORBRIDGE

One of the great joys of academic life is learning from others, and here I have in mind other academics as well as respondents in 'the field' in India. I have been lucky enough to work in several great universities, but I will always look back with affection on the London School of Economics (LSE) and in particular my time in its Department of International Development. Two colleagues, Tim Allen, an anthropologist, and Diana Weinhold, an economist, put on a course each year for our Masters students, which really encouraged everyone to think critically about some of the different methodologies they held dear. And it was fun in my own seminars to see really smart students with an economics background critique books which maintained that the failure of aid project A was evidence in itself of the failure of 'Development'. Of course it wasn't. Generally, the author was describing a principal-agent problem in a small project that mattered hardly at all to long-run income trends, even at a local level. Development is largely about the accumulation and distribution of capital. But it was also exciting to see the table turned, especially when a student with many and sustained experiences in 'the field' challenged the plausibility of some of the data sets that were being taken for granted in large-*n* statistical studies.

For myself, I am deeply grateful for the mixed-methods training which geography provided me in the 1970s, even when I struggled with really advanced statistical techniques and with Hindi classes populated by students who were taking the course as part of their degree. It was not easy (or for me possible) to keep up! Knowing your limitations helps. It also helped enormously that the fieldwork which I pursued after the 1980s was almost always with other colleagues – for example, Sarah

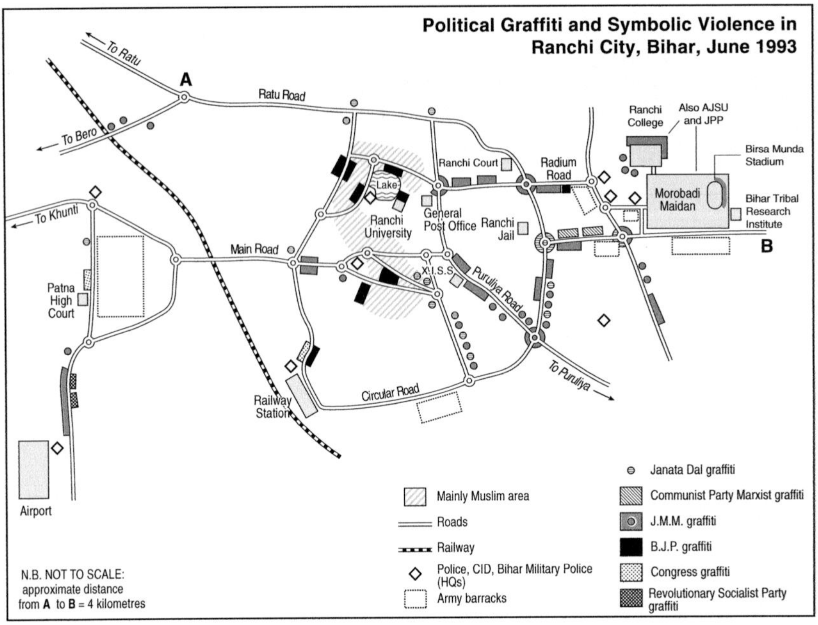

Figure 16.1 Political graffiti and symbolic violence in Ranchi City, Bihar, June 1993 (Stuart Corbridge). A black and white version of this figure will appear in some formats. For the colour version, please refer to the plate section.

Jewitt: with whom, one weekend in 1993, I toured Ranchi City, Bihar, India on a motor scooter, mapping local geographies of graffiti (see Figure 16.1, and note the purple BJP – Hindu nationalist party – graffiti encircling Ranchi's Muslim quarter) or it was carried out as part of a large research team. Working with Manoj Srivastava, Rene Veron and Glyn Williams in eastern India on what became the *Seeing the State* project and book (Corbridge *et al.*, 2005) sharpened my understanding of how team-based fieldwork and rigorous project design could help overcome the classic problem which all fieldworkers face in the human sciences: the possibility that one's field site is exceptional and that no generalisations are permissible. In that project, we worked in five field sites (districts: see Figure 16.2), where we hypothesised that locally prevailing political settlements – with a strong history of casteism, say, or an organised left-wing presence – would significantly affect the ability of local people to access entitlements (work, education, justice) from the state: a hypothesis that was significantly but not fully confirmed.

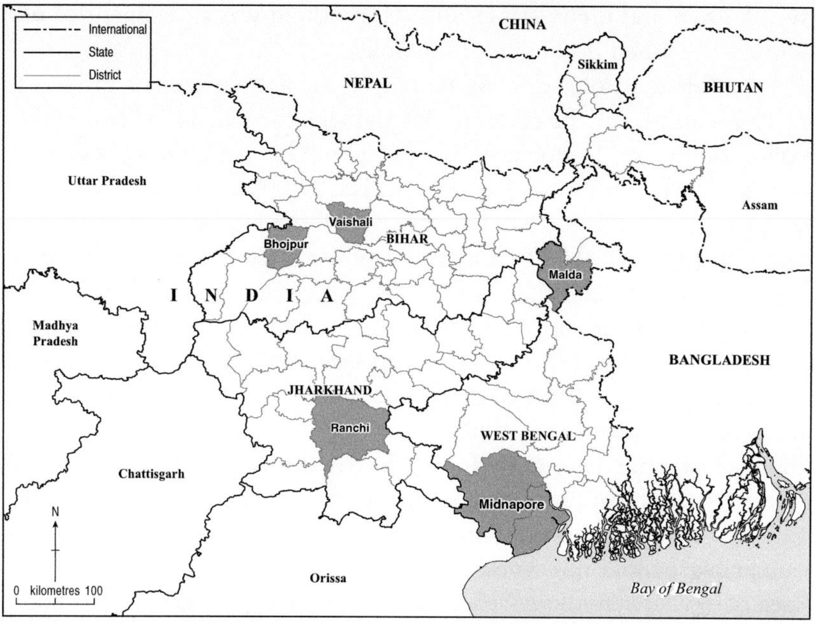

Figure 16.2 Map of the five districts studied in India (Stuart Corbridge)

But I have also learned in India, notably working with my friend and colleague, Sanjay Kumar, that there are extraordinary lessons to be learned from even the smallest $n = 1$: an individual. I'd like to call this thinking 'synechdocal'. Less pretentiously, it is like the stone that is thrown into a pond which causes ripples to spread outwards. The example I like to give is one that came from Sanjay's work as a forest officer, though I have met many people in a similar situation to the respondent who we called Polus B in our paper (Corbridge and Kumar, 2002). Polus, in the 1990s, was an *adivasi* (scheduled tribal) schoolteacher in Ranchi District, Bihar (now Jharkhand). He wanted to chop down 10 jackfruit trees on his homestead land, as he was legally entitled to do. Before he could do so, however, and because of a colonial environmental regime that was meant to protect untutored *adivasis* from unscrupulous forest traders, Polus would have to trek 40 kilometres to the district capital, Ranchi City, to establish his *bona fides*. Not only would such journeys be expensive and multiple, they would bring Polus into contact with various state authorities, from the Forest Department to the police force. Polus knew the drill. 'Cuts' or commissions would be demanded and taken. Nothing too surprising. The rates are generally well known and pretty stable. But always there was an element of risk. Particularly when dealing with the police.

So, what did Polus do? He employed a *dalaal* (a fixer or broker) to make the deals, pay the commissions and take the felled trees to an urban collection centre. In doing so, he recognised he was getting value from the *dalaal*, who would need to pay out Rs. 20,000 for transportation and stumpage costs. But, by the time he paid all the commissions, Polus was left with just Rs. 20,000 himself, out of a market value for his trees of Rs. 100,000 (or Rs. 80,000 net). The *dalaal* ended up with just under Rs. 30.000, with around Rs. 30,000 going to various amins, clerks, forest guards, forest range officers, circle inspectors and check-post personnel.

Not a happy story. But an instructive one. First, this one story brings to life a set of state–society interactions that would be experienced by millions of poor people in India. Each commission paid, each insult offered, each threat made, builds up our account of what anthropologists call the 'everyday state' in India or South Asia. For many poor and vulnerable people the world over, life is lived not so much in a civil society – where are there well-defined rights and responsibilities – but in intensely political societies where connections to the powerful are sparse but greatly prized. And second, Polus's experience, which, to repeat, was widely shared, disinclined many like to him to harvest trees or to invest in the planting of new trees (assets). A forestry regime that was intended to protect locals from avaricious outsiders had led to an unexpected degree of ecological stagnation. Significantly, when the legislation protecting *adivasi* farmers was relaxed, as it was for a short while in the 1990s, tree-cutting took off, but so too did new tree-planting. Landscapes changed as local institutions (the rules of the game) took on a different shape.

References

Corbridge, S. and Kumar, K. (2002). Community, corruption, landscape: tales from the tree trade in eastern India. *Political Geography* 21, 765–788.

Corbridge, S., Williams, G., Srivastava, M. and Veron, R. (2005). *Seeing the State: Governance and Governmentality in India.* Cambridge University Press, Cambridge.

17 • *Fieldwork, Field-Friends and the Paradox of Absence*

DOUGLAS J. DAVIES

Though I have worked on death rites, Sikhs, emotions and the Church of England, it is to my Mormon research, which began in 1969, that I focus this essay on fieldwork (Douglas, 2010, 2015, 2017, 2018). Through participative and archival research in the UK and the USA, through conferences, the supervision and examination of doctoral work and the ensuing friendships, I have found myself networked with scholars and church members in ways hardly imagined by my first-year post-graduate self. This Mormon chapter of my life has been interspersed with, and much influenced by, the other topics just mentioned, each posing intriguing questions of 'fieldwork' and of experience in one 'field' leaking or flowing into, or even occasionally flooding, the others.

Experience is the key to all fieldwork. For, however we define 'fieldwork', our experience of emotional and intellectual shifts plays an important role in an academically engaged life. As an exercise in a free-thinking association of ideas on fieldwork let me leave all caveats of autobiography and reflexivity to your critical judgement as I take a series of cameo-like topics that strike me as theoretically significant. These resemble coloured elements in a kaleidoscope of thought whose symbolic power has grown over a lifetime and whose shifting patterns have prompted new insight.

My fieldwork began in the mind. Within the space of a single day my first postgraduate supervisor changed from the Africanist, Godfrey Lienhardt, to the sociologist of religion, Bryan R. Wilson, whose researchers then covered numerous groups but with nobody engaged with Mormons; that turned out to be my task. As I had no knowledge of the 'Mormons', they appeared as but idea in my mind. This mental dimension grew rapidly, especially when I 'discovered' that the Bodleian Library held a collection of the Mormon journal, *The Millennial Star*, which had begun publication in the UK in 1840, as a vehicle for church news. It ceased as this title only in 1970. Daily, selective, reading through this paper treasury, with subsequent work in US archives, generated images of historical Mormonism, a constant reminder of anthropologist Evans-Pritchard's musings on where history transforms into sociology. Library-archival research was exciting, perhaps in ways that today's non-travel, digital resources have toned down.

Fieldwork continued when I set out to meet Mormons, inverting the normal social contact in which Mormons set out to meet others. I first visited the newly built chapel at Merthyr Tydfil, the town that had been the centre of the Welsh Industrial revolution that had, by 1845, already hosted nearly 500 members following three years of Mormon mission. There I happened to meet, and personally 'click' with, a senior church leader, who introduced me to what would be an ever-expanding circle of acquaintance in Wales, and in Utah. Fieldwork certainly benefits from such sensed affinity with persons and, probably in some ways, also with their worldviews. Here, in Merthyr as in Utah, history pervaded contemporary life, with archival and geographical–social fields strongly interplaying. While fieldwork seemed to take the form of working on 'the history of the Church' it was, essentially, working on the current lives of members.

Amongst the kaleidoscopic colours of those early months, one impressive set came from a Mormon student conference and one small prayer-testimony event, where I was the only non-LDS (Latter-day Saint). Here, I first encountered a highly intensified form of the more usual controlled expression of personal emotion present in local testimony meetings. Practically all cried in this close-bonded togetherness, where they spoke of their certain truth of this movement, its leaders, of Jesus Christ and of Heavenly Father's care. Fieldwork is emotion-work.

On a much larger scale came experiences of major Church conferences, both in the UK and USA, where the visual sight of key church leaders manifested the hierarchal leadership of a group that, over my lifetime, grew from one million to some 15 million members.

And what of the physical 'field' in fieldwork? In the Mormon case, the paradigmatic scene lay in Utah and its Salt Lake City. 'This is the place' had been Brigham Young's historical pronouncement over the Salt Lake Valley, when viewing it as the persecuted Saints migrated west after Joseph Smith was killed in 1844. The Mormon hymn, 'Zion Stands

Figure 17.1 The Church of Jesus Christ of Latter-day Saints, 800 East and Center Street, Orem, Utah (Ben Abbott).

with Hills Surrounded' almost inevitably defines this desert field, and certainly comes to mind in a daylight incoming flight, especially when these hills are snow-capped.

Fieldwork, too, involves a complex interplay of abstract theory and physical–social contexts. One particular complexity of place, and of what is now often captured in terms of 'material culture' and 'lived religion', for a non-Mormon fieldworker, concerns Mormon temples. Unlike those testimony and other events in local churches, which are open to all, and even unlike more private moments in Mormon homes where ordinary life may involve prayer and where domestic artefacts betoken the faith, the temples are only accessible by faith- and practice-tested members. Except, that is, during periods following temple construction or refurbishment and prior to dedication or rededication. I have been to three such events, two in the UK and one in Utah. The reason I focus on the temple, itself distinguished as a most sacred place by Mormons, is to discuss 'absence' in fieldwork; the things we do not see.

Although I have researched the rites and 'secrecy' of temple events, know that illicit filming of them has occurred and have had both members and ex-members speak to me about their diverse temple experiences, I have not experienced them during ritual performance. I have, also, taken the decision not to repeat in my own publications the details of the rites that members themselves might well regard as disrespectful. In this sense, fieldwork involves absence. Though much could be said about this in more theoretical terms I simply highlight 'absence' here as a paradox of knowledge, and as a quandary of interpretation of the privacy of personhood.

Personhood often involves friendship and so does fieldwork. Many fieldworkers will, in their older age, have field-friends. If that term 'field-friend' does not already exist, let me coin it here! (I ignore any potential field-foes that some might also accumulate.) Field-friends are those whose company has involved shared knowledge and some participation in their own worlds. Here emotions are, again, radically important, especially the social emotions of respect and trust. Fieldwork without trust will be deceptive and, ultimately, disfigured. Perhaps I should note here that I specifically avoid the conceptual trap of insider–outsider terminology, which I regard as largely simplistic and unhelpful, despite their frequent discussion in 'methodology'. For there are more than a few insiders who feel outside some time in their life, just as some outsiders may, similarly, feel insiders. For both the emotional and cognitive domains of life easily foster each identity over time. One theoretical query here might ask how insiderhood or outsiderhood pertains to friendship? A further ethical, existential, query would touch on betrayal – itself a key human phenomenon. Just how might description of a field progress to disclosure, and onwards to betrayal?

Time passes; with fieldwork being both engagement in and reflection upon passing time. Fieldwork, as an integration of what some traditions describe as ethnography and ethnology, involves both encyclopaedic and symbolic knowledge, as anthropologist Dan Sperber once categorised this cumulative though far from smooth process of human knowing. Moments of insight make scholarship joyous. Let me pinpoint but two such moments in my Mormon work, one early, another late. The first followed an early 1970s toying with church-sect typologies and came not when reading innumerable *Millennial Star* volumes but when walking past Oxford's Radcliffe Camera building. It dawned on me that Mormon life, centred around the local chapel, was very 'church-like' while temple activities comprised the Mormon 'sect'. This simple thought proved a turning point in my early theoretical work. It then took another 15 years or so, following other work on burial and cremation rites at large, before another dawning moment showed me that Mormonism was, in itself, a vast and often explicit project in death-conquest. I saw that the Prophet Joseph Smith's own enormous experience of grief could be interpreted as the matrix of key LDS ritual practice of genealogy and vicarious temple ritual for the dead. The embodied tingle-factor of these momentary insights, along with just a few others in my intellectual life, have brought the realisation that fieldwork is, in fact, a life-work where the active engagement with other people, times and place interplays with what it would be a shame to describe as the 'passive' periods of integrative reflection, a reflection which often generates insight in parallel with the act of writing. Fieldwork is writing, not only in the idiographic 'field notes' of a day, but in that later nomothetic process framed in monographs. This long-haul flight of data and fancy necessitates trust, both in field-friends and in ourselves, our observation, recall, judgement and theoretical configurations. Fieldwork can be a life-long field-day.

References

Davies, D. J. (2010). *Joseph Smith, Jesus, and Satanic Opposition: Atonement, Evil and the Mormon Vision*. Ashgate, Farnham.

Davies, D. J. (2015). *Mors Britannica: Lifestyle and Death-Style in Britain Today*. Oxford University Press, Oxford.

Davies, D. J. (2017). *Death Ritual and Belief: The Rhetoric of Funerary Rites*. Third, expanded edition. Bloomsbury, London.

Davies, D. J. (2018). Anthropology and theology: fugues of thought and action. In *Theologically Engaged Anthropology*, ed J. D. Lemons. Oxford University Press, Oxford, pp. 194–210.

18 • *Ornithological Fieldwork: Essential and Enjoyable*

ROY DENNIS

Over more than 60 years of working in wildlife conservation, I have regarded fieldwork as the bedrock of my being. It's a quest for knowledge and a chance to pass it on to others. Fieldwork with birds involves much more than the basic identification of species. It is, for example, about knowing their signs – understanding the story told by a pile of feathers on the ground (did the bird moult or was it killed?) and identifying a broken eggshell and knowing how the birds live their lives. The most basic but crucial information comes from fieldwork – how many pairs bred this year; what was their breeding success? That fieldwork knowledge is your currency. No matter who you meet – farmer, forester, policeman, politician, student or teacher – they will recognise your skill and expertise, while you will know that, no matter what they ask, you can answer with clarity, conviction and enthusiasm – if you know the answers. And besides all that, fieldwork is fun.

You do have to work at it, but that was always what I found most alluring. To sally forth as a birdwatcher, with a notebook and binoculars, was an exciting expedition. New places were discovered, old ones re-examined at different seasons or times of day. It was a continuous process of learning, adding sightings and data to your ever-growing bank of knowledge. When and where did I first realise that a common gull's legs were green? I can tell you. It was perched on a post in a muddy inlet on the Hamble River, Hampshire, and I was 14. And then there's the essential encouragement of mentors. My scoutmaster had a copy of *The Popular Handbook of British Birds* (Hollom, 1952) and I learnt a lot more about the common gull that day beyond it having green legs. Locked away in my brain, fieldwork-derived information is precious and enduring.

A Young Person's Perspective

As a young person, I found the physical challenge of fieldwork an exciting part of its magic: climbing tall trees to get to nests, wading river channels to explore islands and early-morning hitchhiking to my favourite birding places. Looking back, we were allowed to do so much, often on our own. Recently, I re-read my early field notebooks – my annual diaries go back to when I was 12 – and found the first references to my bird-ringing activities. I've been a bird-ringer all my life, starting at St Catherine's Point on the Isle of Wight. We used to write to the Elder Brethren of Trinity House and, incredibly, they would give us letters of permission to catch and ring birds at the lighthouse. St Catherine's Lighthouse is 80 feet high, and the keepers would allow us to climb the tower and go outside onto the balcony, at night, to catch birds with butterfly nets as they fluttered in the dazzling rays of the powerful beams, night-flying migrants returning over the Channel from France: garden warblers, whitethroats, redstarts, spotted flycatchers and, once, a quail. I recall one spring night when we caught and ringed more than 50 birds. It was magical and we were adding knowledge to the great database maintained by the British Trust for Ornithology.

There is a tendency to think that it's very difficult to gain a special ability in fieldwork: it's too time-consuming, too tricky and, some would say, too much hard work. But within four years of starting out as a budding birder, I was able to land a job for the summer and autumn months as an assistant warden on Lundy Island. What an experience that was. I jumped at the chance, catching my first seabirds, counting my first migrants, showing visitors new birds and how to identify them and, of course, bird-ringing, often with real excitements, such as on the rainy day when I found the first recorded Baltimore oriole for Britain, wind-blown from North America (Workman, 1963).

Reflections on Fair Isle

Six months later, I was travelling north on the steamer from Aberdeen, for a year as assistant warden at the world-famous Bird Observatory on Fair Isle. Peter Davies was the warden, and he was great at teaching me to become a field ornithologist. And that's how I still think of myself, despite my later interest in large mammals, ecology and rewilding.

Later, I returned as warden in charge of Fair Isle Bird Observatory and now, after 60 years of being enthralled by that island, I am its Honorary

Figure 18.1 Roy Dennis and lynx, 7 July 2001, Oslo (from the author's photo collection). A black and white version of this figure will appear in some formats. For the colour version, please refer to the plate section.

President. But it's the early years which I remember most fondly: counting, ringing and studying the seabirds on the great cliffs; taking small boat trips to hidden beaches to ring black guillemots and shags; scrambling down 400-foot screes (often un-roped) to catch razorbills and guillemots in incredibly smelly jumbles of rock; and the thrill of catching fulmars and puffins in flight with a genuine Faroese fleyg net. All of the ringing data were carefully recorded at the end of the day, contributing to a store of knowledge. And years later, we had the thrill of seeing many of those birds caught again, sometimes decades later, and learning more about them.

Some of our exploits, looking back, were hair-raising and would never be allowed today. We wanted to ring kittiwakes on the great western cliffs but, because of the sea swell, they had always been just out of reach. One day, I realised that if we took a ladder with us, and two people held the ends of the open boat against the base of the cliff, I could climb up and ring the chicks. It was exhilarating and highly dangerous.

Figure 18.2 Roy Dennis and Lapland bunting (from the author's photo collection)

We would make night-time expeditions onto the grassy slopes of the high cliffs to mist-net and ring storm petrels. Adding to the science was important, but so was the experience that people gained. Many of them thanked me in later years for taking them to such exciting places and told me that the experience was what had made them so interested in nature. For some, it had started a life's work as a naturalist.

At the time, you could have said that some of the fieldwork might not have had much point beyond being exciting, especially on Fair Isle. We observed and recorded the daily migrations of birds over years: massive flocks of redwings and fieldfares on their way from Scandinavia to Scotland or, suddenly, one spring day, the first ever Cretschmar's bunting for Britain (Dennis, 1969). And then, a few years ago, scientists – with their present-day ability to analyse huge chunks of data – looked at all the migration counts on the island since 1948 and published papers which help us understand the changing nature of climates and bird populations (Miles *et al.*, 2017). That day-by-day, year-by-year fieldwork has contributed importantly to science in an uncertain age.

Fieldwork is not just about acquiring knowledge: it's a physical and intellectual demonstration of our ability to understand and record nature.

It helps us add to the debate on how we look after planet Earth, and to be honest and proud in discussions with other interest groups, without feeling insecure about our knowledge and our abilities. Any young person with the necessary drive and determination can, as I did long ago, get started on fieldwork. I say, give it a go – and don't let anything prevent you from doing the things you love. The harder it is, the greater and more enduring the rewards.

References

Dennis, R. H. (1969). Cretschmar's bunting on Fair Isle: new to Britain and Ireland. *British Birds* 62, 144–148.

Hollom, P. A. D. (1952). *The Popular Handbook of British Birds*. Witherby, London.

Miles, W. T. S., Bolton, M., Davis, P., *et al.* (2017). Quantifying full phenological event distributions reveals simultaneous advances, temporal stability and delays in spring and autumn migration timing in long-distance migratory birds. *Global Change Biology* 23, 1400–1414.

Workman, W. B. (1963). Baltimore oriole on Lundy, Devon (1958). *British Birds* 56, 52–55.

19 · *Exploration Science on the Shore of the Arctic Ocean: a Personal Experience*

DAVID J. A. EVANS

Upon embarking on my PhD programme in 1984 at the University of Alberta I was aware that I was entering into an arena of exploration science. For I had joined the research team of Professor John England, the pre-eminent and passionate expert on glaciation and sea-level history of the Canadian High Arctic. I had to quickly adapt to the style of field research that had been undertaken by generations of Quaternary geologists in northern Canada, which involved long summer seasons, small teams living frugally in tents, the foot-traversing of vast areas of unexplored terrain with occasional camp moves facilitated by helicopters. My field area was the north-west coast of Ellesmere Island on the shores of the Arctic Ocean, an area visited by less than a handful of scientists in the 1950s and 1960s, getting around by dog sled teams, skis or walking in between their drop off and pick up by a small Piper Cub plane on skis or balloon tyres.

In that pre-Google Earth era I had nothing more than very preliminary 1:250,000 scale maps and 1950 aerial photographs from which to select my 1985 season camp locations. I also had to direct the pilots, who were to fly in first my fuel dump and then me and my two assistants to the field area in the stalwart Canadian bush plane, the De Havilland Twin Otter. My arrival over my field area in mid-June was to prove a little stressful, as most of the snow upon which the Twin Otter had landed on skis two months earlier was still covering the majority of the ground and we were now on tyres! 'That's my landing site', I shouted to the pilot, my head peering through into the cockpit and pointing to a raised marine delta on one of my aerial photographs – 'crikey, it looked an awful lot bigger under the stereoscope than it does right now'. After

Figure 19.1 Camp at Cape Armstrong, north-west Ellesmere Island, Canadian High Arctic, 1987 (David Evans)

an hour of circling and touch-and-go practice runs in a fully loaded plane we were on the ground, a feat of flying that to this day I find astonishing, and the first of many throughout my many trips to the Arctic that have for me put Canadian bush pilots in the league of superhuman; the wheel marks of the plane were imprinted right on the lip of the delta terrace – there was absolutely no room for error! In less than 30 minutes we were unloaded, our SBX11 shortwave radio tested, our rifles unpacked and loaded (we were deep in polar bear country after all) and the Twin Otter was roaring off the delta surface and ascending slowly out of our sea-ice-covered fjord, leaving us in the overwhelming silence of the High Arctic.

What followed in that first field season of three for my PhD research was six weeks of traversing mountainous and fjord terrain, first by cross-country ski, then on foot, sometimes 'making time' by using our skis as skates on the sea ice, wading through ice-cold water in rivers raging with fresh snowmelt, and all the time mapping landforms and sediments on aerial photographs, logging sections and collecting samples for radiocarbon dating. We saw another person only on the three occasions that we received a helicopter to help move our camp to another fjord system, the pilot delivering our mail, some fresh meat and news of the outside world. For this was before the era of satellite phones and our only connection to

Figure 19.2 Taking time to write my notes and enjoy a rest from walking in the arctic sunshine. Phillips Inlet, 1985 (from the author's photo collection). A black and white version of this figure will appear in some formats. For the colour version, please refer to the plate section.

the south was the 'sched', twice a day at 07:00 and 19:00, to ensure we were safe and to glean from us, meteorologists for the summer season, our weather conditions. On one of my most memorable days of so many, we reached the summit of a previously unclimbed mountain, to an astonishing view, across the low fog that clung to the fjords and north coast, to the glacier-covered peaks of the Grant Land Mountains of northern Ellesmere Island. To someone who loved his subject and the great outdoors in equal measure this was truly 'manna from heaven'!

For me, the Canadian High Arctic is not only the most beautiful, unspoilt place on Earth, it is the field experience that constitutes my overwhelmingly evocative backdrop and soundtrack to all I have done and still do in my heavily field-orientated career. For I was carrying out unbridled, curiosity-led research with a remit so broad that I could not fail but discover important things at every turn – there was no critical rationalist agenda. Of course there was a bigger-picture hypothesis to test and it was my role to reconstruct the style of the last glaciation and its impact upon sea-level change for my designated part of Ellesmere Island, as part of a larger research programme (Evans *et al.*, 2014; Evans, 2016). But such was the impact on me of the exploration science experience on

Ellesmere Island that to this day I tell my students that there is nothing inherently evil in indulging in a little exploratory, rather than strictly hypothesis-testing, science now and then; in fact, it positively nurtures the inquisitive mind in us all and prepares us to be receptive to all the signals coded into unexplored environments. When the Twin Otter lifted off from the last of my many camps in 1987, and it slowly turned in the sky for me to look down on the spectacular fjords and calving glacier tongues that lay below, I knew that I had been very privileged to see such a place and was glad that I had made the most it. My field notebooks were full of scribbles and sketches on everything from periglacial patterned ground to avalanching glacier snouts – I had done more than work out the glaciation and sea-level history!

References

Evans, D. J. A. (1990). The last glaciation and relative sea level history of NW Ellesmere Island, Canadian High Arctic. *Journal of Quaternary Science* 5, 67–82.

Evans, D. J. A. (2016). Landscapes at the periphery of glacierization: retrospect and prospect. *Scottish Geographical Journal* 132, 140–163.

Evans, D. J. A. and England, J. (1992). Geomorphological evidence of Holocene climatic change from northwest Ellesmere Island, Canadian High Arctic. *The Holocene* 2, 148–158.

Evans, D. J. A., England, J. H., La Farge, C., *et al.* (2014). Quaternary geology of the Duck Hawk Bluffs, southwest Banks Island, Arctic Canada: a re-investigation of a critical terrestrial type locality for glacial and interglacial events bordering the Arctic Ocean. *Quaternary Science Review* 91, 82–123.

20 · *Only Connect – and Make Records*

ALASTAIR FITTER

I did my PhD under the supervision of Tony Bradshaw, a truly remarkable (and delightful) man who bridged the pure/applied divide in science with enthusiasm and aplomb (Fitter, 2010). He was an evolutionary ecologist who wrote an immensely influential review on phenotypic plasticity (Bradshaw, 1965), at a time when the subject was viewed as being in the province of taxonomists rather than evolutionary biologists. His early work on the evolution of resistance to heavy metals in grasses (e.g. Gregory and Bradshaw, 1965) – still one of the best demonstrations of evolution in action, even if largely ignored because the organisms in question are uncharismatic – led directly to his major contributions to the scientific approach to the reclamation of derelict land. He took me on to work on reclaiming old coal tips in the Lancashire coalfield, north-west England, which already by the late 1960s was largely abandoned.

By its nature, the project was field-based. Of course, there were glasshouse experiments to tease out some of the factors that determined plant growth on these unpromising materials, but the main body of work was a series of field trials on old coal tips around Ince, Wigan and Ashton-in-Makerfield. This was then – and sadly still is – an economically depressed area, at least in part because the landscape was so badly scarred. Some might feel (indeed, I feel) that a rainforest, coral reef or alpine summit would have been more congenial, but the true point of fieldwork is that whatever your ecological system, you can only attain realism in your research by performing it in the environment where the system occurs.

Abandoned coal tips turned out to have their own challenges. They were mainly rather bleak expanses of the spoil that had been dug up –

these had been deep mines – to access the coal seams. It was easy to show that the big problems were, first, a serious lack of nutrients, especially nitrogen and phosphorus, and second, on many of the sites, severe acidity. The acidity arose because the coal was laid down in coastal swamps that were intermittently inundated by the sea, trapping the peat (which turned into coal) beneath sands and muds that were sulphate-rich from the seawater. This sulphate turned into sulphide in anoxic conditions and when the resultant rock was dug up and dumped in spoil heaps, it turned back into sulphuric acid, leading to 'soil' pH values sometimes below 3. A consequential problem was drought: the crushed rock that passed for soil held little water and that problem was exacerbated because the poor plants grew miserable root systems, thanks to the infertility and especially the acidity.

The solution was simple: to add limestone and fertiliser, and ideally to grow lots of clover, and soil began to form. My field plots involved working out the best combinations of these and the result was great patches of green grass growing on the grey spoil wastes. I guess I should have anticipated that these would attract attention. My first attempt worked well and I was proud of the green sward. On one visit to check on progress I was disconcerted to find that a group of travellers had parked their caravan bang on top of it. They were curiously reluctant to move, pointing out, not unreasonably, that they had chosen the nicest spot to stop.

I was not going to let my next set of plots, on the site of the old Outwood colliery, go the same way. I chose a site on the top of a steep-sided but flat-topped tip, inaccessible to caravans or any vehicles, and I set out the plots. A new problem appeared: rabbits. Every starving rabbit in the neighbourhood came to feast on the lush grass that I was providing for them, making measurements of productivity problematic. I erected a substantial fence all around the plots and all the rabbits could do was sit outside and gaze longingly at the grass inside.

On my next visit, I climbed the steep slope and as soon as I got to the top it was clear that all was not well: the grass was grazed to the ground. However, the fence was intact and I could not work out how the rabbits had got in. Eventually I discovered that the staples holding the wire in place had been carefully unpicked at one of the corner posts. Then I discovered a large pile of empty 12-bore cartridge cases on the slope of the tip on the far side and realised that an enterprising local had turned my experiment into a giant rabbit trap. By unpicking the wire and rolling it back, he (I guess it was a he) had allowed the rabbits in; then he had

Figure 20.1 Richard Fitter (1913–2005): naturalist, author and compiler of remarkable data sets (from the author's photo collection)

crept up the slope and let rip with his shotgun, before carefully restoring the wire to allow the grass to regrow. I eventually did get useable data from the experiment, but the main lesson was that reclamation worked and that, long before it was fashionable or even mandated by research funders, I had achieved public engagement with my research.

Most of my subsequent research was field-based or field-inspired, but the best field data set I ever had the privilege of working on was not one I compiled: it was my father's remarkable, near 50-year, sequence of first flowering dates of plants around his home in Oxfordshire. He was a professional naturalist, earning his living by writing pioneering field guides and other natural history books. As part of this work, he kept detailed records of many natural phenomena, but especially of birds and flowering times. I analysed his flowering-time data when he moved house in 2000 and the record came to an enforced stop. There were records for nearly 600 species, and sufficient data for a time series on 385 of them.

The data showed that, in the 1990s, the flowering time had advanced by, on average, 4.5 days compared to the previous four decades; however, that average concealed a remarkable variation, from plants whose first flowering date was now eight weeks earlier to some where it was significantly later (Fitter and Fitter, 2002). The link to warming was clear: a rise of 1 °C in spring advanced flowering across all species by four days. At the time, this was the largest data set to show the impact of climate change on the natural world, and its findings have been reinforced by many subsequent studies. It is obvious that I did the easy bit – the data analysis – while my father's contribution in generating the entire data set was fundamental. Because he was 90 when the paper was published and because the findings had a high profile anyway, we were interviewed on the 'PM' news magazine programme on national radio. He was asked why he had collected the data and he replied that when he was a young naturalist he had read that it was a good idea to make records of what you saw, and so he had started doing that and never stopped.

That is an unsung aspect of fieldwork: the data collected by thousands of naturalists, often with no specific plan in mind, are fundamental to doing good ecology, because all good ecology starts with natural history.

References

Bradshaw, A. D. (1965). Evolutionary significance of phenotypic plasticity in plants. *Advances in Genetics* 13, 115–155.

Fitter, A. H. (2010). Anthony Bradshaw. *Biographical Memoirs of Fellows of the Royal Society* 56, 41–61.

Fitter, A. H. and Fitter, R. S. R. (2002). Rapid changes in flowering time in British plants. *Science* 296, 1689–1691.

Gregory, R. P. G. and Bradshaw, A. D. (1965). Heavy metal tolerance in populations of Agrostis tenuis Sibth. and other grasses. *New Phytologist* 64, 131–143.

Figure 1.2 Why have we brought you here? Andrew Goudie and Tim Burt always used to start their Dorset coast field trip with this simple question. On this photo, Andrew is talking to students at Durdle Dor (Tim Burt). A black and white version of this figure will appear in some formats.

Figure 1.14 Edinburgh students visit Siccar Point, James Hutton's famous unconformity, where a gap of some 60 million years separates the two rock formations. Unconformities are discussed in chapters 13 (Tim Burt) and 25 (Peter Higgins). A black and white version of this figure will appear in some formats.

Figure 1.15 Curiosity about the world around us leads inevitably to patient observation of whatever it is we choose to investigate (Des Thompson). A black and white version of this figure will appear in some formats.

Figure 2.6 Modern fieldwork: raft equipped with acoustic Doppler current profilers and single-beam sonar. With this equipment, continuous data of velocity profile and water depth were obtained in an eight-day traverse covering more than 500 km of Fraser River, British Columbia, including passage through the Fraser Canyon (Rennie *et al.*, 2018). Data were supplemented by channel widths derived from aerial photography (Michael Church). A black and white version of this figure will appear in some formats.

Figure 2.10 Citizen science. (a) Members of the Cascades Butterfly Project, a long-term survey in the North Cascades and Mount Rainier National Parks, Washington State, USA. The objective is to understand the effects of environmental change in alpine meadows, for which butterflies are important indicator species. Photo by National Park Ranger. A black and white version of this figure will appear in some formats.

Figure 5.1 Students learning about human geography at the Olympic Stadium in East London. (Photograph provided by the Field Studies Council.) A black and white version of this figure will appear in some formats.

Figure 9.2 Coring the sediments of Fjell Lake, near Bergen, Norway. The student palaeoecology class watches keenly (Hilary Birks). A black and white version of this figure will appear in some formats.

Figure 10.2 Buxbaumia aphylla (John Birks). A black and white version of this figure will appear in some formats.

Figure 13.1 Jim Hanwell (white shirt) examining the surface of the Vallis Vale unconformity (Tim Burt). A black and white version of this figure will appear in some formats.

Figure 13.2 Jim Hanwell and Tim Burt at the Vallis Vale unconformity (Elizabeth Burt). A black and white version of this figure will appear in some formats.

Figure 12.2 The climatological station at Malham Tarn (54.10° N, 2.16° W, 395 m above mean sea level), photographed on 27 May 2017 (Stephen Burt). The louvred white box is a Stevenson Screen. A black and white version of this figure will appear in some formats.

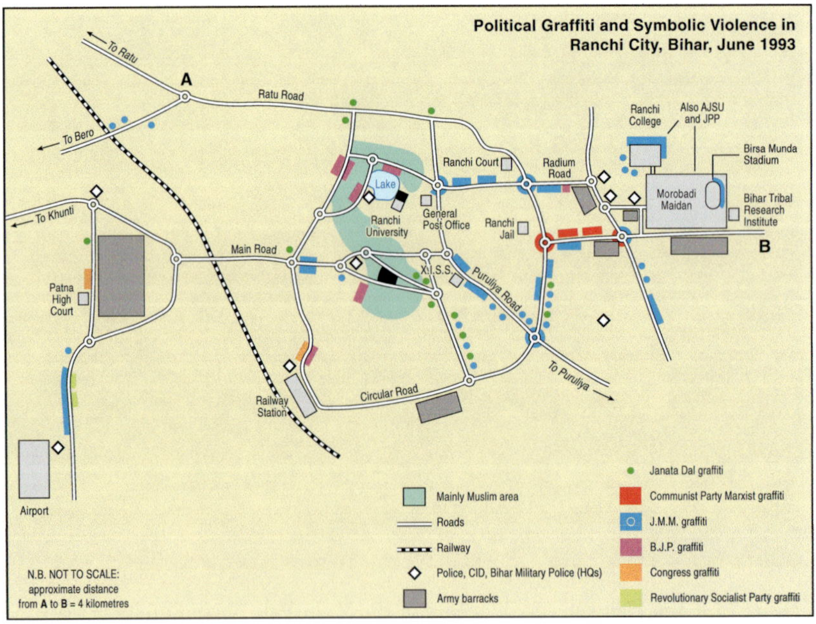

Figure 16.1 Political graffiti and symbolic violence in Ranchi City, Bihar, June 1993 (Stuart Corbridge). A black and white version of this figure will appear in some formats.

Figure 19.2 Taking time to write my notes and enjoy a rest from walking in the arctic sunshine. Phillips Inlet, 1985 (from the author's photo collection). A black and white version of this figure will appear in some formats.

Figure 21.1 David Goode making measurements on the Silver Flowe, 13 August 1965 (Philip Smith). A black and white version of this figure will appear in some formats.

Figure 23.2 Mus musculus domesticus – Valtellina, Italy (Heidi Hauffe). A black and white version of this figure will appear in some formats.

Figure 24.2 Male chacma baboon from our habituated study group feeding in the flooded De Hoop Vlei (Russell Hill). A black and white version of this figure will appear in some formats.

Figure 41.1 A female *Kawanaphila nartee* feasts on the spermatophylax while sperm are transferred from the spermatophore to her sperm storage organs (Leigh Simmons). A black and white version of this figure will appear in some formats.

Figure 23.3 Heidi Hauffe with crow (Heidi Hauffe). A black and white version of this figure will appear in some formats.

Figure 14.3 A number of pelagic anglerfish species, such as the humpback anglerfish (*Melanocetus johnsonii*) shown here, live in the deep below the photic zone. Typically, they have black skin, only producing light in a small organ at the end of a rod pointing up from the forehead. This organ is used as a lure to attract prey. Large gapes, sharp fanglike teeth and widely distensible stomachs enable these fish to hold onto, and swallow, surprisingly large prey (Ingvar Byrkjedal). A black and white version of this figure will appear in some formats.

Figure 37.2 Plecotus auritus, the brown long-eared bat (courtesy of John Altringham). A black and white version of this figure will appear in some formats.

Figure 25.2 Freshwater study – Kinloch Burn, Isle of Rum NNR (Peter Higgins). A black and white version of this figure will appear in some formats.

Figure 25.3 Bog vegetation survey – Kinloch, Isle of Rum NNR (Peter Higgins). A black and white version of this figure will appear in some formats.

Figure 26.1 Holding my breath and measuring the rugosity of a coral colony in Clam Gardens, Lizard Island on the Great Barrier Reef (Emily Husband). A black and white version of this figure will appear in some formats.

Figure 31.1 Devil's bolete (original artwork by Albin Schmalfuss). A black and white version of this figure will appear in some formats.

Figure 38.2 A family of tawny frogmouths aligned and in stick-pose, but their multiple outlines are less concealing (as branch-like) than that of a single bird (Stuart Rae). A black and white version of this figure will appear in some formats.

Figure 32.2 Teaching students on our Blencathra field course (from the author's photo collection). A black and white version of this figure will appear in some formats.

Figure 34.3 The South Drain, which runs along the edge of Westhay Heath, eventually running into the Huntspill River (Stephen Moss). A black and white version of this figure will appear in some formats.

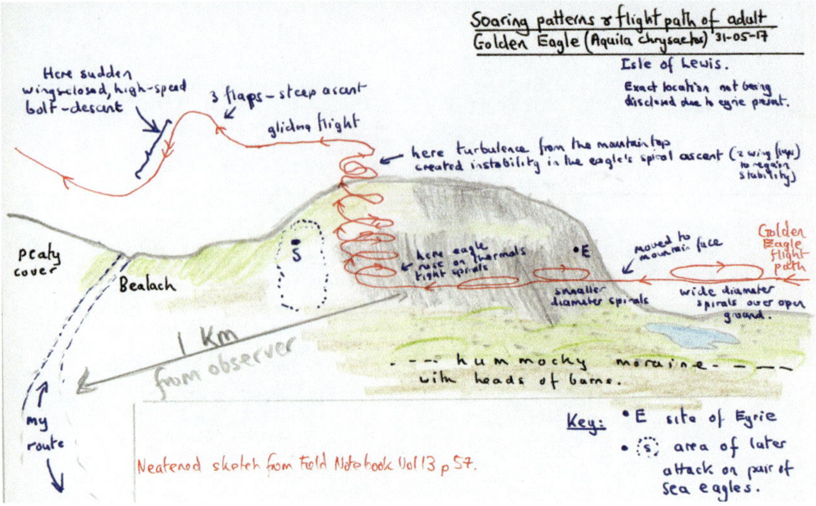

Figure 35.2 Golden eagle territorial behaviour near its eyrie in the Outer Hebrides. (Grid reference withheld for reasons of eyrie security). (Original artwork by Stephen Mott.) A black and white version of this figure will appear in some formats.

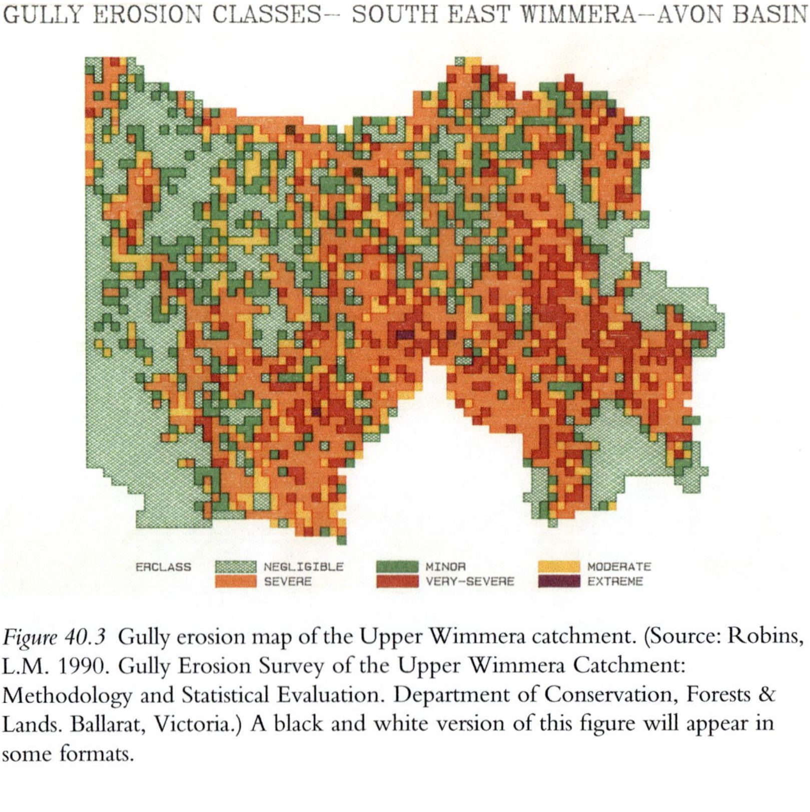

Figure 40.3 Gully erosion map of the Upper Wimmera catchment. (Source: Robins, L.M. 1990. Gully Erosion Survey of the Upper Wimmera Catchment: Methodology and Statistical Evaluation. Department of Conservation, Forests & Lands. Ballarat, Victoria.) A black and white version of this figure will appear in some formats.

Figure 42.3 Hydrological fieldwork means there is no shortcut to getting good data! Here, Dr Paul Gares is about to measure velocity in a rill channel on a tobacco field, during extreme rainfall (Tim Burt). A black and white version of this figure will appear in some formats.

Figure 46.4 Freshwater invertebrate identification using Field Studies Council fold-out charts (Field Studies Council). A black and white version of this figure will appear in some formats.

Figure 47.1 How it all started – my first fossil, found in the back garden at the age of seven: an imprint of a bivalve in a flint pebble derived from the chalk (Upper Cretaceous) (Maurice Tucker). A black and white version of this figure will appear in some formats.

Figure 18.1 Roy Dennis and lynx, 7 July 2001, Oslo (from the author's photo collection). A black and white version of this figure will appear in some formats.

Figure 28.1 Charley Krebs live-trapping a red-backed vole (*Myodes rutilus*) at Kluane Lake in the Yukon. These voles have a three–four-year cycle in this white spruce boreal forest (Charles Krebs). A black and white version of this figure will appear in some formats.

Figure 48.1 Camping in the Grand Canyon on a Colorado River sand bar (Alan Werritty). A black and white version of this figure will appear in some formats.

Figure 49.1 Nature offers us so many learning opportunities, each one building our connection with the Earth (Natalie White). A black and white version of this figure will appear in some formats.

Figure 49.2 The awe and wonder found in our environment helps to strengthen our connections with nature (Natalie White). A black and white version of this figure will appear in some formats.

Figure 50.3 Students looking at leaf litter samples in Slapton Woods (Lewis Winks). A black and white version of this figure will appear in some formats.

Figure 51.5 Professors Jeff Warburton and Tim Davies studying the crust of a dried-out bog pool on Bleaklow in the South Pennines, England. Over the past 10 years there have been transformative advances in understanding peat erosion and restoring bogs at a large scale to secure their carbon stocks (Tim Burt). A black and white version of this figure will appear in some formats.

21 · *Studying Patterned Bogs*

DAVID GOODE

In 1964, I was offered a research studentship at Hull University to investigate pattern formation on bogs of the Silver Flowe National Nature Reserve in south-west Scotland. These bogs exhibit the same striking patterns of pools and ridges that are found on peatlands throughout the boreal zone. Although there were many theories about the origins of such patterns (e.g. Sjors, 1961), I was aware that no one had yet produced a plausible explanation that would apply across the whole range of peatlands – from the extensive so-called string bogs of Scandinavia, Siberia and northern Canada to the blanket bogs of Scotland and even Tierra del Fuego. The patterns posed a great puzzle, and I was totally absorbed by the problem.

How should I go about my investigation? On my first visit to the Silver Flowe I could see that some areas of bog had concentric patterns of pools and ridges. Other areas had extensive pool networks, and some had no patterns at all. I decided to look in detail at a small area that exhibited these differences. I wanted to know how the different kinds of pattern related to the overall morphology, peat depth and gradient of the bog. Then there was the influence of physical hydrology in controlling pattern development and the distribution of plant species. There were also stratigraphical investigations to be done. My aim was to discover the key factors that influenced pattern formation, and it turned into a daunting task.

Academic decisions were one thing, practicalities another. The Silver Flowe lies in an isolated glen in the Galloway Hills, about 14 miles from the nearest village. Fortunately, there was a small cottage very close to my study area, known as the Backhill of Bush. Built as a summer shieling for shepherds but long abandoned, it was now available as a mountain bothy for use by hill walkers. For the next three years this became my second home where I stayed during intensive periods of fieldwork. Sometimes I didn't see another soul for several weeks. There were no mobile phones. And it rained, incessantly, for days on end!

During the first few months there were jobs that required two people, so I persuaded one or two research students to join me in laying out a grid across the study area, surveying the bog surface and measuring peat depths using peat-borer rods. There was much tramping along grid lines with peat depths recorded every 3 m. This resulted in a detailed map of surface contours and the underlying mineral-ground topography. I also wanted to map the surface patterns and decided to use a robot camera, suspended from a meteorological balloon. That was easier said than done. I would inflate the balloon with hydrogen to a diameter of about 2 m, then walk for nearly 1 km with the balloon strapped to my rucksack. It felt somewhat like moonwalking. On the study area I followed the grid lines, hopefully holding the balloon at a constant altitude, taking photographs of each grid square. This required totally calm conditions, which occurred infrequently in the Galloway Hills, and when I did get calm summer weather midges were out in force (I covered exposed bits of my body in soap; even the thinnest layer was sufficient to stop them biting, and I walked around smelling of Imperial Leather!).

One evening I was sitting alone in the bothy, eating a rather good meal, which was cooked on a primus stove. My microscope, which I used for identifying *Sphagnum* mosses, stood on the table beside me. Suddenly, the door opened and an army officer appeared, followed by a troop of young soldiers. He quickly took stock of the situation and announced, 'Ah! Here we have a boffin! Take no notice of the boffin. Go upstairs and find your quarters.' I remember having a short conversation, explaining what I was doing, after which I saw little of the army until they left for their 'exercise' the next morning. I was delighted to have a calm day and was getting on well with the balloon photography when there was a sudden, sharp report. The balloon had burst and the camera in its aluminium frame was coming down fast. I ran. Fortunately, the camera didn't land in a pool and there was no damage, but I was unable to take any more photographs on that visit. Did the army shoot the balloon? I shall never know, but I had my suspicions.

With the photography completed it was clear that pool networks were restricted to the deepest areas of peat, where there were basins in the mineral ground. These were also the flattest areas of the bog. Linear pools aligned parallel to the contours occurred where the bog gradient was less than 1 in 37. On steeper slopes there were no pools. I was beginning to understand that hydrology could be a critical factor. So I installed three continuous water-level recorders on platforms supported by conduit piping that stood on the underlying bedrock. Data from these showed that the range of water-table fluctuation differed considerably between pools and ridges. I wanted to explore this in more detail, along a transect across a series of linear pools and ridges. But bogs are difficult places to measure changes in water levels. The peat surface sinks as you tread on it. I found that my weight within 1.5 m of a tube-well could cause the water level to rise by 3 cm. I needed a device that would allow me to record the water level in a series of tube-wells, without my proximity influencing the local water table. Our departmental technician produced a wonderful instrument that did exactly that. It consisted of an aluminium tube, 2 m long, with a graduated dipstick at one end, and electrodes at the bottom, that could be lowered into a tube-well by turning a knob at the other end. When it touched water, a light was switched on; it was simple and effective. This let me make accurate measurements from outside the critical distance around each tube-well and it also minimised damage to the living surface of the bog.

The results showed very clearly how the water table behaved in a series of ridges and pools. The range of fluctuation was up to three times greater in the ridges than in the pools. I was able to demonstrate that the surface layers of the ridges, to a depth of about 10–15 cm, were highly permeable, with water moving freely through the surface *Sphagnum* layer. Below that, there appeared to be little water movement through the more humified peat. In these lower layers of the ridges, the water level only fell as a result of evapotranspiration. Each ridge was in effect acting as a dam and the width of the pools was directly related to the gradient of the bog surface. Other things emerged from my fieldwork, but understanding the physical hydrology was the key that I was looking for (Goode, 1970, 1973; Boatman *et al.*, 1981).

While out on the bog with my nose down among the *Sphagnum* moss, or recording water levels, I was constantly aware of birds in the local

Figure 21.1 David Goode making measurements on the Silver Flowe, 13 August 1965 (Philip Smith). A black and white version of this figure will appear in some formats. For the colour version, please refer to the plate section.

neighbourhood, by their calls. Dunlins nested on a patch of bog nearby and I would frequently hear their long, drawn-out trill, like a referee's whistle. Peregrine falcons and ravens nested on the high crags above my bog and occasionally I had brief glimpses of golden eagles soaring at the head of the valley. One day, a solitary figure came tramping across the bog and introduced himself as Derek Ratcliffe. He was on his way to check how the peregrines were faring. Derek was equally at home here, as he had written a detailed description of the Silver Flowe in 1958. I remember him asking about the azure hawker (*Aeshna caerulea*) dragonfly, which still occurs here in one of its most southerly localities in Britain.

In 2017, Richard Lindsay invited me to join a small group who were investigating how the vegetation and patterns might have changed since my studies over 50 years ago. We located my original grid lines and a drone was flown to photograph the bog surface in colour and infrared. The bog surface was surveyed with remarkable accuracy to produce a map of the topography. The whole operation was completed in a few hours.

References

Boatman, D. J., Goode, D. A. and Hulme, P. D. (1981). The Silver Flowe III pattern development on Long Loch Bog B and Craigeazle mires. *Journal of Ecology* 69, 897–918.

Goode, D. A. (1970). *Ecological Studies on the Silver Flowe National Nature Reserve.* PhD thesis, University of Hull.

Goode, D. A. (1973). The significance of physical hydrology in the morphological classification of mires. In Procedings of the International Symposium on Peatland Classification, Glasgow, September 1973, International Peat Society.

Sjors, H. (1961). Surface patterns in boreal peatlands. *Endeavour* 20, 217–224.

22 · *Mapping the Rise of the Animals: Cambrian Bodies in the Sirius Pass, North Greenland*

DAVID A. T. HARPER

First some key facts: the Sirius Passet fossil biota is the most remote, least well known and, to date, one of the least diverse of the Cambrian exceptionally preserved assemblages or Lagerstätten (Harper *et al.*, 2019). Following its serendipitous discovery in 1984, by field geologists working for the Geological Survey of Greenland, the locality in the Buen Formation, on the edge of J. P. Koch Fjord in North Greenland has been visited by only seven collecting expeditions, most recently in 2009 and 2011 by multidisciplinary and multinational groups, led by the author when a member of staff in the Natural History Museum of Denmark, University of Copenhagen, and in 2016 and 2017 by a UK–South Korea group. The Sirius Passet Lagerstätte occurs in black mudstones deposited at the shelf–slope break on the Laurentian margin. Although deformed and metamorphosed by a Devonian tectonic event, the Ellesmerian Orogeny, the locality preserves, to a large degree, the original depositional relationships and allows an interpretation of the environmental setting of this early Cambrian ecosystem; thus it is key to understanding the environmental constraints on the Cambrian Explosion. The fossiliferous site is located in the Sirius Pass, beside J. P. Koch Fjord, Peary Land, North Greenland, at 82° 47.59′ N, 42° 13.54′ W (Figure 22.1) and an altitude of 420 m. This remote locality can only be reached by short take-off and landing aircraft that can use a rough airstrip in the valley, 2 km to the west–north-west.

During my first few years as Professor of Palaeontology in the University of Copenhagen, I was fortunate to lead and participate in a number of expeditions in the fjord region of north-east Greenland (Cambrian and Ordovician rocks), the Isua region (Archean terranes

Figure 22.1 The Sirius Passet locality, showing the relationship of the younger fossiliferous, Cambrian shales and siltstones of the Buen Formation (to the left) to the older Upper Precambrian pale limestones of the Portfjeld Formation (to the right) (David Harper)

with some of the evidence of the oldest life forms on Earth) and the Qaersut region (Cretaceous–Paleogene boundary), West Greenland. But without doubt, for a palaeontologist researching early animal life, the locality in the Sirius Pass was the Lord Mayor's Show! This was a key target.

The participants on previous expeditions in the 1980s and 1990s had collected fossil material exclusively from the scree slopes adjacent to the locality and they maintained that there was little more to be had. Moreover, rumours persisted that the airstrip near the locality was already victim to extreme weather conditions and unsuitable to permit even a Twin Otter to land. But the opportunity to collect fresh material from rock exposures, in a constrained stratigraphical framework where diversity and environmental changes could be mapped onto the section, deserved very serious consideration. Moreover, as it contained the earliest of the 'Cambrian Explosion' biotas, much could be learnt and understood from Sirius Passet about how, when and where animal-based communities began to dominate marine life. But we needed better-preserved material and a proper stratigraphical context.

This was the bid presented to the Agouron Institute, Carlsberg Foundation and Geocenter Denmark. All agreed that this was high risk, but the risks were worth taking, and so funding was granted. The 2009 expedition was small-scale and exploratory, with only four participants. We established our route from Copenhagen to Longyearbyen (Svalbard) by a regular SAS flight, and then onto the Danish military base at Station North by Twin Otter. After a short stay on the base and

Figure 22.2 From left to right, colleagues Arne Nielsen, Jakob Vinther and (the back of) Paul Smith, working at the Sirius Passet locality (David Harper)

reunited with our field equipment, we continued by Twin Otter. After a few sweeps around J. P. Koch Fjord with the stall lights flashing, and with some reluctance, the Twin Otter pilots agreed to land. The landing was bumpy and short. If the pilots were to collect us (and our samples), we would have to repair the runway. This we did, each evening filling in the runnels with gravel and indicating the track of the airstrip with large rocks spray-painted red. Not quite the main runway at Kastrup but good enough to impress the returning pilots a few weeks later, who landed without incident. We established our camp by the fjord, a short distance from the airstrip, on a spot frequented by arctic hares, the occasional wolf and a passing herd of musk ox. The locality is about a 45-minute walk up into the mountains; the rocks were beautifully exposed and highly fossiliferous (Figure 22.2).

What did we achieve in our 2009 and 2011 and the subsequent expeditions? Most importantly, we firmly established that fresh rock, hammered out of exposures, yielded the best-preserved fossil material. Moreover, each specimen could be carefully located on a stratigraphic log, and diversity through the section could be tracked and mapped onto palaeoenvironmental data. The taxon count rose to 45 species, with rarefaction curves predicting that there were still more to be discovered. We have been able to reconstruct in some detail the digestive systems and speculate on their contents in a number of animals (Strang *et al.*, 2016). In one species, the nervous and visual systems have been illustrated (Park *et al.*, 2018), and some of the presumed apex predators were probably more placid, sweep-net feeders, indicating that already there was an abundant potential harvest in the water column (Vinther *et al.*, 2014). The modes of preservation of the fossils were more complex and variable than hitherto realised (Topper *et al.*, 2018) and animal life initially diversified at the margins of its environmental tolerance (Hammarlund *et al.*, 2019).

At the core of this exciting research is fieldwork, mapping and collecting the remains of animals from rocks half a billion years old, in one of the most beautiful and remote areas of our planet. What could be more rewarding for a palaeontologist?

References

Hammarlund, E. U., Smith, M. P., Rasmussen, J. A., *et al.* (2019). The Sirius Passet Lagerstätte of North Greenland: a geochemical window on early Cambrian low oxygen environments and ecosystems. *Geobiology* 17, 12–26.

Harper, D. A. T., Hammarlund, E. U., Topper, T. P., *et al.* (2019). The Sirius Passet Lagerstätte of North Greenland: a remote window on the Cambrian Explosion. *Journal of the Geological Society*, https://doi.org/10.1144/jgs2019-043.

Park, T-Y. S., Kihm, J-H., Woo, J., *et al.* (2018). Brain and eyes of *Kerygmachela* reveal protocerebral ancestry of the panarthropod head. *Nature Communications* 9, 1019.

Strang, K. M., Armstrong, H. A. and Harper, D. A. T. (2016). Minerals in the gut: scoping a Cambrian digestive system. *Royal Society Open Science* 3, 160420.

Topper, T. P., Greco, F., Hofmann, A., Beeby, A. and Harper, D. A. T. (2018). Characterization of kerogenous films and taphonomic modes of the Sirius Passet Lagerstätte, Greenland. *Geology* 46, 359–362.

Vinther, J., Stein, M., Longrich, N. R. and Harper, D. A. T. (2014). A suspension-feeding anomalocarid from the Early Cambrian. *Nature* 507, 496–499.

23 · *Evolution in the Cellar: Live-Trapping Wild House Mice in the Italian Alps*

HEIDI C. HAUFFE

The house mouse (*Mus musculus*) has contributed to our knowledge of animal biology for almost 400 years, with a multitude of carefully selected laboratory strains now acting as models for human medical research. Less well known is the role this animal has played in our understanding of speciation; that is, how one species evolves into two or more. Although Charles Darwin speculated about speciation in *The Origin of Species*, the process is still hotly debated.

Let us delve into the science behind this. Specifically, research on the wild house mouse has clarified the role of chromosome fusions in **sympatric speciation**. The house mouse usually has a **karyotype** of 19 pairs of **telocentrics** plus two sex chromosomes (XX or XY, as for humans) or 40 chromosome arms altogether. However, in western Europe, where the house mouse lives almost exclusively with humans and their livestock, a common mutation occurs: the Robertsonian fusion (Figure 23.1). These mutations cause the union of two pairs of telocentrics to form a pair of **metacentrics**. After a few generations, all mice in the same village will have the same fusions: up to nine pairs! Importantly, no genes are lost during the fusion and the number of chromosome arms remains the same. Of course, different villages can have fusions with different combinations of telocentrics. These **chromosomal races** are given names based on the geographical area of discovery. The first such race was discovered in 1969 in the Poschiavo Valley of southern Switzerland (the 'Poschiavo race'). Just 50 years later, more than 100 such races have been described across Europe and north Africa (see Piálek *et al.*, 2005; Giménez *et al.*, 2017).

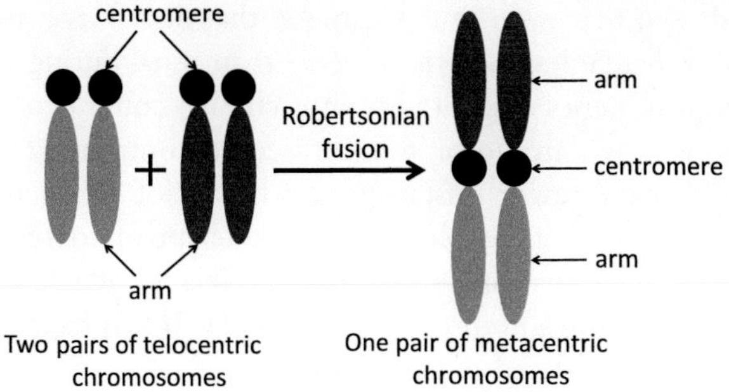

Figure 23.1 Robertsonian fusion: the fusion of two pairs of telocentrics to form a pair of metacentrics

karyotype: the number of chromosomes found in every body cell. The number of chromosomes is often species-specific (humans have 23 pairs plus two sex chromosomes). In the eggs and sperm, the number of chromosomes is half that of the body cells; this halving takes place during the process of cell division called **meiosis**.
sympatric speciation: when two groups of the same species living in the same area stop reproducing with each other and become two species.
telocentric: a chromosome with the centromere at the tip.
metacentric: a chromosome with the centromere in the middle.
chromosomal race: mice in the same village that have Robertosonian fusions between the same pairs of telocentrics.
hybridisation: when mice of two races interbreed; offspring are called **hybrids**.
linkage disequilibrium: when blocks of genes evolve together because they are not broken up during gene exchange, which normally occurs during meiosis.

When two races meet together in the same village, **hybridisation** often occurs. Since the parents have different chromosome fusions, the hybrids have a karyotype carrying the fusions of both parents. As the total number of chromosome 'arms' remains the same, a hybrid mouse looks and acts like any other mouse. However, during **meiosis**, these fusions cannot pair correctly and do not separate properly, so that the gametes (sperm and eggs) have the wrong number of chromosomes – and do not survive. Therefore, a hybrid mouse has fewer offspring than its parents.

We have shown that even in the gametes that do survive, mismatching 'arms' also exchange less genetic material than usual during meiosis: this means groups of genes associated with each race could evolve together. Here, then, is how chromosome fusions could contribute to speciation. Since hybrids have fewer offspring, and blocks of genes are inherited together in them, if these blocks of genes happen to dictate mating preferences of each race, mice of each race should evolve to recognise each other – and stop interbreeding altogether. If this happens, the two races would be considered new species.

Such a process may have occurred in the isolated alpine village of Migiondo, Valtellina, Italy, as noted by Italian colleagues in the early 1980s. At that time, two chromosomal races were found here: the Poschiavo race (with seven pairs of fusions) and the Upper Valtellina race (with eight pairs of fusions). After several years of searching, the authors found no hybrids. Was this a speciation event? This remote village is where my fieldwork started in 1989, with my then DPhil supervisor at the University of Oxford, Professor Jeremy B. Searle (now at Cornell University).

A Breakthrough – Speciation!

After several years of field and laboratory studies, and several postdoctoral and PhD studies, we showed that the semi-sterility of hybrids, as well as **linkage disequilibrium**, were responsible for this apparent speciation event. However, intriguingly, one of these races subsequently disappeared, leading to the conclusion that such sympatric speciation may only occur in small populations with a low influx of new mice. Such conditions are common in many small alpine villages. Hence, we expect that other such 'natural experiments' in speciation may occur.

Although our fieldwork for these studies concentrated on Valtellina, where we eventually identified six chromosomal races (two new to science!), our focus on hybrid zones for this species for over 20 years has taken us to farms and villages across the southern central Alps and Po Valley, from Varese to Lake Garda (e.g. Hauffe *et al.*, 2012). Here, every old village is dotted with tiny farms, hosting a typical menagerie of livestock: two milking cows, a pig or chickens to eat the leftovers and a cage of rabbits. There is usually a feed room with bags of grain and old bread, an underground cellar with homemade cheeses, and stored potatoes and fruit. This is perfect mouse habitat – as long as there are no cats!

Figure 23.2 Mus musculus domesticus – Valtellina, Italy (Heidi Hauffe). A black and white version of this figure will appear in some formats. For the colour version, please refer to the plate section.

Catching Mice, and Taking Samples

Capturing house mice is rather different from trapping non-commensal species. On one hand, it is easier because no permission is needed: house mice are pests and, therefore, not protected (although animal welfare laws still apply). However, driving up to a dairy farm at milking time (just before 16:30 in the Alps, if you want to find a farmer at home) and explaining that you want to trap their mice for three nights and, yes, release them afterwards, gets a fairly sceptical look! By the second day the farmer is more curious, giving advice, showing you where the mice holes are, relating fascinating life stories and, if you are very lucky, offering coffee and indeed phoning relatives in the neighbouring village (where you will find them eagerly waiting!).

We mainly use Longworth and Ugglan live traps, stocked with hay for bedding and baited with chicken feed, chocolate and peanut butter. Traps are placed with the opening flush with walls, between sacks of grain, along the edges of shelves. House mice are curious animals and, if present, will enter the traps after just one or two nights. But they are

Figure 23.3 Heidi Hauffe with crow (Heidi Hauffe). A black and white version of this figure will appear in some formats. For the colour version, please refer to the plate section.

anything but docile: for handling, we use a tall bucket (they can jump more than 50 cm) and thick washing-up gloves to avoid bites. Even then, they are courageous, and will run up a sleeve to leap for freedom. With practice, a mouse can be caught quickly and measurements taken, together with a tiny biopsy from the tail (for cytogenetics and DNA). Mice can be microchipped (for future identification) and released at the trap site. 'Chromosome spreads' need to be prepared from fresh samples, but this can be done in a 'field laboratory' with a small centrifuge, water bath and microscope. DNA samples are stored in ethanol to be sent and processed in laboratories, for further genetic analyses.

Although I have branched out into other rodent-related disciplines (conservation genetics, zoonotic disease and microbiome–helminth–host

Figure 23.4 These small underground cellars used for ripening cheese and storing wine are typical house mouse habitat in northern Italy (Heidi Hauffe)

interactions), fieldwork has remained a critical component of my experimental designs. Every student joining my research group is expected to go out with expert technicians for at least a week to experience the physical effort of live-trapping and the importance of precision sampling.

Live-trapping small mammals is a simple, yet exciting, way of introducing school children to nocturnal biodiversity. But, more than science and education, such fieldwork allows us to reflect on why we started on this path in the first place. For me, simply describing, understanding and conserving our extraordinary natural world forms the impetus. Each year, I set aside a few weeks to indulge in this demanding yet undying passion.

References

Giménez, M. D., Förster, D. W., Jones, E. P., *et al.* (2017). A half-century of studies on a chromosomal hybrid zone of the house mouse. *Journal of Heredity* 108, 25–35; DOI: 10.1093/jhered/esw061.

Hauffe, H. C., Giménez, M. D. and Searle, J. B. (2012). Chromosomal hybrid zones in the house mouse. In *Evolution of the House Mouse. Cambridge Series in Morphology and Molecules*, eds. M, Macholan, S. J. E. Baird, P., Munclinger and J. Pialek. Cambridge University Press, Cambridge, pp. 407–430.

Piálek, J., Hauffe, H. C. and Searle, J. B. (2005). Chromosomal variation in the house mouse: a review. *The Biological Journal of the Linnean Society* 84, 535–563; DOI: 10.1111/j.1095-8312.2005.00454.x.

24 · *Reflections on 'Babooning'*

RUSSELL HILL

I still look back on the fieldwork I undertook for my PhD as one of the most amazing periods of my life and an opportunity I was very fortunate to experience. Arriving in late 1996, I spent 14 months living on De Hoop Nature Reserve in the Western Cape Province of South Africa. South Africa was emerging from the shadow of the apartheid era at that time and, although its legacy was still clearly apparent, I fell in love with the country and its people. While the location of my current field site is now in the opposite corner of the country, my early fieldwork experiences ensured my research has remained situated in the Rainbow Nation ever since.

My PhD research focused on the behaviour and ecology of the chacma baboons living in South Africa's most southerly primate population. De Hoop Nature Reserve is situated approximately 50 km from Cape Agulhas, the southern tip of Africa, with the reserve's extensive coastline giving way to impressive sand dunes and a unique and diverse vegetation type – lowland fynbos. Having grown up close to the coast in Cornwall (UK), the climate and environment seemed familiar to me. Nevertheless, the wildlife, including the seasonal influx of the southern right whale and the resident endangered bontebok and Cape mountain zebra were wondrous additions. As were, of course, the chacma baboons I had come to study.

On reflection, there was a degree of good fortune about my field site that was not apparent to me at the outset. My initial intention had been to study the impact that predators have on primate behaviour, but if leopards were present at De Hoop, they were transient and at very low densities. In my pilot work earlier in 2016, I had visited other sites with

Figure 24.1 Dawn over De Hoop Nature Reserve, South Africa (Russell Hill)

much higher predator densities, which were impractical for that very reason – lions, buffalo and elephants all represent very significant risks when your method of data collection requires you to walk on foot alongside your monkey subjects. Fortunately, De Hoop was much safer from that perspective and by virtue of its location offered up new research questions. Serendipity always seems to make an important contribution to fieldwork!

Day length is an important constraint on behaviour. For species such as chacma baboons that conduct all of their essential activities in the daylight hours, the length of the day restricts the time they have available for foraging and other important activities such as grooming. By darkness, these animals need to have retreated to the safety of their sleeping cliffs. Seasonal variation in day length was thus an important factor for the baboons and in winter, when day lengths dropped below 10 hours and food availability was short, it placed a significant constraint on their foraging and social behaviour, which ultimately limited the size of group they could live in (Hill *et al.*, 2003). In contrast, the longer summer days gave the baboons unparalleled flexibility to respond to high midday temperatures and adopt a flexible thermoregulatory strategy that allowed them to take a siesta on their sleeping cliffs in the middle of the day – something not evident in equatorial populations (Hill, 2005, 2006).

Figure 24.2 Male chacma baboon from our habituated study group feeding in the flooded De Hoop Vlei (Russell Hill). A black and white version of this figure will appear in some formats. For the colour version, please refer to the plate section.

While these insights were important and unanticipated at the outset of my fieldwork, I also had not appreciated quite how the location of my field site would influence my daily schedule. Behavioural data collection on primates traditionally involves habituating the primates to allowing human observers to walk with them and collect observational data on their daily lives, from distances of just a few metres. This approach offers a fascinating window into their lives as you take part in the soap opera of their daily interactions. On arriving at the field site in early summer I was afforded with long days within which I could attempt to make contact with our focal groups, slowly acclimatising them to my presence. Once data collection began, however, this also meant very long days in the field since the protocol required us to follow the monkeys from dawn to dusk, collecting them at their sleeping site as the sun rose and staying with them until they retired again at night. In midsummer, data collection could start as early as 4:30 a.m., and there could be a walk of up to an hour to reach their remote sleeping site beforehand. The final data points could be at 8:30 p.m., resulting in nearly 18-hour days around the summer solstice. Rarely did I work more than a couple of days in succession at these times, despite the benefits to knowing where you had left the monkeys the night before. In contrast, the modest working days of a little over 10 hours in winter meant that five days could comfortably be worked in succession such that data were quickly accumulated.

From memory, I never begrudged the early starts and long days. Some of the most amazing sights I witnessed were at dawn and dusk, and it was a privilege to be out (often alone) at these times. The light at dawn was often far more beautiful than at sunset. I still remember vividly the most intensely coloured sky I have ever seen on one of those dawn walks. The sight of a flock of flamingos against the dawn sky is an equally enduring memory. Likewise, one evening I was waiting for the baboons to move up to their sleeping cliff when a caracal walked slowly through the group, heading out on its nocturnal hunt. It was wonderful to see such a beautiful animal so close (by habituating the baboons, other animals also seemed to lose their fear when we were with them), but I also crossed another candidate off my list of potential primate predators; even relatively small juvenile baboons seemed unperturbed by its presence.

Fieldwork, for me, was a wonderful experience, bringing me so much closer to nature than any of the experiences I have had in other nature-watching contexts, such as safaris and game drives. It was such a defining experience in my own career that I have also looked to support others seeking similar opportunities. As a result, I have been lucky enough to have supervised over 20 research postgraduates, mostly PhD students, undertaking field projects in southern Africa, and I have enjoyed each of their individual journeys. In 2012, I was also fortunate to help establish the Anthropology Field Station, a permanent research base for the Department of Anthropology, and my Primate and Predator Project in the Soutpansberg Mountains of South Africa. While providing a base for long-term data collection, the field station also provides the accommodation to support postgraduate study as well as a variety of internship opportunities for those gaining their first experience of field research. The field station also supports our undergraduate field course. Fieldwork remains the cornerstone of anthropology and yet overseas field courses were essentially absent from undergraduate provision within UK Anthropology departments. Feedback from our students suggests the course is a highlight of their degree. I can trace my commitment to all of these activities to my time at De Hoop. I hope that some of our students will be as inspired by fieldwork as I was and that they look to pursue it to some extent in their subsequent careers.

References

Hill, R. A. (2005). Day length seasonality and the thermal environment. In *Primate Seasonality: Implications for Human Evolution*, eds. D. K. Brockman and C. P. van Schaik. Cambridge University Press, Cambridge, pp. 197–213.

Hill, R. A. (2006). Thermal constraints on activity scheduling and habitat choice in baboons. *American Journal of Physical Anthropology* 129, 242–249; DOI: 10.1002/ajpa.20264.

Hill, R. A., Barrett, L., Gaynor, D., *et al.* (2003). Day length, latitude and behavioural (in)flexibility in baboons (*Papio cynocephalus ursinus*). *Behavioral Ecology and Sociobiology* 53, 278–286.

25 · *Bogs, Birds and Bones: Interdisciplinary Fieldwork on the Isle of Rum National Nature Reserve*

PETER HIGGINS

An Invitation

This is a tale of a 40-year relationship between a college programme, its staff and students, and the Isle of Rum National Nature Reserve (NNR). In 1972, the then Director of the Nature Conservancy Council for Scotland (NCCS), Dr John Morton Boyd, wrote to outdoor education staff at Moray House College of Education in Edinburgh, inviting them to bring students to an island that the Nature Conservancy had bought from the owner, Lady Monica Bullough, in 1957. The 26,000 acres of hills and bogs that comprise the Isle of Rum (or Rhum as it was then called) had been bought for the nation (at £1 per acre), and came with its own castle. Neville Crowther, a lecturer, botanist and field naturalist who taught on the one-term outdoor activities programme for teachers, accepted the invitation enthusiastically, and the first ever educational field course on the island took place in the spring of 1974. The fit was perfect: the mountains, wild-land and coast of the island provided opportunities for outdoor adventures such as mountaineering on the famous Rum Cuillin; the diverse habitats were ideal for field studies, and the landscape provided rich opportunities to discuss the cultural heritage and consequences of centuries of upland management for sheep-grazing and deer-stalking.

'Why is What Where?' A Year-Long Learning Project

Every year in spring, a member of staff (occasionally two) and student groups have spent a fortnight on the island. The course has changed and so has the institution – Moray House is now the School of Education of Edinburgh University, and the students are members of two MSc programmes – 'Outdoor Education' and 'Outdoor Environmental Education'; the annual intake is 30 to 40 students, with an even gender split and about half from overseas.

While the Rum 'Ecology and Field Studies' course is the final taught part of the MSc programmes, it has a very close link with the first of the year – 'Interpreting the Landscape' – an introduction to the nature of landscapes, and how as educators we might best approach teaching this. The course uses a wide range of teaching approaches, with inputs from specialist staff in the School of Education and beyond, and a geological field trip to important sites close to Edinburgh, notably the famous 'Hutton's Unconformity' at Siccar Point, one of the most significant geological sites in the world (Edinburgh Geological Society, 2018).

Directly after the end of the academic course (in September), the students and core staff make a one-week canoe-camping descent of the River Spey in the Cairngorms, honing practical skills as they go, and also using the Spey and its catchment to further develop concepts and skills in 'interpreting the landscape'.

Throughout the year, staff frequently revisit themes studied, and the consistent intent is to help students think about the landscape, and to inspire questions on origins, functions and perceptions: so '*why is what where?*' is a deliberate attempt to provoke and generalise to anything (what) in any place (where), followed by more specific questions such as 'why is it like that?', 'what do you think/feel about that?', etc. This intent continues when the students arrive by ferry to Rum. There are no laboratory facilities, and accommodation is basic, so the course maintains its long-standing tradition of being taught almost entirely outdoors. As the 'culminating' course of the year it picks up and develops the themes outlined above, facilitated by the same staff member teaching both academic courses and leading the Spey canoe descent.

The Story of Rum: a Context for the Course

Many books have been written on the geology, ecology and human history of Rum (e.g. Clutton-Brock and Ball, 1987; Magnusson, 1997; Love, 2002). By way of context for fieldwork it is important to note the following:

- The geology of Rum is almost exclusively Torridonian sandstone, ultrabasic and granitic igneous rocks (Goodenough and Bradwell, 2004).

- The 9,000+ years of human history is significant – from one of the oldest settlements in Scotland, through periods of population clearances (1826, 1828), to private ownership for sheep and as a sporting estate (Rixson, 2001). Its current, permanent population numbers no more than 30 inhabitants.
- It has been publicly owned since 1957 – managed for conservation and research by Scottish Natural Heritage (SNH) (2009). Access was heavily restricted by the owners until the late 1980s but, more recently, public access and recreation have been embraced as a key function of the NNR.
- Its geological, ecological and human heritage is designated under a wide range of UK and international regulations and agreements. Of considerable prominence are its closely studied red deer and the globally significant, massive nesting colonies of the Manx shearwaters, which are found in burrows high on the Rum Cuillin.
- The impact of historically high populations of sheep and deer has been significant, degrading the soils and preventing tree regeneration.

An Island Context for Experiential Constructivist Learning

This context above is important, as the MSc programmes are for generalists – and the purpose of this interdisciplinary field course is to play to the curiosity of students, and to help them develop a passion and skills for understanding and communicating the values of the environment in their future careers. Most are unfamiliar with basic ecological concepts and principles, and many will not see particular relevance of technical fieldwork in their careers. But here, the island context, its natural boundary (shoreline) and interaction of the terrestrial and marine ecosystems, offers a chance to simplify and exemplify the wonder of ecology and our part in influencing environmental change. As the students are 'apprentice' outdoor educators, pedagogical practices are important, and so the course is intended to model these – through experiential constructivist approaches (e.g. Taber, 2014). An example of this follows, but in addition to the habitats mentioned below, students are also introduced to a range of others (e.g. seashore) and field study techniques, and have the opportunity to design their own 'study' day. They also accompany staff on a common bird census in an area that was originally open moor, but was planted with a range of tree species in 1978 (the area has been monitored since 1972, providing data on population changes as the trees have matured).

Rum Puzzles

What follows is an example of a sequence of 'puzzles' the students are introduced to. For the first of these they are told that if they think they know the explanation/answer, to keep it to themselves! They may share it with the tutor but not their classmates until the end of the course.

The course always begins with a local 'guided walk', during which the students are introduced to the dominant wet-heath and bog vegetation, and one plant in particular – the bog asphodel (*Narthicium ossifragum*). The name of the plant is debated – the '*ossi*' bit and the '*fragum*' bit. What has this plant to do with breaking bones (of sheep)? The answer is not provided – the students are left with the puzzle – 'is *Narthecium ossifragum* innocent or guilty?'

While the weather dictates the sequence of activities, the following always feature in the week, as students are introduced to further (related) puzzles (all of which they are asked to think about over the week rather than answer at the time):

- Freshwater habitat surveys – samples invariably contain no molluscs, except occasional tiny freshwater limpets (*Ancylus fluviatilis*). The students are asked if they have seen any other molluscs – and asked to think why there are so few or none.
- A visit to the deer study area at Kilmory, and discussions of deer behaviour include the observation that they gnaw shed antlers – why would they do so?
- A night-time visit to the Manx shearwater nesting burrows on the mountain Hallival – where, amongst other discussions, the students are presented with the puzzle that red deer have been observed eating shearwater carcases (Furness, 1988).
- Sessions on geology remind the students of the main rock types.
- Discussions on human history include the 'clearances' of the human population (Love, 2002) in 1826 and 1828; the introduction of sheep; and the implications of the export over time, of large numbers of deer and sheep carcases (flesh and bones) from the island.
- Towards the end of the week, students are asked if they have seen any snails, in water or on land – generally they have not, as they are scarce. This is usually the final puzzle.

All of these are drawn together in a concluding discussion where the students invariably 'join up the dots', as they realise the lack of calcium on

Figure 25.1 Rocky shore study – Loch Scresort, Isle of Rum NNR (Peter Higgins)

the island is implicated in all of the above, and that it is absent in local rock types, except in a few localised places such as Allt nam Ba (Hughes, 1960[1]).

Late in the week, students also have a session on the open sea, the exploitation of fish stocks and the effects of climate change on ocean warming and increasing atmospheric CO_2 on acidification. This includes a discussion of the impact on diatoms and other species with calciferous structures. The dawning realisation is that the only significant source of calcium imports to the island is from seabird faeces, notably from the large breeding population of shearwaters. This was a problem for those who tried to farm the land in the nineteenth and twentieth centuries too, and the students are reminded that they have seen the evidence of this in the disused lime kiln by the old pier.

Finally, the students are asked to state their verdict on *Narthicium ossifragum* (innocent or guilty?). The ensuing discussion considers the plant's preference for boggy areas, and perhaps that the visibility of the bright yellow flowers ensured that it got the blame when sheep broke

[1] Hughes (1960, p. 384) reports that remarkably, in 1897, in this isolated location, Archibald Geikie recorded the presence of 'a much-altered grey and white limestone or marble' (Geikie, 1897, p. 351).

Figure 25.2 Freshwater study – Kinloch Burn, Isle of Rum NNR (Peter Higgins). A black and white version of this figure will appear in some formats. For the colour version, please refer to the plate section.

their legs (through having brittle, calcium deficient bones)! However, the real culprit is of course the calcium-poor soils (Mabey, 1996).

A Relationship with a Particular Place

There must be something in a relationship that lasts over 40 years – especially when it involves study outdoors in one of the wettest parts of Scotland! But the relationship is strong, partly because the two key staff who have taught this course (Neville Crowther, 1972–1995 and Peter Higgins, 1992–present) have each spent in total well over a year of their lives on the island with students, have been involved in independent research and monitoring and wrote the island's environmental education plan for Scottish Natural Heritage about the island's environmental education plan (Higgins *et al.*, 2001). However, the relationship that students develop is the important issue, and invariably those who took the course decades ago have vivid memories of it.

Figure 25.3 Bog vegetation survey – Kinloch, Isle of Rum NNR (Peter Higgins). A black and white version of this figure will appear in some formats. For the colour version, please refer to the plate section.

Nonetheless, there is a bigger 'place' that this course, and the MSc programmes as a whole, intends to foster a strong, perhaps intimate, relationship with the planet. As part of the course, through lively discussions, students are also engaged with important contemporary global environmental issues. There is a focus on the biogeochemical cycles that we and all species are part of. By engaging with the complexity of the natural world, developing a connection with it and understanding the consequences of our actions, we can develop an ethic of care. This relates not just to the students on the programme, but to the students they will eventually work with – to help them care for place – when place is the planet (Higgins, 2010).

Acknowledgements

An apprentice must acknowledge the master-craftsman. Neville Crowther grasped the opportunity, developed the course and inspired generations of students both on and off the island to appreciate and care for the natural world. He taught me most of what I know of that island, and much that I have forgotten; he continues to be an inspiration. Thanks also go to the students I have shared this unique course with, and to SNH staff and others on the island for continuing to welcome us and enthusiastically sharing its wonders.

References

Clutton-Brock, T. and Ball, M. (1987). *Rhum: The Natural History of an Island.* Edinburgh University Press, Edinburgh.

Edinburgh Geological Society (2018). *Hutton's Unconformity at Siccar Point.* See www.edinburghgeolsoc.org/edinburghs-geology/huttons-unconformity/ (accessed July 2019).

Furness, R. (1988). Predation on ground-nesting seabirds by island populations of red deer *Cervus elaphus* and sheep *Ovis*. *Journal of Zoology* 216, 565–573.

Geikie, A. (1897). *The Ancient Volcanoes of Great Britain, Volume 2.* Macmillan, London.

Goodenough, K. and Bradwell, T. (2004). *Rum and the Small Isles: A Landscape Fashioned by Geology.* Scottish Natural Heritage, Perth.

Higgins, P. (2010). Pedagogy for 'global intimacy'. In *Education in a Global Space: Research and Practice in Initial Teacher Education*, eds. T. Wiseley, I. Barr and B. King. Scotdec, Edinburgh, pp. 180–188.

Higgins, P., Crowther, N., Nicol, R. and Meldrum, G. (2001). *Rùm National Nature Reserve: Environmental Education Plan.* Scottish Natural Heritage, Perth.

Hughes, C. (1960). An occurrence of tillyite-bearing limestone in the Isle of Rhum, Inner Hebrides. *Geological Magazine* 97, 384-388.

Love, J. (2002). *Rùm: A Landscape without Figures.* Birlinn, Edinburgh.

Mabey, R. (1996). *Flora Brittanica.* Sinclair-Stevenson, London.

Magnusson, M. (1997). *Rùm: Nature's Island.* Luath Press, Edinburgh.

Rixson, D. (2001). *The Small Isles: Canna, Rum, Eigg and Muck.* Birlinn, Edinburgh.

Scottish Natural Heritage (2009). *The Story of Rum National Nature Reserve.* Scottish Natural Heritage, Perth.

Taber, K. (2014). *Student Thinking and Learning in Science: Perspectives on the Nature and Development of Learners' Ideas.* Routledge, London.

26 · *Exploring World(s) Down Under*

EMILY HUSBAND

Tropical ecosystems and, in particular, coral reefs have always, albeit somewhat subconsciously, inspired my love of geography ever since I studied them for my geography GCSE high-school examination in 2011. The colours, biodiversity and complexity of life on the reefs, to me, seem like a wondrous, secret (calcareous) city functioning beneath the ocean's surface, which only a lucky few humans may be privy to on rare occasions. Despite occupying just 0.2 per cent of the oceans, coral reefs are home to one-third of all marine species, owing to the capacity of key reef structure-building corals to provide numerous hiding places, niches and attachment sites for coral larvae. However, as in many ecosystems, recent global and local disturbances have caused huge changes in coral communities, structural degradation of coral reefs and a subsequent loss of fish life and biodiversity living among the calcareous structures.

For my undergraduate dissertation I completed desk-based research, using only remotely sensed data to examine the ocean temperatures that cause coral bleaching. Consequently, being informed that my Masters project would involve a somewhat 'compulsory' field trip to the Great Barrier Reef was quite a pleasant shock! My project aimed to develop a globally applicable and novel technique for assessing the structural complexity of shallow-water coral ecosystems. More specifically, it involved building a database of genera-specific coral 'rugosity' data and applying this to remotely sensed imagery or video transects of the ecosystems so that they can more easily be monitored over large scales. Rugosity is a metric often used to monitor reef structural complexity and it describes the roughness or irregularity of a coral reef surface. In the field this is measured using the rugosity index – a ratio between the planar diameter

Figure 26.1 Holding my breath and measuring the rugosity of a coral colony in Clam Gardens, Lizard Island on the Great Barrier Reef (Emily Husband).
A black and white version of this figure will appear in some formats. For the colour version, please refer to the plate section.

of a coral colony and its respective contour length. Thus, my fieldwork had two parts of data collection: measuring the rugosity of corals and collecting aerial and video imagery: easy . . .

The first two and a half weeks of my fieldwork were spent on Lizard Island, a granite continental island an hour's flight, on a tiny, bumpy five-seater plane, from Cairns. The island was completely different to any environment I had worked in during the course of my undergraduate degree, and undoubtedly much warmer! I spent time mastering a variety of new skills which are not commonplace to learn at school, namely snorkelling, fishing and, most importantly, measuring carbonate structures underwater. My daily commute was much altered too: a five-minute walk along a beach followed by a 10-minute motor boat drive through tropical waters to reach my office, one of the many fringing reefs or reef platforms that surround the island. Here I spent my days measuring the rugosity of as many corals as I could whilst willing myself not to jump out of the water and back to the boat any time I saw one of the sharks, rays or jellyfish that were often within eyesight. Being informed that the research station had had only one major shark-related incident in 30 years of operation did NOTHING to calm my fear of this awe-inspiring environment.

Unfortunately, for a good proportion of our stay on Lizard, torrential rain and stormy weather hit the island and prevented any *in situ* data collection. Although this was to be expected, having arrived in Australia during cyclone season, it was incredibly frustrating not to be able to get out into the water or collect any aerial data of the reefs. Such a sense of 'failure' that came with this was something I had not experienced previously with fieldwork. Even during the worst weather on previous field trips, I would usually be able to wrap myself up in waterproofs and get out into the field. On a positive note, spending prolonged time in the research station's library and laboratories allowed me to gain great insight into the breadth of other coral research and help with some of the other science being undertaken on the island.

The second half of my research took place on Heron Island. Despite being a fraction of the size of Lizard, this island really captured my geographical excitement and imagination, not only for highlighting the vast differences in coral reef environments (Heron Island is a platform reef), but also for teaching me more about tropical environments than that which is just applicable to coral. Within two hours of stepping off the boat I had witnessed baby turtles hatching and scrambling towards the sea, female turtles tirelessly pulling themselves up the beach to nest, the dusk playfulness of baby sharks and rays, a full lunar eclipse and I had walked in the steps of Sir David Attenborough through the island's noddy and shearwater seabird-dominated forest. I spent the majority of the evening and indeed the first day on the island with my eyes wide open, trying to take everything in about this incredible environment, so far removed from anywhere I had ever been before. A completely humbling moment came one morning at 6 a.m. when, having gone out into the ocean to make the most of a morning low tide, two female turtles were still on the beach laying. Whilst popping up for breath between measuring corals I watched the turtles intently, observing their calm nature, before seeing them slink back, exhausted, into the sea.

The tidal range at Heron Island meant that during low tide I had to lie on the seabed, face to face with the sea slugs, to measure the corals,

Figure 26.2 A slightly easier environment in which to measure corals: low tide on Heron Island, Great Barrier Reef . . . before the rain came! (Mike Salter)

making for extremely efficient coral measuring. Unfortunately, we were once again treated to three days of storms with rapid currents, which prevented extensive data collection and any drone-derived imagery, further highlighting just how unpredictable fieldwork in the tropics can be. The daily occurrence of turtle hatchings, however, did lift my spirits in between time spent inside, identifying each and every coral I had measured.

This fieldwork taught me so much more about tropical ecosystems, the ocean and, in fact, the world, than just that related to my project. In the run-up to and planning of my fieldwork, I had been so focused on preparation and learning about coral rugosity, that everything else came as a complete surprise. As expected, large areas of the coral reefs I saw were dead, and many were dominated by coral rubble and algae that had colonised the carbonate skeletons. Many researchers who had worked on the islands for years were obviously saddened by the change they had witnessed in these ecosystems over their careers, and it was also eye-opening for me to realise that I may never see pristine reefs at Lizard or Heron Island. However, if we can better monitor the changes in these ecosystems with methods such as those my research is aiming to develop, we may stand a better chance of identifying the reefs most at risk from future disturbance and preserving the structural complexity of the reefs we have left.

Reference

Husband, E. H. (2019). *Coral Colony-Scale Rugosity Metrics and Applications for Assessing Temporal Trends in the Structural Complexity of Coral Reefs*. Masters Thesis, University of Exeter.

27 · *Experiments by Nature: Strength in Realism*

CHRISTIAN KÖRNER

How do plants respond to environmental conditions? A common approach to finding an answer is to do an experiment in which the plant's environment is varied. Experiments with plants can be simulated on a computer, performed in a greenhouse or growth cabinet, undertaken through field manipulations or conducted by observing what nature has come up with along environmental gradients or contrasts. The last of these, which I call 'experiments by nature', are the strongest in terms of realism, because they yield a long-term 'ground truth' beyond all initial effects of manipulative experiments. But, they are the most difficult to match into standardised protocols that permit statistics (Körner, 2018). One may, for example, have 1,000 mini-pots with *Arabidopsis* (rockcress) in a common garden, but whatever plant study one replicates on Kilimanjaro is pseudo-replicated because there is only one Kilimanjaro ($n = 1$), while all the garden pots are great from a purely statistical point of view.

Field observations that are designed in ways to account for clearly defined environmental contrasts or gradients can yield responses by plants or plant communities that have been established over long periods, sometimes millennia, and thus offer 'nature's' answer to certain habitat conditions. Unlike most manipulative experiments in gardens or greenhouses, plants (and soil animals and microbes) have 'selected' their position (their habitat) and arrived at some sort of steady state that an artificial system would never arrive at within a realistic study period. We can test climatic warming (elevation, latitude), alternative snowmelt regimes, changes in moisture, nutrients and even carbon dioxide (CO_2)

Figure 27.1 Natural CO_2 springs in Tuscany, Italy, permit the study of growth responses of trees to elevated CO_2 for decades. Here the site is at Rapolano (Chritian Körner; see Hättenschwiler *et al.*, 1997).

concentration (natural CO_2 springs) across such sharp, but 'old' gradients, and 'harvest' the results. Such fieldwork is using an observation design that is not dissimilar to the experimental designs with artificial, disequilibrium systems, yet adequate replication becomes a challenge. Here, I give three examples based on timely research questions, and the answers found through fieldwork.

Does Elevated CO_2 Stimulate Tree Growth?

Thousands of years of geological CO_2 release exposes trees in the immediate surroundings of so-called CO_2 springs to a broad spectrum of CO_2 concentration for their entire life. Close to such (replicated) springs, pure CO_2 kills all life, but further away, mixtures of air with CO_2 can be identified, and the growth rings of trees enable us to explore their life's history. This way it can be shown that growth in young trees in open land, following a disturbance and associated nutrient overflow (e.g. coppicing, clear cutting) is stimulated by a doubling of CO_2 concentration but, after 25 years, as the canopy closed, the effect disappeared (Hättenschwiler *et al.*, 1997; Figure 27.1). Who would pay for 25 years of CO_2 enrichment in as close as we can get to a natural setting? The results complement and validate those from manipulative field experiments with undisturbed natural forests that were artificially exposed to elevated CO_2 concentrations, and all together they showed that elevated CO_2 has no effect on tree growth, once a steady-state nutrient cycle has been

Figure 27.2 Bird cliffs in Svalbard at close to 80° N exemplify an experiment by nature to test plant nutrition–temperature interactions in one of the coldest places on Earth with plant life (Christian Körner)

established (Bader *et al.*, 2013; Klein *et al.*, 2016; Ellsworth *et al.*, 2017). The simple answer is: it needs more than carbon to grow a plant (Körner, 2006).

Does an Extremely Low Temperature Prevent Plants from Responding to Soil Nutrients?

Here is nature's answer. At ca. 10 degrees south of the north pole, the non-ice-covered terrain is Arctic desert in Svalbard. The growing season lasts six–seven weeks of Arctic 'summer', with air temperatures hardly exceeding 0 °C during 24 hours of daylight. This is a truly deserted world, with only tiny traces of plant life in the coastal plains. Yet, cliffs used by millions of summer-nesting birds are packed with guano and provide plant nutrient saturation under otherwise high-arctic life conditions. These slopes are covered by knee-deep, super-lush, tall herb fields, composed of *Koenigia islandica*, *Oxyria digyna*, *Poa alpina* and a few other taxa (Figure 27.2). Again, this field observation validates experimental trials with nutrient addition at 2,500 m elevation in the Alps, where both pioneer glacier forefield and late successional vegetation show dramatic sensitivity to nutrient addition (Schäppi and Körner, 1996; Heer and Körner, 2002). In other words, low temperatures do not prevent plants from utilising extra nutrients.

Are Drought Conditions Slowing Plant Growth Because Plants Run Out of Carbohydrates?

It is widely held that water shortage causes stomata (leaf pores) to close and thus prevents plants from photosynthesising; hence, there is no growth. Field data can disprove this paradigm and point at a fundamental misconception in our traditional view of the plant world. It turns out that water shortage is affecting the delicate meristems (where plant growth takes place) long before it affects stomata and photosynthesis. As a result, during the early part of drought, plants accumulate (rather than lose) photosynthates in the form of carbon reserves (e.g. starch; Körner, 2003). When the drought becomes severe, the plant's demand for fresh carbon becomes almost zero (no growth) and the strongly diminished rates of maintenance respiration can be fuelled from reserves for several months in a severe drought. In other words, when trees die from drought they die with a full or half-full 'stomach'. These field observations have been validated by experiments under controlled conditions (Muller *et al.*, 2011). So, plants die from drought because of tissue dehydration and not because of carbon shortage.

My Take-Home Message

Fieldwork introduces us to the highly needed realism in plant research, and it both challenges and greatly advances the theories of plant life. Crucially, comparative plant ecology – a term coined by Phil Grime and colleagues in Sheffield, UK - is the 'alpha and omega' of a functional approach to understanding plant responses to the environment (Rorison, 1987; Grime *et al.*, 1988).

References

Bader, M. K. F., Leuzinger, S., Keel, S. G., *et al.* (2013). Central European hardwood trees in a high-CO_2 future: synthesis of an 8-year forest canopy CO_2 enrichment project. *Journal of Ecology* 101, 1509–1519.

Ellsworth, D. S., Anderson, I. C., Crous, K. Y., *et al.* (2017). Elevated CO_2 does not increase eucalypt forest productivity on a low-phosphorus soil. *Nature Climate Change* 7, 279–283.

Grime, J. P., Hodgson, J. G. and Hunt, R. (1988). *Comparative Plant Ecology*. Unwin Hyman, London.

Hättenschwiler, S., Miglietta, F., Raschi, A. and Körner, C. (1997). Thirty years of *in situ* tree growth under elevated CO_2: a model for future forest responses? *Global Change Biology* 3, 436–471.

Heer, C. and Körner, C. (2002). High elevation pioneer plants are sensitive to mineral nutrient addition. *Basic and Applied Ecology* 3, 39–47.

Klein, T., Bader, M. K. F., Leuzinger, S., *et al.* (2016). Growth and carbon relations of mature *Picea abies* trees under 5 years of free-air CO_2 enrichment. *Journal of Ecology* 104, 1720–1733.

Körner, C. (2003). Carbon limitation in trees. *Journal of Ecology* 91, 4–17.

Körner, C. (2006). Plant CO_2 responses: an issue of definition, time and resource supply. *New Phytologist* 172, 393–411.

Körner, C. (2013). Plant-environment interactions. *Strasburger's Plant Sciences,* eds. C. Korner, J. W. Kadereit, G. Neuhaus and U. Sonnewald. Springer, Berlin.

Körner, C. (2018). Concepts in empirical plant ecology. *Plant Ecology and Diversity* 11, 405–428; DOI: 10.1080/17550874.2018.1540021.

Muller, B., Pantin, F., Genard, M., *et al.* (2011). Water deficits uncouple growth from photosynthesis, increase C content, and modify the relationships between C and growth in sink organs. *Journal of Experimental Botany* 62, 1715–1729.

Rorison, I. H. (1987). Mineral nutrition in time and space. *New Phytolologist* 106, 79–92.

Schäppi, B. and Körner, C. (1996). Growth responses of an alpine grassland to elevated CO_2. *Oecologia* 105, 43–52.

28 · *Big Problems – Small Animals*

CHARLES J. KREBS

Few mammals have inspired more fieldwork by ecologists than lemmings and voles. Who has not heard of the tale of lemmings jumping off cliffs or of hoards of voles attacking crops of wheat or barley? The base of ecologists' interest in these small animals is that they contradict our assumed *balance of nature* idea, drilled into us from an early age. Lemmings of the tundra, and voles of grasslands and forests, often fluctuate in three–four-year cycles of abundance, and the challenge to the field biologist has enduringly been to understand why these cycles occur.

What Drives Lemming Cycles?

In 1959, I went to Baker Lake, a small settlement in the central part of northern Canada, to find out why lemmings fluctuated so much in abundance. Two very popular ideas were suggested to explain lemming cycles. Either lemmings ran out of food under the snow in the winter or they were hunted to near extinction by predators such as weasels and snowy owls. To my surprise neither of these ideas held water. There was always ample food available, even at high lemming density, and there were too few predators to make inroads on the lemming numbers.

Dennis Chitty at Oxford suggested a third possible reason for lemming and vole population changes – aggressive social behaviour. We looked at wounds on the skins of dead lemmings and found that bite marks were very common, and once males reached sexual maturity bite marks greatly increased. If social aggression was a possible driver of population changes, it had to operate in two ways, through direct killing of one lemming by another, or by reducing reproduction. Female lemmings have litters of

Figure 28.1 Charley Krebs live-trapping a red-backed vole (*Myodes rutilus*) at Kluane Lake in the Yukon. These voles have a three–four-year cycle in this white spruce boreal forest (Charles Krebs). A black and white version of this figure will appear in some formats. For the colour version, please refer to the plate section.

five to seven young every three weeks when they are breeding full speed. We found that lemmings could breed under the snow in winter, at a time of year when one would think breeding was impossible. But lemmings had not read the textbook, and consequently there was consternation in 'lemming country' when we saw very few lemmings in the autumn when the snow came, and yet seven months later hundreds of animals emerged from the melting spring snow. However, breeding under the snow does not occur every winter and so another biological mystery had to be investigated. We were left, after four years of research, with the discovery that lemming cycles were not a simple biological problem!

Turning to Meadow Voles, Red Grouse, and a Field Experiment

It was clear that lemming research was difficult because of winter weather and the remoteness of the tundra areas of northern Canada. So, in 1962,

I began eight years of work in the USA, in grasslands that contained meadow voles which also had three–four-year cycles. There was an idea that cycles required bad winters and snow, but the field voles of California and Indiana turned out to have cycles in areas where it virtually never snowed!

At this time in my career I discovered the research on red grouse in Scotland, led by Adam Watson and David Jenkins, based near Aberdeen. The Scottish research had focused on territorial behaviour in red grouse, and we wondered if meadow voles might operate like 'flightless' red grouse. We thought that perhaps territorial behaviour would give rise to many surplus animals that moved off into poor habitats and died from food shortage or predation. But no one knew how many voles were moving about and not being territorial.

So, we began a removal experiment in grassland and removed about 80 per cent of the field voles from one hectare of grassland by live-trapping every two weeks throughout the year. The results were another great surprise because it was like emptying the ocean! As fast as we removed resident voles, new, untagged ones moved in to occupy the habitat. Removing 100 per cent of the adult animals every two weeks was not enough to keep the population down, and nearly 2,000 voles were removed from this grassland over one year, with almost no detectable effect on population size. So, the message became 'study dispersal movements and territoriality'.

There was always a problem in thinking that aggressive social behaviour could cause population cycles because changes in both reproduction and mortality were involved. High mortality could readily be due to predators, of which there were many, but also from infanticide of young voles at an early age. But how could we explain reproductive changes? The idea that stress might be the answer came from the laboratory work of John Christian and David E. Davis in the 1950s and 1960s. They used laboratory mice in small enclosures in the laboratory, and found that strife between individual mice led to stress and hormonal imbalances that shut down female reproduction. But these studies were done on very high-density laboratory populations, and we wondered if the same mechanism could possibly work in field populations that lived at very low densities.

Snowshoe Hare Cycles: a New Beginning

But we were not to find the answer to this question on meadow voles because, in 1970, I moved to the University of British Columbia, and we began to study the snowshoe hare cycle in northern Canada. The snowshoe hare has a nine–ten-year cycle that occurs across the boreal forests of northern Canada and Alaska and into Siberia. All the same, ideas of what might cause cycles had appeared in the literature on snowshoe hares, so we began anew to investigate the role of food, predators and social behaviour in possibly causing 10-year cycles. No one in their right mind would study the snowshoe hare cycle since it took a decade to get a single set of observations, and experiments were, to say the least, challenging! But a university professor has the freedom to take on long-term studies, and in 1973 we began a study of snowshoe hares around Kluane Lake in the Yukon.

The first question was whether hares ran short of good food in the winter and starvation was a primary cause of the decline? We added artificial food to three populations of hares in winter to see if we could stop the cycle. Nothing happened, and we began to think that food was not a factor producing population changes in hares. In collaboration with Scott Gilbert, Tony Sinclair, Rudy Boonstra, Stan Boutin and Jamie Smith we then began a second set of 10-year experiments, this time manipulating both food and predation. We could exclude many important hare predators with an electric fence, which prevented Canada lynx and coyotes from catching hares. The result was a bonanza of hares inside the electric fence, and we realised that predation was critical to determining how hare populations rose and fell. In contrast to the cycles of lemmings and meadow voles, social behaviour in snowshoe hares was almost non-existent. There were no clear territories in hares, and no records of any infanticide to produce changes in mortality rates. Predation was the key to hare population dynamics (Krebs, 2011, 2013).

But there was still a problem in explaining reproductive changes in hares. Like meadow voles and lemmings, hare reproduction collapsed by 50 per cent at high density and during the decline of the cycle. Rudy Boonstra had the bright idea that perhaps stress was again the critical agent, but this time the stress arose from predator chases of hares, since often the hares escape their predators. Fortunately, around 2000, it became possible to measure directly the stress corticosteroids in individual hares from chemical residues in their faecal pellets. Rudy and his student, Michael Sheriff, showed that stress in hares reduced their reproduction and it also carried over to their offspring. Stressed females gave rise to stressed offspring, so there was a maternal effect by which stress was passed on to the next generation (Krebs *et al.*, 2018).

All of this has involved a great deal of fieldwork by many people. I was fortunate to lead many of these studies and to collaborate with many devoted graduate students, technicians and fellow scientists to get to where we are today. The role of stress, whether arising from predator chases or social interactions, is now an active and key area of study, and more surprises will no doubt come again. If anyone thinks that the natural world is a simple system, they have only to study the history of population cycles.

Acknowledgements

I am deeply indebted to my mentors the late Dennis Chitty and Adam Watson for their influence on my work, and through me on to my students and colleagues. We need much more field research before we will have a general understanding of ecological systems.

References

Krebs, C. J. (2011). Of lemmings and snowshoe hares: the ecology of northern Canada. *Proceeding of the Royal Society of London, Series B* 278, 481–489.

Krebs, C. J. (2013). *Population Fluctuations in Rodents*. University of Chicago Press, Chicago, IL.

Krebs, C. J., Boonstra, R. and Boutin, S. (2018). Using experimentation to understand the 10-year snowshoe hare cycle in the boreal forest of North America. *Journal of Animal Ecology* 87, 87–100.

29 · *Soil Survey: a Field-Based Science*

ALLAN LILLY

Although I spent much of my pre-school and school years wandering around derelict industrial sites in Scotland's central belt, I was also exposed to rural life on a regular basis during the school holidays, when I attended Lanark Auction market with my father. I had developed an interest in geography at school and decided to study it at the University of Strathclyde's Geography Department. It was there that my interest in soils grew and was nurtured by Donald Davidson but it was a visit to the department by John Bibby, one of the senior soil surveyors in the Soil Survey of Scotland, that finally opened my eyes to the possibility of combining geography and soil science by joining the Soil Survey. Being able to wander the Scottish countryside, far from the industrial waste-land, and map soils seemed like the ideal job. Indeed, Bob Glentworth, Head of the Soil Survey of Scotland, often told his field staff that they were lucky to be getting paid to enjoy themselves digging holes and mapping soils. So, in 1980, armed with a degree in geography and biology and with some knowledge of hydrology, geomorphology, ped-ology, biogeography and plant science, I applied for and was accepted as a junior soil surveyor in the Soil Survey Department of the Macaulay Institute for Soil Research. Thus, my training in how to survey soils began.

The Soil Survey of Scotland

By the time I joined the Soil Survey of Scotland, it was already 40 years old and was in the process of completing the national 1:250,000 recon-naissance scale soil map of Scotland, but there was still plenty of surveying to be done. It was R. S. Smith, the Director of the Illinois Soil Survey, who said in 1928 '... I cannot conceive of the time when knowledge of soils will be complete. Our expectation is that our successors will build on what has been done, as we are building on the work of our predecessors.' Sadly, this wasn't the case for the Soil Survey of Scotland, but I had 10 good years of making soil and land capability maps, which are still being used today. We are now finding new ways to make the information available to new generations of environmental scientists and indeed anyone interested in soil, whether as a user of the land or as someone with an interest in how the soil has shaped Scotland and its people, or simply just to wonder at the variety of soils we have. As C. F. Marbut, Director of the Soil Survey Division of the US Department of Agriculture said in 1920, 'Probably more harm has been done to the science by the almost universal attempts to look upon the soil merely as a producer of crops rather than as a natural body worth in and for itself of all the study that can be devoted to it, than most men realize.'

Why Map Soils?

As well as the aesthetics of soils, as described by Marbut, soil can be considered as a finite resource and a soil survey is a means of characterising the amount and quality of land and its potential productivity, and, increasingly, quantifying and qualifying the ecosystem services that a healthy soil provides. Soil maps also provide the means to extrapolate experimental results derived from one area to another with similar soils, without the need for further experimentation at that site. Fundamentally, soil surveying is a field-based science where the surveyor conceptualises the process of soil formation for a specific area and then uses auger borings and inspection holes to validate (or refute) that conceptual model. It is not a reductive science whereby the system is broken down into component parts and investigated, rather it is deductive whereby the surveyor uses a conceptual understanding of soil processes, bringing together elements of hydrology, geomorphology, glacial and land-use history and biology.

Surveying and Field Mapping

The Soil Survey of Scotland began the mapping of Scotland's soil resource in 1938, and by 1947 was charged with the systematic mapping of this resource. The initial surveys were in connection with

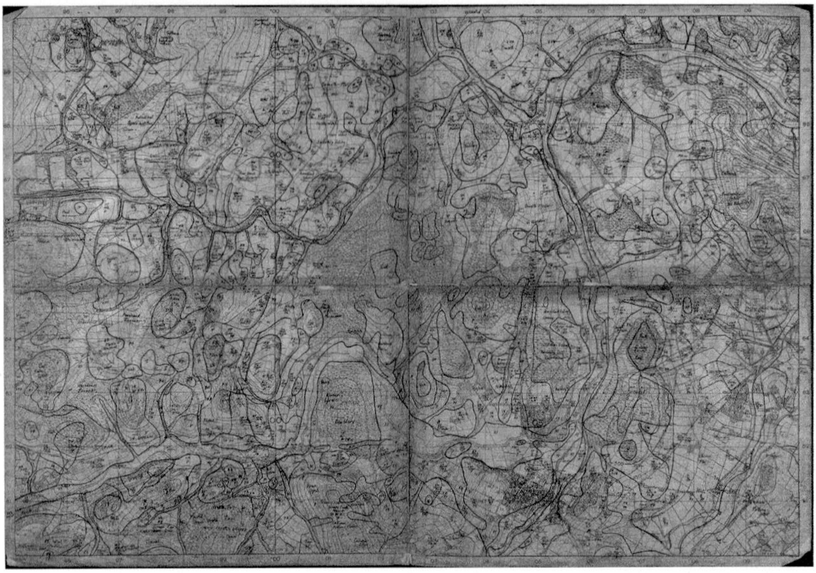

Figure 29.1 1:25,000-scale GSGS topographic map used to record observations and to map soils

afforestation, but at the onset of World War II, the focus switched to agricultural land and increased food production. The mapping and classification of soils in Scotland owes much to the influence of Russian soil scientists; indeed, the first soil surveyor at the Macaulay Institute (Sandy Muir) spent time in the Dokuchaev Institute in Moscow and learned to survey in Crimea. A further Russian influence came to Scotland via the USA and Canada through Bob Glentworth, a former head of the Soil Survey of Scotland. He had worked in Manitoba where soils were mapped based on their parent material and hydrologic conditions, a system still used in Scotland today to distinguish 'Soil Series'.

The early surveyors used 'wartime' Geographical Section, General Staff (GSGS) editions of topographic maps as field sheets. As they traversed the landscape, they dug inspection pits (soil augers were of little use in the stony Scottish soils) and meticulously recorded the location of these in the field, making brief notes in their notebook. Later (often in the evening after a day's surveying) they transferred their notes to the field sheets, using abbreviated shorthand descriptions of parent material, hydrologic class and often soil texture and thickness of the topsoil, using fine-nibbed (0.5-mm) Rotring pens (Figure 29.1).

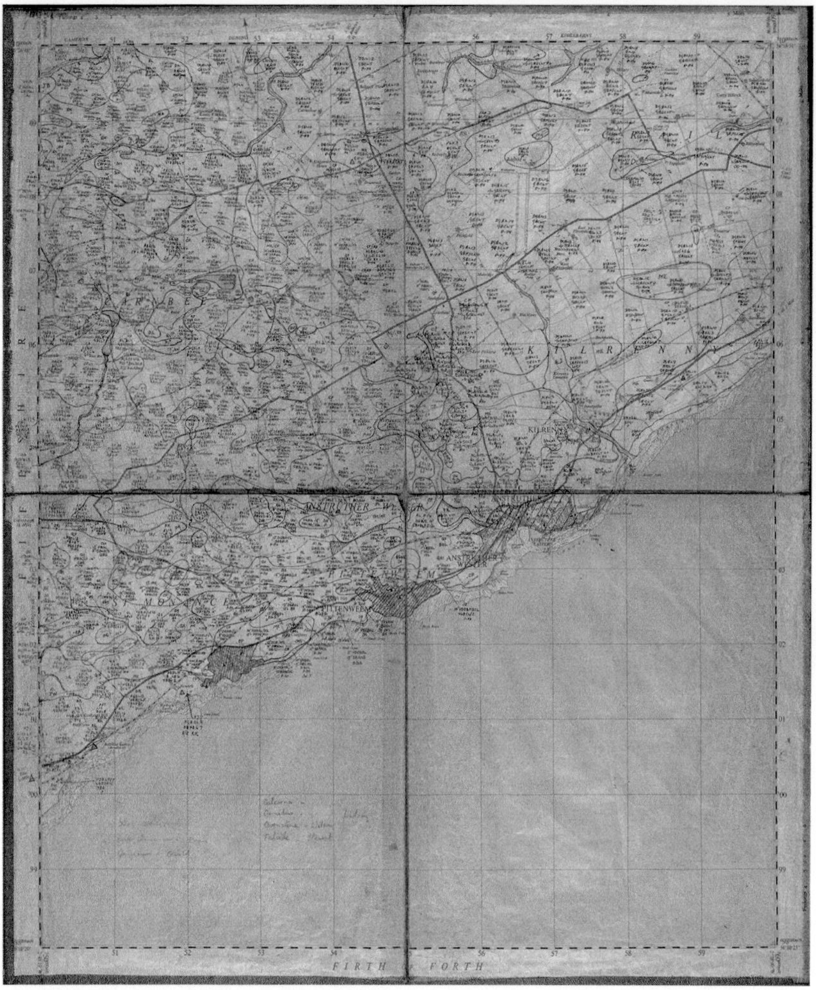

Figure 29.2 1:25,000-scale Ordnance Survey topographic map used to record observations and to map soils

These GSGS maps were based on the Cassini projection system but, later, 1:25,000 Ordnance Survey maps were used as the field sheets of choice (Figure 29.2). These maps were based on the transverse Mercator projection system and they have the same national grid system that is still currently in use in Great Britain. By the 1960s, black and white aerial photographs, generally around 1:24,000 scale, were increasingly being used in the field. These photographs partially overlapped such that, when viewed through a stereoscope, they gave the impression of a three-dimensional image. A trick that surveyors developed in the field was to relax both eyes so that they were looking at the same point on each of the overlapping photographs with different eyes and, with a bit of practice and moving the photograph slightly, they could see the image in three dimensions in the field! China graph pencils were used to mark the locations of inspection pits on the photograph in the field.

Whether using GSGS, Ordnance Survey maps or air photographs, the surveying process was the same. The surveyor would dig inspection holes to verify or refute his (they were all male) conceptual model or to confirm that it was the same or different from the previous one. They worked systematically over the land, digging at a density dictated by the complexity of the landscape and they drew lines on the map to signify a change in soil type using topography, land use and, in semi-natural soils, the relationship between soils and vegetation as a guide. This technique was known as 'free survey'. In all cases, field notes were generally transferred to the field sheet in the evening, ready for the next day's surveying.

The field season normally lasted from the beginning of April to the end of September, at which point the sun became too low in the sky to illuminate the bottom of the inspection pits. During the winter months, these boundaries between different soil types identified in the field were 'fixed' by being inked onto clean copies of field sheets. When air photographs were introduced, the boundaries sketched in the field were redrawn with the help of a stereoscope to view the relief in three dimensions, and changes in soil type that related to changes in topography could be accurately mapped. However, irrespective of the scale of the mapping, there was always a compromise between the heterogeneity present in the field and what could be realistically shown on a paper map. Different surveyors had widely different interpretations of what could or should be shown on the maps.

The good days of late September–early October were often used to go back to areas and dig a full-sized profile pit, describe and sample the soil layer to a depth of around 1 m in order to characterise the soils of the area. These profile descriptions along with the National Soil Inventory of Scotland form the majority of the 15,000 soil profiles held within the Scottish Soil Database.

The systematic soil survey of Scotland effectively came to an end in 1986. During the time that the Soil Survey of Scotland was active, approximately 95 per cent of the cultivated land in Scotland was mapped at a scale of 1:25,000 (for publication at 1: 63,360[1] or 1:50,000 scale) and all of the country at a reconnaissance scale of 1:250,000.[2]

Use of the Data

Since its beginnings in the late 1930s, the surveyors of the Soil Survey of Scotland have amassed nearly 15,000 soil profile descriptions (around 11,000 with analytical data) and this database underpins much of the current research in Digital Soil Mapping (Poggio and Gimona, 2014) and in providing data for national scale modelling (Boorman *et al.*, 1995; Smith *et al.*, 2010). There have been a wide range of scientific papers published that made use of the data collected (Lilly and Baggaley, 2013; Toberman *et al.*, 2015; Wylie *et al.*, 2016; Tipping *et al.*, 2017). There have also been a range of maps produced to help manage land more effectively, which use the original soil maps as a basis (for example LCA,[3] Risk maps[4]). The data collected also contributed to the production of the first global map of soil organic matter.[5] Alongside the mapping and profile data, we have around 50,000 dried and preserved soil samples in our National Soils Archive. These samples are invaluable 'time capsules', which can be used to assess changes in soil properties through time and allow modern researchers to extract more information from soils using new analytical techniques.

Many of the uses that current researchers are putting the data and archived material to would be totally bewildering to the early surveyors, but it is a testament to the thoroughness and dedication of the surveyors who laboured in the field and moors of Scotland to map the soils and collect data that we are still using the maps and data today and finding new ways to analyse soils to better understand our natural environment.

[1] https://maps.nls.uk/series/soils/

[2] www.hutton.ac.uk/learning/natural-resource-datasets/soilshutton/soils-maps-scotland

[3] http://soils.environment.gov.scot/maps/capability-maps/national-scale-land-capability-for-agriculture/

[4] http://soils.environment.gov.scot/maps/risk-maps/map-of-topsoil-compaction-risk-partial-cover/

[5] www.fao.org/global-soil-partnership/pillars-action/4-information-and-data/global-soil-organic-carbon-gsoc-map/en/

Acknowledgement

This work and the work of the Soil Survey of Scotland was funded by the Rural & Environment Science & Analytical Services Division of the Scottish Government and its successors. I would like to thank all my Soil Survey of Scotland colleagues (past and present) and to acknowledge the work of Robert (Bob) Grant, former Head of the Soil Survey of Scotland, which provided much of the basis of the early history of soil surveying in Scotland.

References

Boorman, D. B., Hollis, J. M. and Lilly, A. (1995). *Hydrology of Soil Types: A Hydrologically-Based Classification of the Soils of the United Kingdom. Institute of Hydrology Report No.126.* Institute of Hydrology, Wallingford.

Lilly, A. and Baggaley, N. J. (2013). The potential for Scottish cultivated topsoils to lose or gain soil organic carbon. *Soil Use and Management* 29, 39–47.

Poggio, L. and Gimona, A. (2014). National scale 3D modelling of soil organic carbon stocks with uncertainty propagation: an example from Scotland. *Geoderma* 232–234, 284–299; DOI: 10.1016/j.geoderma. 2014.05.0040016-7061.

Smith, J. U., Gottschalk, P., Bellarby, J., *et al.* (2010). Estimating changes in national soil carbon stocks using ECOSSE. Part II. Application. *Climate Research* 45, 193–205.

Tipping, E., Davies, J. A. C., Henrys, P. A., *et al.* (2017). Long-term increases in soil carbon due to ecosystem fertilization by atmospheric nitrogen deposition demonstrated by regional-scale modelling and observations. *Scientific Reports* 7, 1890; DOI:10.1038/s41598–017-02002-w.

Toberman, H., Tipping, E. W., Boyle, J. F., *et al.* (2015). Dependence of ombrotrophic peat nitrogen on phosphorus and climate. *Biogeochemistry* 125, 11–20.

Wylie, C. E., Shaw, D. J., Fordyce, F. M., *et al.* (2016). Equine grass sickness in Scotland: a case control study of environmental geochemical risk factors. *Equine Veterinary Journal* 48, 779–785.

30 · *A Travelling Ethnography of Urban Technologies*

ANDRÉS LUQUE-AYALA

After years of working with indigenous peoples in the Amazon, drawing on ethnographic approaches, I refocused my research on urban technologies. I came to be fascinated by the political implications of transformations in urban infrastructures (e.g. water networks, waste collection or transport systems). As it became clear to me that technology is a powerful political force shaping the social dimensions of cities, I grew interested in the politics embedded (and often hidden) in both technological change and its materiality. Technology for me was not a passive agent, but one with an agency of its own. Its material configuration plays a fundamental role in establishing and maintaining social relations (i.e. it is relational) – this is precisely what is at stake in its political power.

With these ideas in mind, I headed to Mumbai, in India, and São Paulo, in Brazil, hoping to engage ethnographically with an infrastructural object that, in my view, had significant potential for environmental, social and political transformations in cities: the humble solar hot water (SHW) system. This is an energy technology that, in replacing electricity or gas-based systems for heating water (e.g. electric showers or gas boilers), helps in lowering greenhouse gas emissions. Given its affordability, it is often used in social housing or as a mechanism to increase energy access for the poor, signalling implications for social justice. But as I started fieldwork, as a newcomer in the field, there were still important unresolved methodological questions. How do you examine technology from a social science perspective? How do you uncover and account for its inherent agency? Is it possible to use an ethnographic approach to do research on urban infrastructures?

Figure 30.1 Fieldwork on the roof of buildings, here at a workshop learning how to build do-it-yourself SHW systems in São Paulo, Brazil (Andrés Luque-Ayala)

'Tracing the Life' of a Technological Object

Drawing on the combined methodological and theoretical outlook of actor–network theory (ANT) scholars, I set out to 'trace the life' of SHW systems. I aimed for a field-based understanding of how the materiality and relations of this technology make up the social world, to be achieved by 'latching on' and following SHW systems. I looked at the things that 'stick' to them and the 'translations' that occur as they move from vision and imagination to material presence and practice. In ANT, 'the social' is traced through material means, 'follow[ing] the actors in their weaving through things' (Latour, 2005, pp. 68–69). By closely following human or non-human agents and describing in detail the relationships amongst them, tracing the life of a technological object offered productive opportunities for engaging with the integration of social and technological domains (Law, 1991). This allows for greater attention to the material dimensions of such processes while challenging us to think about the agency of these material elements, and the possibility of them having a voice: energy itself, electricity, water, solar collectors, water tanks, showers and pipes.

Following technology as a research practice meant an exhausting schedule of meetings, interviews, site visits, travels and journeys. In each city I interviewed users, manufacturers and dealers of SHW systems; I visited new and old housing projects using them; I shadowed the work of environmental activists promoting their use; I kept detailed photographic records of every SHW system installation I encountered; I attended training events on their manufacture, installation and standards; I participated in industry conferences put together by each city's renewable energy community; I accompanied installers of SHW systems in their work and, at times, I installed the systems myself; I climbed onto roofs and, there and then, asked about pipes and joints, broken glass and bits of plastic lying around. Interviewing users of the SHW systems would involve asking whether they would mind showing me the bathroom where the water (heated by the sun) is used. My request would be welcomed with excitement, as the user – often a proud owner of an SHW system – would excitedly explain the arrangement of water faucets and opened the tap in the hope that, despite a rainy day, hot water would still be available! Then, as hot water came out, a look of satisfaction would follow; a sense that they knew a trick no one else knows.

Dwelling versus Travel: a Methodological Approach Informed by Ethnographic Sensitivities

Over time, this multi-sited and multi-methods research became a 'travelling ethnography' of urban technologies (cf. Clifford, 1997). I followed SHW systems in their different incarnations, from their conception (as objects endowed with the power to offer a technological solution to an energy problem) to their implementation and use. Fieldwork occurred at many sites and took different shapes, following commodity chains, examining manufacturing processes and visiting points of use (Marcus, 1998). It required interviewing experts and outsiders, aficionados and lay people, users and non-users. Such multiplicity of sites for ethnographic inquiry offered a window into 'decisions about encoding and standardisation, tinkering and tailoring activities ... and the observation and deconstruction of [social and political] decisions carried into infrastructural forms' (Leigh-Star, 1999, p. 382).

Using a methodological approach informed by ethnographic sensitivities, the fieldwork generated an abundant data set and rich narrative accounts. As a multi-sited ethnography of an urban process, it required constantly travelling across the city (visiting sites of visioning, manufacture, delivery, assembly and use) and across cities (tracing how dynamics occurring in one city influence another and how a seemingly similar urban process plays out in different urban contexts). Tracing the life of SHW systems, in practice, required criss-crossing Mumbai and São Paulo on a daily basis as well as visiting other cities to see the sites of inspiration and imagination (such as the national capitals, where policy guidelines affecting many cities are developed), or to understand the material origin of the technology (such as at manufacturing sites). Taking cue from Marcus's multi-sited ethnographic practice (Marcus, 1995) and Clifford's travelling ethnographies (Clifford, 1997), this methodological approach acknowledges the itinerant nature of urban technology, from government offices to the sites of advocacy and from factory to dwelling. While in the past ethnography 'privileged relations of dwelling over relations of travel', travelling ethnographies decentre the fieldwork 'site' as the primary locale of ethnographic research and instead 'take place', in the words of Clifford, 'in worldly, contingent relations of travel, not in controlled sites of research' (Clifford, 1997, p. 22 and p. 68). As described by Marcus (1998, p. 90), it is a form of research that

> is designed around chains, paths, threads, conjunctions, or juxtapositions of locations in which the ethnographer establishes some form of literal, physical presence, with an explicit, posited logic of association or connection among sites that in fact defines the argument of the ethnography.

As fieldwork practice, it responds to the multi-sited nature of any ethnography aimed at uncovering the manner in which global processes – such as the contemporary transformation of energy systems – play out in local domains.

But there are important caveats to be made. As a trained anthropologist, I am mindful of using the term ethnography lightly. A 'travelling ethnography of urban technologies' does not claim to embody the ethnographic purity of classical ethnographic texts (e.g. Malinowski, 1922; Mead, 1935). Instead, it sees social science research as a non-linear and flexible process, best described as a 'contingent construction' (Crang and Cook, 2007, p. 207). This form of field research acknowledges the practical need to vary the emphasis amongst methods, and readily takes advantage of the unexpected data-gathering opportunities that emerge in the field (Crang and Cook, 2007). Yet, it is a methodological approach that does not come free of burdens and anxieties. During fieldwork, adopting practices of 'moving to and from, in and out, [and] passing through' (Clifford, 1997, p. 67) imposed an extenuating schedule of activities and travels. My memories of the everyday in Brazil and India are more related to traversing the city by public transport than to interacting with my object of study; the research fieldwork, in practice, was more about developing and maintaining a database of contacts (via telephone calls and mobile phone text messages) that could lead to interviews (in offices, living rooms and, occasionally, factories) than about dwelling in the observation practices required by an ethnography. Time constraints for fieldwork in each site and city inevitably resulted in methodological anxieties around the 'attenuation of the power of the fieldwork' (Marcus, 1995). This required balancing the ethnographic 'dwelling' of participatory observation with the copious but sometimes antiseptic and place-detached narrative accounts resulting from interviews.

This travelling ethnography of infrastructure enabled an interrogation of complex processes that are usually taken for granted. It provided entry points for political, epistemological and ontological insights as much as more practical policy recommendations. It revealed how infrastructure has defined spatial and temporal reaches that transcend a single event or one-site practice, uncovering how infrastructure 'sink[s] into other structures, social arrangements and technologies' (Leigh-Star, 1999, p. 381). Paraphrasing Susan Leigh-Star, who speaks of an 'ethnography of infrastructure', this ethnography of urban technologies revealed how infrastructure is embedded in social processes, supporting them in invisible ways.

References

Clifford, J. (1997). *Routes: Travel and Translation in the Late Twentieth Century*. Harvard University Press, Cambridge, MA.

Crang, M. and Cook, I. (2007). *Doing Ethnographies*. Sage, London.

Latour, B. (2005). *Reassembling the Social: An Introduction to Actor-Network Theory*. Clarendon, Oxford.

Law, J. (1991). Introduction: monsters, machines and sociotechnical relations. In *A Sociology of Monsters: Essays on Power, Technology and Domination*, ed. J. Law. Routledge, London, pp. 1–23.

Leigh-Star, S. (1999). The ethnography of infrastructure. *American Behavioral Scientist* 43, 377–391.

Malinowski, B. (1922). *Argonauts of the Western Pacific*. Routledge & Kegan Paul, London.

Marcus, G. E. (1995). Ethnography in/of the world system: the emergence of multi-sited ethnography. *Annual Review of Anthropology* 24, 95–117.

Marcus, G. E. (1998). *Ethnography through Thick and Thin*. Princeton University Press, Princeton, NJ.

Mead, M. (1935). *Sex and Temperament*. William Morrow and Company, New York, NY.

31 · *My Date with the Devil*

PETER MARREN

I can't say I have ever had much of a passion for proper scientific fieldwork. In fact, on about the only occasion I ever did any, I disliked the experience very much. I was about 23, and, wearying of ever finding a full-time job in nature conservation, I got myself enrolled as a temporary member of staff at a northern university where I spent about four months on a cold, treeless fell, counting gentians. It was very boring and it drove me to the bottle. I made the discovery that the teeniest, most drivelling drop of scientific information requires days and days on your knees out there under the clouds with your quadrat frame, and weeks and weeks more of statistical skulduggery back in the lab.

Apparently, all this counting eventually helped to establish the admittedly fairly interesting fact that sheep are not the great enemy they are cracked up to be. Despite being hoovered up by sheep every season before they can ripen seed, the gentian still manages to spread itself about with the help of little creeping stems, like strawberries. They don't need to set seed. They don't even need to put forth their starry blue flower, which Sarah Raven likened to a wayward blossom stuck down the end of a cigarette holder. So, they will probably survive overgrazing and all the other problems we throw at them. But, to a thoughtful person, shivering out there on the fell, this very resilience seemed to beg some broader questions about existence. Is it enough just to sit there vegetating, on a bare, bleating hillside, without purpose, without even much of a role in the ecosystem? (For nothing else seems to depend on the gentian's continued existence.) I thought about that, as I too sat there, counting and clicking, and recording the results on graph paper, and in my special scientific notebook. I never did become Dr Marren PhD.

Figure 31.1 Devil's bolete (original artwork by Albin Schmalfuss). A black and white version of this figure will appear in some formats. For the colour version, please refer to the plate section.

My idea of fieldwork is less scientific data analysis as chasing down a species. In that regard, I want to introduce you to a charismatic organism in whose company I once spent a few congenial weeks. Its name is the Devil's bolete or *Rubroboletus satanas*, 'the red boletus of Satan', or the Devil's own toadstool. The real Devil is red and has a really serious body odour problem, both characteristics which are shared by his toadstool. Othmar Lenz, who first described *R. satanas* in the early nineteenth century, claimed he was made ill by the terrible pong 'emanating' from the type specimen in front of him. I don't suppose he was tempted to eat it, but others have been, and the effect was always interesting. Projectile vomiting for 'up to six hours' is usual, sometimes accompanied by 'bloody diarrhoea'. You would think, wouldn't you, that they might have been warned by the name *satanas*? When the bolete's toxic principal was finally located, extracted and fed to mice, the latter immediately dropped dead. It is wonderfully evil.

It looks the part too. In shape, the Devil's bolete has been compared to 'a badly made dumpling', with a great fat stalk like a turnip or Chianti bottle. The bright red pores, which slowly fade to orange, are tucked in under the cap, and so, seen from above, the fungus resembles a ball of putty that has been mildly beaten with a toffee hammer. Sometimes you find several such balls, all squeezed together like a clover leaf. Unfortunately, you need to turn the fungus over to see the red pores underneath, and that great fat stalk with its net of red veins. Another colourful surprise happens if you cut the flesh with a penknife and watch it turn from yellow to a pretty shade of blue. You can get all the colours of the rainbow out of *R. satanas*.

The Devil's bolete entered my life in 1997 when I was asked by Plantlife to investigate its status in Britain, in part by checking records and herbarium specimens, but also by going out and looking for it. It had been added to the Species Action Plan list as a potentially endangered species, in need of protection. But although the Devil's bolete was famous, it seemed that hardly anyone had seen it in the flesh, and there were only two records on the database. Was it pegging out? And, if so, could anything be done to preserve it?

Of course, you can't just go for a walk and expect to find species like this, even on recorded sites. Unlike flowers, fungi produce their fruit bodies – their only visible, recordable part – only when conditions are right for a truly remarkable spurt of growth. The Devil's bolete, which can swell to the size of a football (though tennis-ball size is more normal), probably requires rather exceptional conditions to fruit well in Britain. Like all boletes, it is spongy and soft when mature, and the fruit bodies only last a few days. Your chances of being there at the exact moment it chooses to heave itself above ground are not great.

So, what do you do? What I did was get in touch with everyone who took a field-orientated interest in boletes. That was never a very large number of people, and back then it consisted of about 20 enthusiasts, led by Alan Hill, who maintained a database of records and much else besides. Alan took me to the one place he knew where, with luck, you could expect to find the Devil. We were lucky in that the weather that August was just right for big boletes: still, hot, sweaty sunshine following on from heavy rain. And there it was, on a steep bank overlooking a lake: bulges of fungal putty, protruding from the earth among the snaking roots of mature beech trees. By lying down and peering up at the fruit bodies, you could make out glints of red from the pores (that's a phrase characteristic of James Bond villains: 'Drax's eyes glinted redly at Bond'). Strangely enough, they lacked any detectable smell; perhaps the appalling stink of carrion described by others is confined to specimens that have started to rot. These were still young, with a curiously dense, tough texture, but with a bit of give, almost like rubber. You could imagine them bouncing. So, there I was, standing in a fungal garden of a near-legendary species – and with a view to match: an arm of Capability Brown's great lake, with Blenheim Palace in the background. It was as though the Devil in the guise of a poisonous toadstool had chosen to lurk inside one of the most gloriously Baroque visions of the English landscape.

We counted the fruit bodies, and made a map of the site, and a lot of notes. Even without knowing it, you could see from the way the caps were distributed among the roots that there must be a connection between fungus and tree. Like most boletes, *R. satanas* is a mycorrhizal species, which infects tree roots and draws from them water and dissolved sugar in exchange for certain trace elements released from the soil by the fungus. It depends on trees, especially beech, and presumably also depends on a sufficiency of calcium, for the Devil's bolete is nearly always found on chalk or limestone, in this case Oxfordshire's low escarpment of 'coral rag'. What else? It seems our species seems nearly always to occur in bolete 'hotspots' where, for whatever reason, rare species cluster together.

I found it in other places, too, always after a tip-off from my generous band of bolete-hunters. These were usually less romantic locations than the beech bank at Blenheim. One was on a former hedge line close to a public toilet. Another was by a road. But bit by bit, in this entirely empirical way, an impression of a much-misunderstood species was building up. One that occurs on clay overlying limestone in warm, sheltered places; one that seems to prefer mature beech on surprisingly dry sunny banks, but which will also fruit under oak, and even among rockrose on open downland. One that is more likely to be found in August than October, which may be one reason why mycologists have missed it. A species that is undoubtedly rare, in the sense of being rarely spotted, but which is quite widespread in southern and western England.

A few months after my brief acquaintance with the Devil-in-the-field, they put it on a stamp. It was not a very good image, and might even have been misattributed, but it was still a happy moment. The result of my work was that, after due deliberation, the Devil's bolete was removed from the growing list of obscure species with Action Plans. From a database of just two records, there are now more than 40 known sites for the species. It is still listed as 'Vulnerable', although to what, and by whom, I really could not say. Targeted conservation is, in this case, not very meaningful. The Devil's bolete will probably have to take its chance in this our collapsing natural world. Perhaps the Devil will look after its own. For my part, the survey tickled my curiosity and pleasured my aesthetic sense. It did not stoke a passion for fieldwork, but it was certainly more fun than counting gentians. I felt I had contributed a tiny grain of knowledge to the long beach of conservation science. The bolete also makes a useful example of the unwisdom of making policies without knowledge, bricks without straw. In our business we should always, always doubt what nobody is sure about!

32 · *Peregrinations through the Heathlands and Moorlands of Britain: an Applied Plant Ecologist's Tale*

ROB MARRS

My earliest interests in fieldwork were spawned from a schoolboy passion for hill walking on trips organised by my two biology teachers, Ella Bryce and Sandy Robertson. While biology was not part of the deal, there was obviously a subliminal message sinking in. I then had the privilege of doing biology at the University of Stirling in Scotland, where again my enthusiasm was stoked by official field courses to Millport (where I met my wife) and Loch Rannoch, and in between times, unofficial ones run by the Students' Biological Society to Durness in Sutherland. Students don't seem to organise their own field courses now, unfortunately. This overall experience must have worked because I gradually transferred from a biochemist to a card-carrying plant ecologist; eventually doing a PhD in physiological ecology. The PhD was meant to be centred on a garden experiment, but I soon expanded this to do fieldwork. What I really liked was that you could study in great places and get paid to do it. This really did open up a career option.

The PhD was great. I did single-handed fieldwork in Donside, north Wales, Cumbria, central Scotland and in Galway, and along with my rather eccentric supervisor, John Proctor, two large-scale surveys of the Lizard peninsula, in Cornwall, south-west England (Marrs and Proctor, 1978, 1979). Often, I was collecting plants and soils, but in Ireland and on the Lizard, assessments of all the plants growing in the community were made. My first Lizard trip was in the very hot summer of 1976. I recall a train trip to Preston where I met up with John and we set forth to Lizard Point to survey the heaths on the Cornish serpentine. We travelled in his motorhome, converted from an old ambulance, along with his wife, Sue, and baby daughter, Katy. We had two great weeks and two things impressed me: first, the fact that there was a very serious drought effect which became much more obvious as you moved south (it is always more impressive to see what you know in real life, much more impressive than on the telly); and second, the fabulous plant biodiversity of the Lizard heaths.

Since then I have developed a career in plant ecology, but always involving fieldwork. After a postdoctoral position working in land reclamation, again in Cornwall, I was given my dream job working at Monks Wood Research Station where I was told to run a research contract on scrub and bracken control for 50 per cent of the time and use the other 50 per cent to do 'something interesting'. I joined what had been the 'Lowlands and Grassland Heath Section' and almost everyone there were serious field biologists – Terry Wells, Caroline Sargent, Owen Mountford, Lena Ward, Brian Davis and Eric Duffey – really inspiring colleagues. There was free travel using company cars, so I managed to spend at least six months of the year working in the field. Most of my work was done in Suffolk where the weather is usually great in summer – I adopted a casual attitude, 'if it was wet today, I went tomorrow'. As my career progressed I was pushed rather reluctantly into management, and quietly taken aside and told I did too much fieldwork! However, my timesheets were always spot-on; the number of management meetings I 'was at' on a Breckland heath beggared belief.

Throughout, I have essentially only ever done three things in the field – surveys, experiments and teaching – and I love doing them all. What I have learned is that it is always essential to have a properly designed, rigorous sampling plan – if you don't, you won't be able to enjoy the wonder of seeing your hard-won field results turn into proper scientific output. Here I ascribe to the 8P-maxim of the British Army's parachute regiment – **P**roper, **P**rior **P**lanning and **P**reparation **P**revents **P**iss-**P**oor **P**erformance. This may appear boring and regulates what you do in the field, but it always helps get clean data for analysis. Planning is key.

My Surveys

I started with the Irish and Cornish PhD trips but thereafter surveyed vegetation growth and both soil and ecosystem development in a massive survey of restored land on the Cornish China Clay tips (Roberts *et al.*, 1981).

This was not onerous in terms of assessing biodiversity, it was merely collecting plant and soils sampled from the restored sites using a structured survey design. I then surveyed all heathland sites in Breckland (Norfolk) for heather performance and threats from scrub and bracken encroachment. This experience led me into two really interesting projects. The first was a survey of Scottish native pinewoods, carried out with colleagues Mike Sykes and Brian Mitchell, for the Nature Conservancy Council. This was a year-long study of these fabulous woods with the sole aim of planning an experiment to assess the effects of red deer grazing. I learned a huge amount from working in the field with Mike (the forester) and Brian (the zoologist). On one trip around Deeside and Speyside, we were accompanied by Bob Bunce, another colleague who had already done a survey of the pinewood vegetation. Bob was inspirational in getting his knowledge of Scottish native pinewoods across to a beginner like me; his enthusiasm for proper statistically rigorous sampling and analysis mirrored my own and it was fantastic to see his pinewoods' classification in action. The second was a survey of the impact of grazing on upland ecosystems, led jointly with another colleague, David Welch. Here we had teams distributed across Britain looking at vegetation inside and outside enclosures; I did East Anglia, the North York Moors and the North Pennines. That was a brilliant exposure to countryside that I didn't know.

As my career progressed I have done less survey work personally but I have supervised a national study of bracken response to control treatments, led by Mike Le Duc (Le Duc *et al.*, 1997), and my students have used surveys carried out using Bob Bunce's methods to study the flora of British woodlands (Corney *et al.*, 2006), Scottish 'soft coasts' (e.g. Lewis *et al.*, 2013) and national change via the Countryside Survey (see also Bunce *et al.*, 2019). Analysis of other people's data is much less fun than collecting your own but it is still fulfilling when you have some idea of the systems being studied.

My Experiments

I love experimenting in plant ecology; this is real science, developing a hypothesis, an experimental test of that hypothesis, collecting results, a statistical analysis of the results and then acceptance or refutation of the hypothesis. I started with glasshouse experiments, growing plants in different soils; the fieldwork here was just collecting plants and soils. I only started working on serious field experimentation when I went to

Figure 32.1 With honours student, Erin Stoll, sampling vegetation and soils in the long-term burning and grazing experiment on Hard Hill at Moor House, North Pennines. The experiment was set up two years after I was born! (from the author's photo collection).

Monks Wood, where such experiments were common practice for Terry Wells and Brian Davis. I have studied both scrub and bracken control, heathland restoration, acidification of soils, grazing and moorland burning (e.g. Marrs *et al.*, 1986, 2018; Milligan *et. al.*, 2016, 2018). Throughout, I have followed the example of Sandy Watt (ex Cambridge University) and Terry Wells, focusing on long-term studies (e.g. Marrs *et al.*, 1986). One set of experiments in Breckland lasted for 18 years, another set in the Peak District, Derbyshire, have been going for 25 years and counting, and I have also collected and analysed data from long-term studies at Moor House. Most of these experiments are very large-scale and you can see them from space. When doing fieldwork in such long-term studies, it is rare these days for a single person to do it all; you must consider yourself a time traveller akin to the BBC time traveller Dr Who. You do your part and then you regenerate into someone else who takes over (I am Tom Baker in Moor House 'Dr Who' terms). As I reach formal retirement I have made a career decision to return to active fieldwork (I always did a few days per year, but now I do more). This year, I have spent three days at Moor House collecting soil samples (easy work but a long and hard carry) and another, showing people the experiments; a day at my experiments in the Peak District, checking they were OK, plus two days measuring bracken density and bracken frond leaf length in 198 different quadrats. This might appear boring, but I absolutely love it. Last year I had the privilege of helping my former student, Simon Smart, do some Botanical Society of Britain and Ireland (BSBI) recording on Jura. I had never done this type of fieldwork before but really enjoyed legging it over the moorland, even if my rather elderly legs complained a bit.

My Teaching

I moved to the University of Liverpool in 1991 and immediately became involved in teaching plant ecology. In my view you cannot teach ecology without field experience. At first, we ran practical days in our introductory module, which, to put it bluntly, weren't very good. They were on Wednesday afternoons, then transferred to Saturdays, but always in March and April when there isn't much to see and the weather is more often than not poor. My colleague, the late Brian Moss, was a kindred spirit and we conspired to change this to two residential courses at the end of the spring term. Eventually, the management was worn down, and we were allowed to run mirrored field courses; Brian led one at the Field Studies Council (FSC) centre at Slapton and I led the other at the FSC's Dale Fort field centre. Our aim was not to teach anything in particular, but to inspire students in ecology and the wonder of being outdoors. We succeeded, as our module got rave reviews from the students. The courses have continued in more or less the same form although venues have changed. It is brilliant to see students, often initially rather fearful of the fieldwork experience, opening their eyes and developing ecological skills.

I also got involved in year-three honours field courses, which was brilliant fun, run first by Mike Johnson and then Brian Moss. Originally, they were sited in the Peak District, but after the foot-and-mouth disease outbreak in 2001 we moved it to FSC Blencathra, English Lake District. Brian put most of the Blencathra course together and it was an exceptional one, bringing in geology, climate, land-use history, water chemistry, catchment nutrient budgets and a 'public inquiry', together with ethics and philosophy. Along the way and following our instinct for field-led teaching, we developed a Masters course and ran up to four field courses each year (which was totally uneconomic, but nobody

Figure 32.2 Teaching students on our Blencathra field course (from the author's photo collection). A black and white version of this figure will appear in some formats. For the colour version, please refer to the plate section.

showed us the accounts, so we carried on until told to stop). This was the best teaching I ever did: we ran courses in north Wales, on the Norfolk Broads in cabin cruisers, FSC Orielton, in Pembrokeshire, Wales, and one on degraded land in Cornwall. Our students really appreciated the exposure to the variety of field case studies. Our current Masters course has just one field course (the accountants caught up with me) at the Scottish Centre for Ecology and the Natural Environment (SCENE), on Loch Lomond, in Scotland's first National Park. I designed and run this course single-handedly and I love it. Working in God's own county within western Atlantic oakwoods was absolute paradise.

So, I reckon I am an over-aged schoolboy who has had a dream career working in the 'Sweetie Shop of Ecological Life', carrying out ecological research outdoors. Sometimes it is hard work, but it is always fun, even when wet. People do not dissolve in H_2O! Implementing experimental procedures can often involve very physical work and I have broken the second law of fieldwork so often it is not true. This law states 'never do fieldwork with geologists, soil scientists or physical geographers': you end up helping to carry their samples. I often collect soils and heavy plant samples and so far, my back is intact, but I often find I work alone! The first law of course is 'Never get separated from your lunch'.

References

Bunce, R. G. H., Wood, C. and Smart, S. (2019). The ecology of British Upland landscapes. II. The influence of policy on the current character of the uplands and the potential for change. *Journal of Landscape Ecology* 11. 140–154.

Corney, P. M., Le Duc, M. G., Smart, S. M., *et al.* (2006). Relationships between field-layer species composition and environmental drivers in British semi-natural woodlands. *Journal of Ecology* 94, 383–401.

Le Duc, M. G., Pakeman, R. J., Bärtsch, S. and Marrs, R. H. (1997). Factors affecting vegetation succession following bracken control. In *Species Dispersal and Land Use Processes*, eds. M. Moss and P. Dennis, Proceedings of the 6th conference of IALE. IALE, Belfast, pp. 265–272.

Lewis, R. J., Marrs, R. H. and Pakeman, R. J. (2013). Inferring temporal shifts in land-use intensity from functional response traits and functional diversity patterns: a study of Scotland's Machair grassland. *Oikos* 123, 334–344.

Marrs, R. H. and Proctor, J. (1978). Chemical and ecological studies of heath plants and soils of the Lizard Peninsula, Cornwall. *Journal of Ecology* 66, 417–432.

Marrs, R. H. and Proctor, J. (1979). Vegetation and soil studies of the enclosed heathlands of the Lizard Peninsula, Cornwall. *Vegetatio* 41, 121–128.

Marrs, R. H., Rawes, M. Robinson, J. S. and Poppitt, S. D. (1986). Long-term studies of vegetation change at Moor House NNR: guide to recording methods and database. Merlewood Research & Development Paper 109. ITE, Grange-over-Sands.

Marrs, R. H., Marsland, E-L., Lingard, R., *et al.* (2018). Experimental evidence for sustained carbon sequestration in fire-managed, peat moorlands. *Nature Geoscience* 12, 108–112.

Milligan, G., Cox, E. S., Alday, J. G., *et al.* (2016). The effectiveness of old and new strategies for the long-term control of *Pteridium aquilinum*, an 8-year test. *Weed Research* 56, 247–257.

Milligan, G., Booth K. E., Cox, E. S., *et al.* (2018). Changes to ecosystem properties brought changing the dominant species: impact of *Pteridium aquilinum*-control and heathland restoration treatments on selected soil properties. *Journal of Environmental Management,* 207, 1–9.

Roberts, R. D. Marrs, R. H., Skeffington, R. A. and Bradshaw, A. D. (1981). Ecosystem development on naturally-colonized china clay wastes. I. Vegetation change and overall accumulation of organic matter and nutrients. *Journal of Ecology* 69, 153–161.

33 · *The Maimai Catchment New Zealand*

JEFFREY MCDONNELL

I knew that I would study geology from a very young age – collecting and studying rocks, canoe trips in northern Ontario, winter survival courses: I was ready. But in the summer of my second year at the University of Toronto, I took a summer job as a geophysical technician in a three-person fly camp in the central Yukon. A helicopter would deposit us 'in the middle of nowhere' for 7–10 days and we would do reconnaissance geophysics with proton magnetometers and electromagnetic induction on lines cut into the boreal forest by First Nations sawyers. My first day alone in the bush I had a grizzly bear encounter with a narrow escape. The summer proved to be a punishing series of such encounters. I returned to Toronto to change my major to physical geography and to seek refuge in a kinder, gentler fieldwork: hydrology. Something closer to civilisation. Fieldwork without grizzly bears.

Fast forward a few more years and an opportunity came to do a PhD in hydrology in New Zealand. A bear-free zone. New Zealand is to a field hydrologist as Scotland is to a whiskey aficionado. My site was the Maimai watershed on the west coast of the South Island (Figure 33.1). The mere act of getting there – on the four-hour drive from Christchurch – was inspiring; over the Southern Alps with their greywacke erosional hillslopes, past pockets of peculiar karst landforms and some of the world's most spectacular braided rivers. Each awe-inspiring. Each motivating to a geomorphologically trained young hydrologist, always asking 'why' are things the way they are?

But Maimai was the true geophysical nirvana. Hills carved into early Pleistocene conglomerate – the series of two dozen small headwater catchments were perfect replicas of large watersheds, sculpted into

Figure 33.1 The Maimai catchment (Jeff McDonnell)

miniature forms. The combination of >2,000 mm of rain per year falling every three days, on average, and the soft, weakly lithified conglomerate made for a perfect mix of hydrogeomorphic action. Landslides abounded. For studying rainfall-runoff events, one never had to wait more than a week for 'new data' to stream in. I laid down a kilometre's worth of cable, connecting my tensiometers, water-level recorders and mini-weirs across one of the headwater catchments, and collected data for a year.

As a young hydrologist, what I learned at Maimai, first and foremost, was how responsive a catchment can be to storm rainfall. Even with no detectable overland flow, the storm hydrograph at the catchment outlet rose sharply like an urban stream with peak flows that would rival any suburban catchment. How this worked and where this water came from was a real mystery. And as a PhD student, this 'what the heck?' feeling was a terrific motivator and inspiration to learn. To coax explanation from the landscape, I dug small pits on the steep hillslopes and reactivated existing ones by other researchers to see how the timing of their activation related to the stream channel. I took samples for stable-isotope analysis at a lab on the North Island and also for chloride and electric conductivity measurements that I could make myself at the University of Canterbury.

I felt a deep connection to the hydrologists who had worked at Maimai before me: Paul Mosley, Amy Pearce, Mike Sklash and others. They had published some of the most important work in the field – and even in the pre-internet age, I was acutely aware of their findings back in Ontario. This was mostly because their work at times contradicted each other. There was an argument over the source and pathway of the waters

Figure 33.2 The author revisits the Maimai catchment (from the author's photo collection)

that comprised the storm hydrograph; the early work by Mosley suggesting it was likely preferential flow of rain water through soil pipes and macropores; the latter work by Pearce and Sklash showing, with stable-isotope analysis of the source waters, that such a process was impossible and that the flashy storm hydrographs were made up of stored water, pushed out by displacement. My work was intent on getting to the bottom of this mystery by testing objectively the two hypotheses with new data and measurement approaches. What I found was that they were both right in some ways and wrong in others. Macropores conveyed old water to the stream. Like lifting a full garden hose in summer and seeing water gush out, so was the Maimai watershed during rain events – almost an instantaneous pressure propagation through the landscape. But like the garden hose when connected to the faucet, the time taken for the cold water to be detected at the hose outlet was long – linked to the velocity of the water molecule. Stream hydrographs at Maimai – despite their flashiness – were composed of water that was several months old.

This pressure–velocity relation and its link to macroporous soils proved fascinating to me – and still does. Now, 30 years later, I continue to scratch my head on this problem with my PhD students and postdocs. While we've branched out to many other catchments around the world, we keep coming back to Maimai – perhaps the most studied of all headwater catchments – as our reference point for ideas and field tests. At Maimai we've since done experiments stripping away the soil to expose the bedrock beneath – to quantify its permeability, microtopographic relief and how this controls the release of hillslope water. We've age-dated the deep bedrock groundwater to understand its connection to the fast responses in the soil above. All this has helped us understand catchment in other parts of the world.

What I've learned most about field discovery is that it is an iterative process. One that builds on past work and past people's findings. 'What the heck' questions still abound and I have found Maimai to be my hydrological compass for attacking them. Looking back, fieldwork at Maimai shaped me via the hours sitting in the rain with time to think (and without looking over my shoulder for bears!). For new students coming into the geophysical sciences, I encourage them to find their own 'Maimai': a place of discovery that inspires. Where questions and puzzles and paradoxes can sustain one for many decades.

34 · *'Writing in the Field': the Importance of a Local Patch*

STEPHEN MOSS

'What's your field?' is a question familiar to many professionals and academics. The inquirer is probably anticipating a response along the lines of 'environmental history', 'I'm an ornithologist', or 'I work for the BBC'.

The answer they probably wouldn't expect is 'my local patch, in a hidden corner of the Avalon Marshes, in Somerset'. But that's how I would respond. That's because 'my field' is more or less just that: a reedbed, bordered by trees, watercourses and grassy meadows. And it's a place I visit as often as I can, to get my regular fix of nature.

I came late to finding a local patch. Despite having been a keen birder since childhood, it wasn't until I reached my early thirties, when I moved back to west London, that I found one. One fine July day, my car broke down on the way to work. While waiting for the AA breakdown service to arrive, I discovered that I was right next to Lonsdale Road Reservoir, alongside the Boat Race course by the banks of the River Thames. After that first accidental visit, I would go there once or twice a week, and systematically record the birds I saw, as the seasons of the year unfolded.

Later, I moved a few miles down the road, and adopted another disused reservoir, near Kempton Park racecourse, as my new patch. Then, a dozen years ago, we relocated to Somerset, where I plan to live out the rest of my years. Here, I have broadened my horizons, with no fewer than three local patches: my garden and its immediate surroundings; that quiet spot on the Avalon Marshes; and my coastal patch, where three rivers meet – the Huntspill, Parrett and Brue.

You'll notice I say 'my patch'; for that is how I think of these places, even though I now share them with my friend and colleague Graeme.

Both of us appreciate the value of getting to know these places really well. This allows us to become attuned not just to the obvious changes in the seasons, but also to the subtler changes from year to year, week to week and – during the spring rush – day to day.

In doing so, we are following in a long and distinguished tradition of 'patchwatchers'. They range from Gilbert White of Selborne, via the poet John Clare, to the godfather of birding, Ian Wallace, whose accounts of the birds he finds in a superficially undistinguished corner of Staffordshire are an inspiration to us all.

The point of 'patchwork' is that it connects us with nature in a unique and very special way. In the wise words of Jeremy Mynott, having a local patch allows us to 'relish in the ordinary and the particular'; thereby 'we are in effect making ourselves more at home in the natural world'.

Visits to my local patch follow more or less the same pattern. Along with my binoculars, I carry a digital voice recorder, on which I make notes of everything I see or hear. This is not just a list of species and numbers, but also what the birds are doing, the first dates of spring arrivals and, of course, all the other wildlife I come across. They act as an aide-memoire, and also a source of great pleasure when I browse accounts from previous visits, going back over the years.

I then take things one step further. Like Wordsworth, who coined the phrase 'emotion recollected in tranquillity', I write about my experiences. This started about 25 years ago, when I began a regular monthly 'Birdwatch' column in the *Guardian* newspaper and decided to recall and write about the highlights of a lifetime's birding, going all the way back to my early childhood.

A few years later, I collected these columns together in a book called *This Birding Life* (Moss, 2006), putting them into chronological order. Rereading them, I felt an immense sense of satisfaction. Like the author Nick Hornby, whose bestselling book, *Fever Pitch* (Hornby, 1992) documented his obsession with watching Arsenal football club, and in doing so revealed the deep continuity between the present and the past, the story they tell is that of my own lifetime's consuming passion for birds.

It's not just about the birds, though. There are also memories of the friends I made, the places I visited and my relationship with my late mother, who – despite never really sharing my interest in birds – always encouraged me to pursue my hobby, when I might easily have lost interest and given up.

When I moved from London to Somerset, I began to document my new-found home in books such as *A Sky Full of Starlings: The Diary of a*

Figure 34.1 A wooded drove running through Westhay Heath, a Somerset Wildlife Trust reserve (Stephen Moss)

Birding Year (Moss, 2010) and *Wild Hares and Hummingbirds: The Natural History of an English Village* (Moss, 2012). I deliberately focused on the importance of the local rather than the global; the immediate, not the abstract; and the commonplace, instead of the rare. By concentrating on my home patch, I also developed my understanding of the crucial importance of place and reinforced and deepened my connection with the natural world.

Now I am trying to pass on my passion for birds and wildlife, and the importance of the local, to others. I lead short tours for Somerset Birdwatching Holidays, and gain great pleasure from showing visitors from elsewhere in Britain the birds of this wetland wonderland. And when I am out with my students at Bath Spa University, who are studying for an MA in Travel and Nature Writing, I focus strongly on the importance of the concept of 'writing in the field'.

To say this is about the process of writing is something of a misnomer. To me, when 'out in the field', we need to focus not so much on the words, as on the way we perceive our surroundings. Seeing, hearing, smelling, touching, tasting and feeling are just as important, if not more so, than putting pen to paper. As a wise old fisherman once said: 'Fishing ain't about catching; if it were, it would be called "catching"'.

Figure 34.2 Westhay Heath, a Somerset Wildlife Trust reserve covering 26 hectares on the edge of the Avalon Marshes (Stephen Moss)

Figure 34.3 The South Drain, which runs along the edge of Westhay Heath, eventually running into the Huntspill River (Stephen Moss). A black and white version of this figure will appear in some formats. For the colour version, please refer to the plate section.

For me, writing in the field is a state of mind. You need to drop your guard, forget what you know and simply absorb what is happening all around you. You have to free your mind from preconceptions, and remember that, however familiar a place may seem, it always has the chance of surprising you. That's one of the many joys of a local patch.

References

Hornby, N. (1992). *Fever Pitch*. Victor Gollancz Ltd, London.

Moss, S. (2006). *This Birding Life: The Best of the "Guardian's" Birdwatch*. Aurum Press, London.

Moss, S. (2010). *A Sky Full of Starlings: The Diary of a Birding Year*. Aurum Press, London.

Moss, S. (2012). *Wild Hares and Hummingbirds: The Natural History of an English Village*. Vintage, London.

35 · *Looking but Not Seeing: How Sketching in the Field Improves Observational Skills in Science*

STEPHEN MOTT

Sketching in the field serves many purposes and the type of sketch should reflect these. It is of value in most types of fieldwork.

There has been a decline in the use of field sketching as a skill for the fieldworker. This may be because teaching the skill has become a casualty of ever-increasing pressure from other parts of the course or curriculum. Field sketching vastly improves observational skills, especially of detail, whatever scale or type of sketching is adopted (e.g. Moseley, 1981; Coe *et al.*, 2010). With the rise of digital photography, it is all too easy to take many photos and use them in place of a field sketch.

I often begin a field course with a sketching exercise. The task may be structured, i.e. I might give a list of features that they should look for, draw and label, depending on the age and the level of achievement of the students. I include this task as part of the familiarisation and orientation work on the first day. I choose the site with care so that it includes features that we examine in detail later. Once they have completed the field sketch and labelled it, almost all the students begin to observe rather than simply look. They remark on features that they did not pick up at first glance. Acknowledging the positive response to their progress is gratifying and encouraging for both student and field tutor.

Observing what is seen and recording it in a field sketch allows the fieldworker to clarify detail, pay attention to areas which arouse interest or questions and seek out any possible relationships between features on the field sketch. Often, sketching encourages the geologist to move back from a particular rock exposure to see the 'bigger picture', which can then be drawn and annotated. After this, a detailed sketch of a particular rock exposure can be made. Field sketches should include a date, a relevant grid reference, annotations and shading. The sketch should include a scale and annotations appropriate to the scientific discipline. Basic drawing skills are all that is needed and students who feel they are poor at drawing need not be discouraged. The weather may affect the amount that can be done.

Sometimes the fieldworker comes with an agenda, which might subconsciously influence what is drawn. This is where the discipline of field sketching is important. A field sketch must be an impartial account of what is there, even if there are features which suggest a conflict with a research agenda. Such agendas are of course natural and understandable but this must not be allowed to influence the accuracy of drawing what is actually there. The impartiality of the field sketch applies to all scientific disciplines.

Two Examples of Field Sketches

Figure 35.1 shows a fair copy of two sketches from my geology field notebook of a hypabyssal igneous intrusion (Bell, 1997). These sketches were made when I discovered an exposure which I did not at first understand. I drew the exposure from a short distance away and then walked over it, filling in the details. There is a sketch map of part of the exposure (top) and a section of another. This compelled me to look at every part of the exposure in a structured way so that I did not become overwhelmed by it. Note the questions which were raised by drawing the sketch.

Field sketching skills are and should be transferable between scientific disciplines. Figure 35.2 shows the behaviour of a golden eagle near its eyrie.

Figure 35.2 is an interesting drawing, for it also serves to provide evidence of eagle behaviour which may assist with later research, such as employing satellite tagging to determine the fate and movements of birds (e.g. Whitfield and Fielding, 2017).

In conclusion, field sketches improve observational skills. They provide a means of gaining perspective at the beginning of fieldwork, and can serve as a tool to provide direction and focus for the next steps in fieldwork. Coincidentally, they may provide evidence for other fieldworkers to use. I suggest the sketch should go beyond the limitations imposed by the expectations and agenda of the fieldworker.

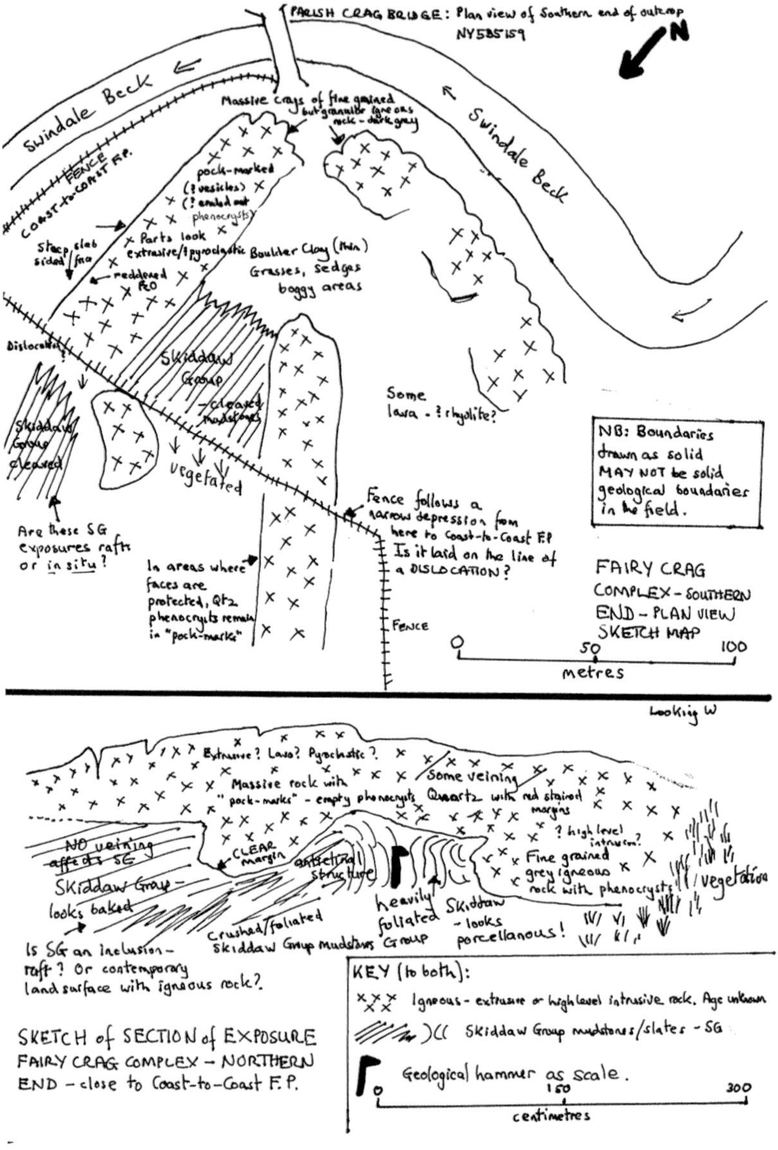

Figure 35.1 Sketch map and section of part of the geology of Fairy Crag Complex, Parish Crag Bridge, Bampton near Shap, Cumbria. Grid reference: NY535160 (original artwork by Stephen Mott).

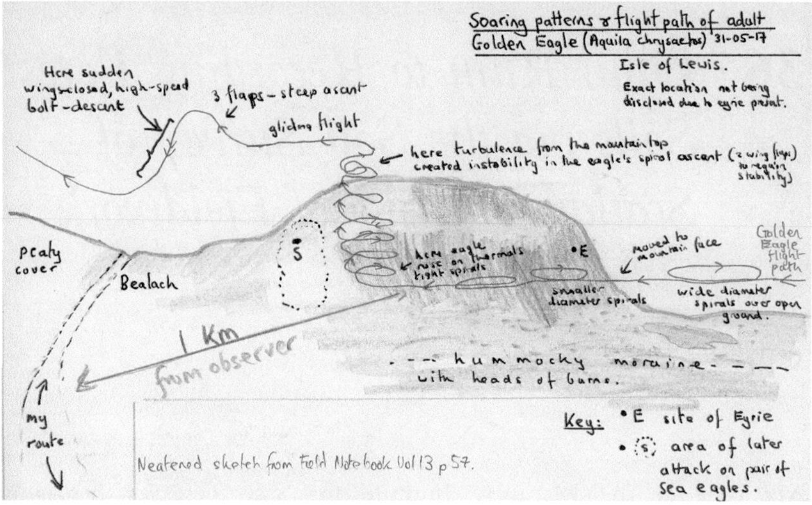

Figure 35.2 Golden eagle territorial behaviour near its eyrie in the Outer Hebrides. (Grid reference withheld for reasons of eyrie security.) (Original artwork by Stephen Mott.) A black and white version of this figure will appear in some formats. For the colour version, please refer to the plate section.

Acknowledgement

I am grateful to Dr Diarmad Campbell, Chief Geologist Scotland at the British Geological Survey for providing time and resources to help me interpret the exposure at Fairy Crag.

References

Bell, A. M. (1997). *The Skiddaw Group and Its Contact with Surrounding Rocks in the Bampton Inlier, Cumbria: BGS Technical Report WA/97/9*. British Geological Survey, Keyworth.

Coe, A. L., ed. (2010). *Geological Field Techniques*. Wiley-Blackwell, Oxford.

Moseley, F. (1981). *Methods in Field Geology*. W. H. Freeman, New York, NY.

Whitfield, D. P. and Fielding, A. H. (2017). *Analyses of the Fates of Satellite Tracked Golden Eagles in Scotland: Scottish Natural Heritage Commissioned Report No. 982*. Scottish Natural Heritage, Perth.

36 · *From Rum to Recording Forest Soils via the Soil Survey of Scotland: a Life of Fieldwork*

ANDREW J. NOLAN

My time at the Macaulay Institute for Soil Research began in 1979 following an application to join the Soil Survey of Scotland as a soil surveyor. I was advised by the then Secretary not to get married, as this wouldn't be at all compatible with survey life! Previously, I had spent the summer of 1977 camping on the Island of Rum, carrying out geological mapping and plant ecology studies as an undergraduate, so had some insights into a fieldwork career. This was in the days prior to the new pier (opened in 2004; before then passengers and goods used to be transferred from the ferry to a flit boat, the *Rhouma*, before landing at an older jetty in Loch Scresort).

I had visited the Highlands and Islands many times in my teens, on Scout camps and expeditions, notably hiking around the northern part of the Island of Jura to observe the Gulf of Corryvreckan whirlpool and the stunning north-west coast, where the striking quartzite cobble raised beaches left a lasting impression, stirring thoughts of Earth processes operating over great timescales. Fortunately, I was able to study geology at a time when the theory of plate tectonics had just provided integration of the many phenomena that earth scientists had observed hitherto. This gave great impetus to the subject. Up to that point, my interest in the natural world had been inspired by authors such as Henry Williamson, Frank Fraser Darling and Mike Tomkies.

On Rum you are immersed in a great igneous cauldron of layered olivine and feldspar crystals, where you can almost feel the shifting pulses of magma and observe the graded settling of different mineral compositions in sedimentary sequence. Here, the volcanic activity during the Tertiary period really comes alive as you systematically map and record

Figure 36.1 Soil surveyors from the James Hutton Institute, Aberdeen, examining a peaty podzol, Leanachan Forest, Lochaber, 2015 (Andrew Nolan)

the bedrock exposures. There were also other moments to contemplate, such as stepping from the deer-grazed wet heath into native woodland plantations, where the luxuriant vegetation, birdsong and invertebrate life were in stark contrast, revealing how much more wildlife some of these areas could hold. Time to explore and wonder, good training for spending weeks out surveying soils and vegetation in remote locations, more often than not on your own. Such places have inspired generations of fieldworkers in the natural sciences.

My main training was in the Golspie Office, under colleagues Don Futty and Willie Towers during production of the 1:250,000 Soil Map of Scotland (with data gathered during 1978–1982). I was impressed by the methodical and systematic approach, vowing to put the principles learned into practice. The survey area included Wester Ross, and magical long summer days were spent amid striking landscapes, mapping and recording soil types and sampling for the National Soils Inventory of Scotland (Nolan and Robertson, 1987). I remember taking photos of numerous erratic boulders of Lewisian gneiss lying on exposed Torridonian Sandstone bedrock, again an indication of the forces that had

Figure 36.2 Richard Hewison, Vegetation Ecologist, James Hutton Institute, recording long-term changes in permanent plots, Island of Rum, 2008 (Andrew Nolan)

created such magnificent natural scenery. No doubt, a GPS (global positioning system) unit would have been very handy then, for now we have small hand-held devices with all the Ordinance Survey map data for the UK on a chip the size of your little fingernail, which can tell where you are to 1-m accuracy. And now you may have a smartphone in your pocket with which you can send a photo or video clip to anywhere in the world in a fraction of a second.

Following completion of the Map of Scotland, the Soil Survey of Scotland recommenced systematic mapping of the agricultural lowlands. However, changing times and funding meant that only three more map sheets were completed before the curtain came down and we moved on to other research projects, notably recording long-term vegetation changes in the Environmentally Sensitive Areas in Scotland (Nolan *et al.*, 2007). Headage payments prevailed, with elevated numbers of domestic livestock compared to the present day, while exclosure cages were used to demonstrate the effects of wild herbivores, enhancing the prospects for woodland regeneration. On the machair of the Outer

Hebrides, maintaining the traditional management practices that had created such patterns of diversity in the first place was an elusive aim. Further extensive fieldwork followed, with research on the designing and testing of sampling strategies for the rapid assessment of grazing and trampling impacts on upland habitats to facilitate sustainable management by the deer management groups.

However, often things come full circle and in the new millennium there has been a greater focus on, and funding for, soil research from the point of view of sustainable goals for human activity, adaptation to climate change and biodiversity and soil carbon management. As a result, once again we have been carrying out detailed soil mapping at a scale of 1:10,000 to improve the resolution of soils data across the National Forest Estate (Nolan *et al.*, 2014). Vitally, this has also provided an opportunity for the training and mentoring of a new generation of fieldworkers, equipped with soil survey skills.

As things move forward with developments in modelling and digital soil mapping, I am left reflecting on our current approach to such fieldwork. The forest soils mapping programme is still being carried out by systematic survey on the ground, essentially using a 'free survey' approach, where the frequency of soil inspection pits depends on the variability and predictability of the soil patterns in any particular area. Knowledge and confidence is acquired in a conceptual and predictive model of the patterns on the ground, using experience gained from previous surveys. This enables the identification of the main soil types and their distribution and the drawing up of a polygon map of soil types. Essentially, this is a classic physiographic approach to mapping, with patterns of soil variation inevitably very closely linked to geology, landform, slope, altitude, nature of the parent material, hydrology and vegetation. On areas of open hill, patterns of semi-natural vegetation may enable reasonably accurate interpretation, even using more remote survey methods. However, most of the current soil mapping is being carried out in existing forests where such patterns have been masked or long since disappeared. Hence, on the ground, survey work is more relevant than ever, especially as many of the soil phases and their implications for forest management relate to soil profile and pedological features such as soil texture and rooting depth.

A question still remains as to the accuracy of the field survey data and the difficulties of classification set against natural variation. In other words, to what extent do we focus on mapping the boundary as opposed to defining the population of soils within the polygon? It would be

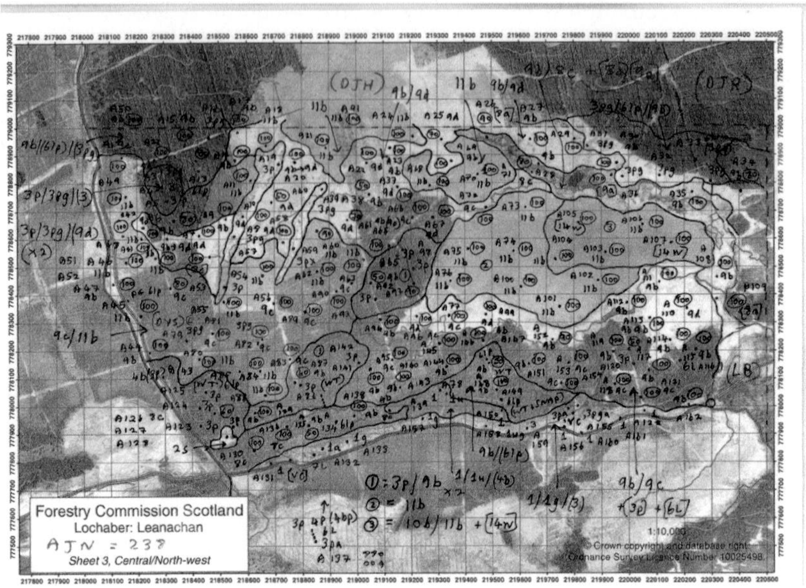

Figure 36.3 Mapping forest soils and peatland types on aerial image, Leanachan Forest, Lochaber, 2015 (Andrew Nolan)

easier for office-based land managers and particularly those using geographic information systems (GIS) if this were more clear-cut, whereas anyone who has spent time out in the field knows that this is very often not the case and that variation is inherent at whatever scale you care to look. Grid surveys provide more objective data on the population within a particular polygon and for intensively surveyed sites can provide the point data to enable the polygons to be constructed with more mathematical confidence. However, for every polygon surveyed, this would require an inordinate effort to generate statistically valid samples. Hence, I conclude that the experience and ability of the soil surveyor will be required for some time to come to make meaningful sense of the patterns of natural variation and to construct a soil map that is fit for purpose.

References

Nolan, A. J. and Robertson, J. S. (1987). Regional trends in dry and moist Scottish moorland vegetation in relation to climate, soils and other ecological factors. *Journal of Ecology* 75, 1145–1157.

Nolan, A. J., Cummins, R. P., Scott, D., *et al.* (2007). *Monitoring Environmentally Sensitive Areas in Scotland. Volume 2: The Breadalbane ESA Final Report, 1995–2004.* Scottish Executive Environment and Rural Affairs Department, Edinburgh.

Nolan, A. J., Rayner, B., Kennedy, A., *et al.* (2014). Sustainable forestry: the imperative of soil mapping in forest resource inventory, modelling and management in Scotland, UK. Presented at the 20th World Congress of Soil Science, Soils Embrace Life and Universe, Jeju, South Korea, 8–13 June.

37 · *In Praise of Bat Detectors*

KIRSTY PARK

When I tell people I work on bats, the most typical response is 'how cool!'. I suspect part of the reason for this (other than the fact that bats are, indeed, very cool) is that to most people they are they are largely unseen, unheard and relatively little known, even though they are amongst our commonest mammals (20 per cent of all mammals are bats). At most, people may catch fleeting glances of them flying around their garden or streetlights at night.

The last 20 years have seen great advances in wildlife filming, which brings all manner of organisms closer to people through their televisions, and there is now some astonishingly good footage of bats. However, there is nothing like observing something directly yourself – one great pleasure in working on bats is introducing them to people, either through showing people bats up close or hearing them for the first time on a bat detector. My first encounter with a bat was when I was about 15 – it had flown into the kitchen the night before and my mother found it the next day in the sink. After we'd dried it out (unaccountably, she thought it was a dead leaf and tried to wash it away), we called the local bat group and, from the moment it took off from my hands, I was entranced. As luck would have it, I then chose to do my undergraduate degree at Leeds University where bat expert John Altringham was (then) a lecturer, and I got my first introduction to studying bats up close. Over the next three years I spent many happy days and evenings in the North York Moors National Park checking bat boxes (Figure 37.1) and recording bat calls with a team of university enthusiasts and a fantastic group of Forest Enterprise wildlife rangers.

Figure 37.1 The author inspecting bat boxes in the North Yorkshire Moors in the early 1990s (from the author's photo collection)

While you need a licence to check bat roosts or catch them in the wild, detectors are a tool that anyone can use without disturbing them. It is not an exaggeration to say that the invention and development of bat detectors has revolutionised the way that we interact with bats, both in terms of research and public engagement. While it was discovered that bats navigated using their ears as far back as 1790, it wasn't until the 1930s that the first detector was made, and it was found that bats use ultrasound (frequencies higher than human hearing) for this purpose. The researcher who coined the term echolocation was Don Griffin and he relates in his book, *Listening in the Dark* (Griffin, 1958), that for the first bat detector they needed 'a station wagon to carry the necessary apparatus, which included a gasoline engine-driven AC generator, oscilloscope, modified movie camera, and a parabolic horn'. We now have detectors less than 5 × 5 cm, which you can plug into your phone, and different types of detectors are available for a range of purposes and prices. While none of them are very cheap, they are now affordable for the enthusiast and enable researchers to collect far more data than could have been envisaged even 40 years ago (to the point now where the analysis and storage of calls can be a major challenge).

The affordability of detectors has also had a big impact on how enthusiasts and researchers engage the public – bat walks with detectors are now a staple part of public engagement activities and have proved very popular. Detectors and volunteer surveyors are also two essential elements in monitoring populations in many countries; in the UK, the Bat Conservation Trust (www.bats.org.uk) offers training and support to thousands of volunteers each year, which enables population trends to be generated for 11 of our 18 resident species.

Using bat detectors, we can identify species, or at least narrow them down to broad groups. We can also quantify relative levels of activity through counting 'bat passes' (a continuous string of echolocation calls). Sometimes, we can hear when an insectivorous bat has homed in on its prey through feeding buzzes, and we can also hear bats interacting with each other through social calls. This means we can find out which species favour particular habitats for feeding and how they move through the landscape between their roosts and foraging grounds. We can use bat activity to assess the impact of human activity that might be damaging to their populations, or habitat management designed to protect them. In my research group at the University of Stirling, we've been using bat detectors to assess the efficacy of agri-environment schemes such as hedgerow or water margin management, both on bats and their insect prey (Fuentes-Montemayor *et al.*, 2011). We've examined the consequences of timber operations in plantations on bat activity (Kirkpatrick *et al.*, 2017), quantified the factors influencing the use of urban woodlands by bats (Lintott *et al.*, 2015), and assessed bats' responses to small wind turbines (Minderman *et al.*, 2012; Tatchley, 2016). One current study (the WrEN project; www.wren-project.com) is focused on the value of secondary (i.e. planted) woodlands for wildlife, including bats, and how to prioritise actions to restore woodland ecological networks (Watts *et al.*, 2016). Bat detectors (and some fantastically hard-working students and field assistants) enable us to survey a large number of sites simultaneously – to date, we've surveyed 133 secondary and ancient woodlands in England and Scotland and have recorded nearly a quarter of a million bat passes from six species or species groups, over 2,500 survey hours. This work has highlighted the importance of species mobility in determining the relative importance of local versus landscape-level attributes (Fuentes-Montemayor *et al.*, 2017); for example, while lower-mobility species such as *Plecotus auritus* (Figure 37.2) were most strongly influenced by local habitat quality, the amount of woodland in the surrounding landscape, and how well connected the focal patch was, became increasingly important for higher-mobility species such as *Myotis* spp. This was particularly the case in more intensive and homogeneous agricultural landscapes.

Figure 37.2 Plecotus auritus, the brown long-eared bat (courtesy of John Altringham). A black and white version of this figure will appear in some formats. For the colour version, please refer to the plate section.

There are limitations to detectors, however – we cannot use bat activity to measure absolute population sizes (although there is work currently underway to try and achieve this), and because of similarities between the calls of some species it can be very hard to be certain about correctly identifying some bats to species. Where I live in Scotland, species identification is less of a problem since we have relatively few of them, but two of my PhD students working in southern Africa had to deal with many more species but much less information about how to identify them from their calls (Cooper-Bohannon *et al.*, 2016). It is also not possible to distinguish between males and females on the basis of their echolocation calls – this affects our interpretation of detector data when gender plays a role in habitat selection, as we found for one of our commonest bat species, *Pipistrellus pygmaeus* (Lintott *et al.*, 2014).

So, while bat detectors are just one tool in the armoury of equipment and techniques needed to study echolocating bats, they are undoubtedly one of the most accessible and transformative and have enabled us to uncover much about bats that we simply would not have been able to without them.

Acknowledgements

All the work from the University of Stirling cited above comes from former and current PhD students and Research Fellows. In order of appearance these are: Elisa Fuentes-Montemayor, Paul Lintott, Jeroen Minderman, Rachael Cooper-Bohannon, Cerian Tatchley, Luci Kirkpatrick and Helen Taylor-Boyd. The WrEN project is a collaboration between the University of Stirling, Forest Research and Natural England Thanks are owed to many undergraduate, Masters students and field assistants who have contributed to our research.

References

Cooper-Bohannon, R., Rebelo, H., Jones, G., *et al.* (2016). Predicting bat distributions and diversity hotspots in southern Africa. *Hystrix: Italian Journal of Mammalogy* 27; DOI: https://doi.org/10.4404/hystrix-27.1–11722.

Fuentes-Montemayor, E., Goulson, D. and Park, K. J. (2011). Pipistrelle bats and their prey do not benefit from four widely applied agri-environment management prescriptions. *Biological Conservation* 144, 2233–2246.

Fuentes-Montemayor, E., Watts, K., Macgregor, N. A., Lopez-Gallego, Z. and Park, K. J. (2017). Species mobility and landscape context determine the importance of local and landscape-level attributes: insights from a large-scale natural experiment. *Ecological Applications* 27, 1541–1554.

Griffin, D. (1958). *Listening in the Dark: Acoustic Orientation of Bats and Men*. Yale University Press, New Haven, CT.

Kirkpatrick, L., Oldfield, I. F. and Park, K. (2017). Responses of bats to clear fell harvesting in Sitka spruce plantations, and implications for wind turbine installation. *Forest Ecology and Management* 395, 1–8.

Lintott, P. R., Bunnefeld, N., Fuentes-Montemayor, E., *et al.* (2014). City life makes females fussy: sex differences in habitat use of temperate bats in urban areas. *Royal Society Open Science*; DOI: 10.1098/rsos.140200.

Lintott, P. R., Bunnefeld, N., Minderman, J., *et al.* (2015). Differential responses to woodland character and landscape context by cryptic bats in urban environments. *PLoS One* 10, e0126850; DOI: 10.1371/journal.pone.0126850.

Minderman, J., Pendlebury, C. J., Pearce-Higgins, J. W. and Park, K. J. (2012). Experimental evidence for the effect of small wind turbine proximity and operation on bird and bat activity. *PLoS One* 7, e41177; DOI: 10.1371/journal.pone.

Tatchley, C. (2016). *Wildlife Impacts of and Public Attitudes towards Small Wind Turbines*. PhD thesis, University of Stirling.

Watts, K., Fuentes-Montemayor, E., Macgregor, N. A., *et al.* (2016). Using historic woodland creation to construct a long-term, large-scale natural experiment: the WrEN project. *Ecology and Evolution* 6, 3012–3025.

38 · *In Search of Tawny Frogmouths*

STUART RAE

When I moved to Australia from Scotland, I didn't know what I would do professionally or privately. I had spent most of my life studying birds, specialising in difficult-to-find species in difficult terrain, notably golden eagle, dotterel and rock ptarmigan in the Highlands. As I went for walks around my new home in Canberra, the bush capital, I was soon familiar with most of the local species even if most were new to me.

I like a challenge but couldn't settle on a local species to study. The only one I was having difficulty finding was the nocturnal tawny frogmouth. But then I found a few and noticed a behavioural trait they had – sitting in the sunny side of trees on winter days. A quick read through the literature revealed that this had only recently been described scientifically, so that topic was off the list to explore. Yet the literature was sparse and most accounts of other aspects of their behaviour were anecdotal; little detailed work had been done.

Beginning to Understand Frogmouths through Searching

Frogmouths are difficult to find, relying upon camouflage and stillness to avoid detection, and perching on dead branches which they resemble. They reminded me of the ptarmigan, the species I studied for my PhD. Ptarmigans rely on camouflage, and sit close to look-alike rocks. The similarity between the two is so strong I now jokingly refer to frogmouths as 'ptree ptarmigans'.

Every species offers something tantalising to study. So I approached the frogmouth dilemma methodically. They are nocturnal birds, obligatory tree-roosting during the day, usually in the same perch the whole time. I decided to search every tree in three study areas, all 1,200 hectares of them. I could only search so many trees each day, and birds can shift trees between days. However, if I restricted my searches to within the breeding season, when the birds were tied to a nest site, either on the nest or very close, I could search an entire block of woodland in a study area each day before moving on.

I reckoned I had about two months to cover as much terrain as possible between the nests being built and the chicks leaving the nest. After that, any birds could move into an area I had searched on a previous day, and I would miss them. I reckoned I had some leeway with this time as not all birds would be at the same stage in their breeding schedule. However, for the same reason, I would have to check each area in which I found no birds at least twice, to be certain of absence.

The positives were encouraging. Once I found a pair with a nest, I could move on. Gradually, I worked my way through the woods, one block at a time. It took me two breeding seasons to complete my search, but the return for effort has been terrific. The birds are long-lived, so the same birds can be in the same territories for several years, and they usually nest close to the previous year's nest site, if not on the exact branch. Some territories have been abandoned over the years, with new territories established in other places, so I still search between each nest site for new pairs every year. I never take anything for granted.

Developing a Long-Term Study

The study is now 10 years old and I monitor over 50 territories per annum. I have 515 nest records to date (a national record). There are now reams of data on their breeding behaviour, and the scientific papers are beginning to appear (some written with my son; Rae, 2009, 2017; Rae and Rae, 2014). A long-term study of any species is probably the most satisfying form of field research, for even if I am collecting data on one topic I amass disproportionately more each year. Later on, when I have other questions, there is a good chance I will have key data, covering several years, already. I am currently assessing the effects of weather on frogmouth breeding success (Rae, 2017), and long-term data are essential.

This is tough work, but there is a feeling of completion in finding every pair of tawny frogmouths in the study areas and logging their breeding success. I don't like to leave gaps and continue to do multiple checks if any birds are missing. They might be difficult to find, but when

Figure 38.1 A fledgling tawny frogmouth aligns itself with one branch while its mother aligns with another, their plumages and poses each merging into the wood (Stuart Rae)

Figure 38.2 A family of tawny frogmouths aligned and in stick-pose, but their multiple outlines are less concealing (as branch-like) than that of a single bird (Stuart Rae). A black and white version of this figure will appear in some formats. For the colour version, please refer to the plate section.

I do find them it is deeply rewarding. Following individuals from hatching to fledglings, and even recognising many by their plumage and behaviour towards me, is fantastic. When I approach them, and they do not react defensively, I regard their acceptance of me as a form of trust. That, I believe, is a sign of good fieldwork. And yes, I am an obsessive field naturalist – obsessed in searching, observing and recording what I see.

References

Rae, S. (2009). Comparisons between nesting densities of tawny frogmouths (*Podargus strigoides*) in open- and closed-canopy woodlands. *Emu* 109, 327–330.

Rae, S. (2017). Breeding success of tawny frogmouths in relation to rainfall. *Canberra Bird Notes* 42, 140–145.

Rae, S. and Rae, D. (2014). Orientation of tawny frogmouth (*Podargus strigoides*) nests and their position on branches optimises thermoregulation and cryptic concealment. *Australian Journal of Zoology* 61, 469–474.

39 · *Don't Just Sit There Reading . . .*

JANE M. REID

'Write about inspiring fieldwork'; that was the instruction. Easy, of course. It's always a pleasure to write about one's enthusiasms. Yet also rather ironic. For me, the point of fieldwork is that it brings insights and emotions that I cannot possibly get from words on a page, not even words crafted by the very best writers. That isn't surprising. They haven't seen what I have seen. Nobody has.

So, fieldwork. Often defined in hard scientific and educational terms: research and learning objectives, methods, meticulous checks and balances – surely the antithesis of 'inspiring'. Yet, in practice, fieldwork assaults the senses and emotions. A battle between triumph and disaster, fluctuating from sublime to truly foul and back again, via the ridiculous. There I was, near the end of my PhD, somehow running a 5-km race as part of Alaskan 4th of July celebrations, in hiking boots and full thermals, into a force-six blizzard, alongside the still-frozen Arctic Ocean, in between sessions of searching for pectoral sandpiper nests on the icy tundra. One runner was far ahead of me, the others were far behind. It was slow-going, so I had plenty of time to ponder what combination of enthusiasm, curiosity, chance and craziness had taken me to that situation. I might still have won the race had I not paused to watch a phalarope sheltering on a roadside pool. I could not resist, it was too beautiful to miss. I doubt that moment will be repeated, for anyone, ever. And later, there the sandpipers were, singing from the top of my tent pole in the bright midnight sunshine, their eggs toasty warm in the arctic landscape, in a hidden nest cup that nobody else could find, filled with lichens carefully selected to insulate from the gradually melting snow (Reid et al., 2002).

Figure 39.1 Jane Reid with a European shag - resident or migrant?

By then, I was already long addicted to birds. In my memory, it was a short-eared owl on Orkney that really did it, when I was eight. I stared at it from 5 metres away, and it stared back. But my parents say I was already addicted, hence why I was outside looking for short-eared owls. Yet it took many more years before I encountered ecology and fieldwork as sciences, and I realised that my enthusiasm could be more than just that. I dropped biology in school, it was so boring, so utterly inferior to what I saw outside. Aged 18, I was a budding theoretical chemist. But gradually I realised that theoretical chemistry might not be the wisest career choice for one who felt happiest outside, while being outside could in fact become useful ecological science.

I moved gradually to population ecology: the science of understanding how populations respond to environmental variation and change, whether in the short term or in the longer term, due to natural selection and ongoing evolution. Critically important questions, but extremely difficult to answer. Fieldwork has a central role to play; it is the only way we can truly understand either the environmental changes or the population responses. Since populations are made up of individuals, population responses ultimately depend on how the environment affects individuals. To understand this, we need to follow the fortunes of free-living individuals over successive seasons and years.

Mandarte is a small island off the west Canadian coast that is of no particular significance, except that it is home to a small population of song sparrows, which has been studied intensively since 1975. These sparrows are all individually colour-ringed as chicks, before they leave their nest, then monitored through the rest of their lives. Since Mandarte is relatively isolated, the local sparrows risk inbreeding with relatives. This, as any animal breeder knows, can have unfortunate consequences. Inbreeding is an increasing problem in wild populations as natural habitats become fragmented, restricting natural movements. My mission was to ascertain whether more inbred sparrows, those whose parents had been closely related, had weaker immune systems, potentially making them vulnerable to infection. The challenge was to safely catch individual sparrows, vaccinate them to fire their immune system, then catch them again 10 days later to measure their immune response. Easier said than done, especially on a rocky island with no power or water, and with one eye always on the small boat that was our lifeline and means of escape. A laboratory study would have been so much easier, but that would have entirely defeated the point. It was a logistics and mist-netting operation of military precision. We emerged two weeks later, exhausted, filthy, scratched and nettled, outwitted by some sparrows but victorious over others, precious blood samples safely in the freezer. Evidence that inbreeding can indeed compromise immune responses in wild animals; we do need to be careful how our activities shatter and fragment the natural landscape (Reid *et al.*, 2007).

Of course, many animals are very good at moving around the landscape, including through seasonal migration, which might ultimately be the best defence against environmental change. Indeed, many animals are much better at moving than us scientists; we tend to be stuck in our place and see what comes to us. But that cannot suffice. So, these days, my main obsession is shag-spotting. With much help from enthusiastic bird-ringers, thousands of European shags have recently been colour-ringed along North Sea coasts, allowing individuals to be identified up to 200 m away through a telescope. For a species that was thought to be 'resident', these shags turn out to be surprisingly mobile through the year. The challenge is to find each individual in winter, and in the summer breeding season, and see how an individual's choice of winter location affects its survival and breeding success. So, almost every weekend for the last nine years, I have been out there, along the rugged Scottish coastline, shag-spotting. I read the tides and weather to predict where the birds will be; each day is a direct test of what I know and what I don't. I'm out there, battling the elements that challenge me and the birds. I forget the university and the work I am supposed to be doing, it's irrelevant. A severe winter storm; death and destruction all around. Then there it was, one of my ringed shags – I knew it well. It flew through the spray and tumult, landed in front of me and shook off the water, iridescent plumage shining in the winter sunlight. What a remarkable bird, surviving amidst the true harshest forces of nature. Now, our hard-won data show that individual shags that stay resident at their breeding colony throughout the year rear more chicks than individuals that migrate away for the winter (Grist *et al.*, 2017). Yet, individuals that stay resident risk death in winter storms. It is an annual game of 'chicken': who will stay, who won't? Will the residents be wiped out as climate change causes more extreme storms, changing the winter distribution and creating a new migratory species? Only our fieldwork, and more time, will tell.

So, fieldwork. Essential to identify the right questions, and to obtain the right answers. Yet so often it is undervalued, and it is hard to combine with almost any career. We need to take care of that. So, don't just sit there reading. It's much better live. Go outside and discover for yourself.

References

Grist, H., Daunt, F., Wanless, S., *et al.* (2017). Reproductive performance of resident and migrant males, females and pairs in a partially migratory bird. *Journal of Animal Ecology* 86, 1010–1021.

Reid, J. M., Cresswell, W., Holt, S., *et al.* (2002). Heat loss and nest scrape design in pectoral sandpipers. *Functional Ecology* 16, 305–312.

Reid, J. M., Arcese, P., Keller, L. F., *et al.* (2007). Inbreeding effects on immune response in song sparrows. *Proceedings of the Royal Society B* 274, 697–706.

40 · *Fieldwork in the Australian Bush: If It Doesn't Kill You, It'll Convert You*

LISA ROBINS

My first experience of fieldwork as a Catchment Project Officer set the trajectory of my career as one focused on integrated catchment management (ICM), coupled with building the capacity of communities to enable their effective engagement in water and land-use decision-making processes. Over a two-year period, I was to work with a small team and a community-based Catchment Co-ordinating Group (CCG) charged with developing the nation's first regional-scale ICM Strategy in the vast, mostly low-rainfall, agricultural catchment of the Wimmera River (30,000 km^2) in the State of Victoria, in Australia's south-east.

Starting in the heat of the summer of 1990, I was tasked with assessing the condition of Crown Land parcels abutting the Wimmera River, from its headwaters at Mt Cole to the regional town of Horsham, about 175 km downstream, and some of its tributaries. Following this, I was to conduct a survey of gully erosion across the Upper Wimmera River Catchment (almost 3,200 km^2). As a recent graduate from the city, the nature and scale of these endeavours seemed daunting.

The river surveillance was to be conducted over a three-month period, combining first-hand observations, together with aerial photographic and topographic/parish map interpretation. This exercise was marked by near misses with venomous snakes, serious dehydration and narrowly avoiding setting wheat crops alight by driving through long grass to access the river. I don't know what the Occupational Health and Safety standards were at the time, but I was without any means of communication and not as well versed as I ought to have been in the threats and risks posed by the Australian bush.

Figure 40.1 The author (wearing Victorian Department of Conservation, Forests and Lands uniform) setting off in January 1990 on a three-month process to assess the condition of Crown Land parcels abutting the Wimmera River from its headwaters to the regional town of Horsham (from the author's photo collection).

The survey of gully erosion used aerial photographic interpretation, together with ground-truthing. Using a 1:100,000 topographic map spread across the lounge-room floor of my small (snake-free) dwelling, I laboriously transferred grid lines to aerial photographs using chinagraph pens, to create 3,177 cells, and recorded the density of existing gullies in each as either negligible, minor, moderate, severe, very severe or extreme. Out in the field, ground-truthing was carried out for 17 cells, in order to determine the actual versus expected gully erosion present. I calculated the cumulative length of gully erosion as 3,180 km for 3,177 km^2; a ratio of 1 km of gully to 1 km^2 of land, with total repair costs of almost $23 million (1990 Australian dollars).

Both the river and gully surveys served as a crash course in developing my knowledge of local flora and fauna; the hallmarks of the landscape, including the back roads and farm gates (and how to open and shut them); the vagaries of the bush; and, most of all, the appalling state of the river and much of the land in the upper catchment, with extensive land and water salinisation. I had come to understand why it was that the

Figure 40.2 Gully-prone hill country of the Upper Wimmera River Catchment (Lisa Robins)

catchment was in desperate need of an ICM strategy. My fieldwork was a prelude to a lengthy process of engaging in diverse conversations with the range of stakeholder interests about how to manage more sustainably the land and water resources of the catchment – river frontage management, steep hill country management, water quality, environmental flows, pest plant and animal control and more. Community engagement and empowerment was enabled through the establishment of a 'forum group' in each of the Upper, Mid, Lower and Plains Wimmera, each of which nominated representatives to the overarching Wimmera CCG. A discussion paper series addressing each of these issues was used to convey the key messages arising from scientific research to inform community decision-making. The process culminated with the launch of the Wimmera River Integrated Catchment Management Strategy on the banks of the river in Horsham, in 1992.

Now, 35 years later, I can see flaws in both my fieldwork and in the strategy itself, although I haven't forgotten some of the constraints that existed at the time – that computers were rudimentary, that there was no internet or email, that geographic information systems (GIS) was a new and quite primitive technology and that library resources were 200 km away in Melbourne. There were also social constraints – as the first

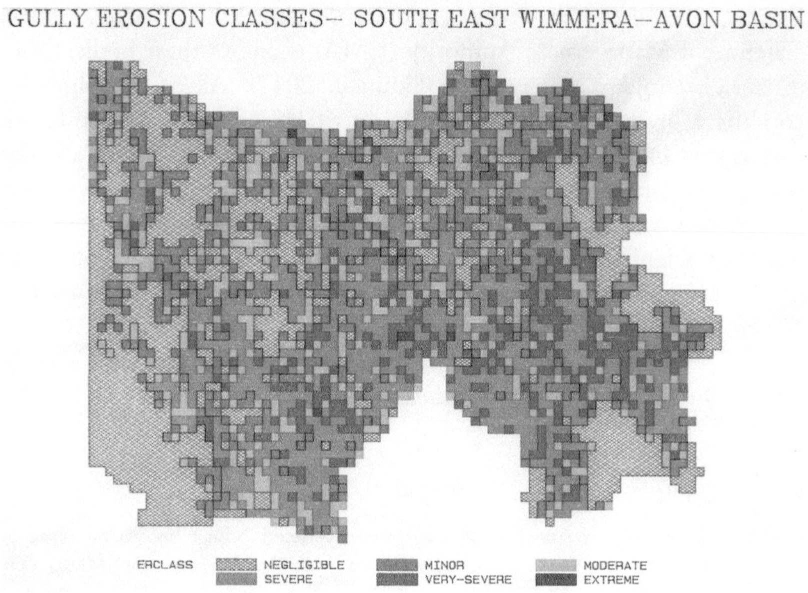

Figure 40.3 Gully erosion map of the Upper Wimmera catchment. (Source: Robins, L.M. 1990. Gully Erosion Survey of the Upper Wimmera Catchment: Methodology and Statistical Evaluation. Department of Conservation, Forests & Lands. Ballarat, Victoria.) A black and white version of this figure will appear in some formats. For the colour version, please refer to the plate section.

female scientist in that country office, I was criticised by a female co-worker for 'wearin' them shorts and doin' a man's job'. However, what was unable to be achieved in terms of state-of-the-art science (or gender equality) was made up for in how we went about 'strategic planning', with genuine, long-term community engagement and empowerment through the suite of community-led committees, together with holding town-hall meetings, field trips and other modes of engagement and communication. To this end, I support the words of General Eisenhower, when he declared, 'In preparing for battle I have always found that plans are useless, but that planning is essential.'

'The Wimmera experience' converted me to the merits of ICM, and I went on to replicate the same process in Melbourne's Yarra River Catchment under the umbrella of 'YarraCare'. Since that time, Australia has made great strides in ICM and put in place a designated network of 56 regional NRM organisations, which cover the entire nation and are governed by community-based boards of management with significant responsibilities to deliver more strategic NRM outcomes (Robins and

Dovers, 2007a, 2007b; Robins and Kanowski, 2011). The Wimmera Catchment Management Authority (CMA) is one of these bodies (Victorian Catchment Management Council, 2017). All 56 organisations conduct a broad range of fieldwork activities, independently and with partners, in line with their agreed strategic plans, and they play a major role in information and knowledge management. They work with and support farmers and rural communities, urban dwellers, indigenous communities, scientists, academics, school kids, teachers, local councils, businesses and government agencies, including research organisations. The Wimmera CMA, the precursor of which was the Wimmera CCG, is now one of the most advanced in its NRM planning and implementation in the nation (see: www.wcma.vic.gov.au).

References

Robins, L. and Dovers, S. (2007a). Community-based NRM boards of management: are they up to the task? *Australasian Journal of Environmental Management* 14, 111–122.

Robins, L. and Dovers, S. (2007b). NRM regions in Australia: the 'haves' and the 'have nots'. *Geographical Research* 45, 273–290.

Robins, L. and Kanowski, P. (2011). 'Crying for our Country': eight ways in which 'Caring for our Country' has undermined Australia's regional model for natural resource management. *Australasian Journal of Environmental Management* 18, 88–108.

Victorian Catchment Management Council (2017). *Celebrating 20 Years: Integrated Catchment Management in Victoria 1997–2017.* Victoria State Government, Environment, Land, Water and Planning, Melbourne.

41 · *Field Studies of Behaviour and Life-Changing Events*

LEIGH W. SIMMONS

My field experiences have been life changing. I grew up in the field. My parents bought a plot of land in the early 1960s, on the edge of the ballast cliffs at Erith in Kent. The land was probably sold for a pinch, as likely or not anything built on top might one day find itself at the bottom of a landslide. They built a bungalow on the widest patch, to which we moved when I was four. The strip of land to the east of the house was deemed too narrow to build upon and was abandoned to run wild. That was my first field site. Although my mother forbad me to go down to the 'lower site', no doubt for fear of me toppling over the unfenced cliff, I nonetheless spent my childhood there. Heavily wooded with English elm, sadly lost to Dutch elm disease in the 1970s, and rich with beds of nettle and bramble, I spent long summer days capturing butterflies: small tortoiseshells, peacocks, red admirals and comma being the most common. I would lie in the long grasses for hours, listening to the stridulating grasshoppers. To this day I recall the thrill of capturing a large male stag beetle, and the screams of my mother when she found it in my shorts pocket that night. I kept a bird list and explored the sand cliffs, searching for nests, and I am ashamed to say I added many eggs to my collection. My time on the lower site developed within me a passion for the field, for natural history, and it kindled the biologist I am today.

I enrolled at the University of Nottingham in 1980, to study zoology. I was not particularly engaged, it must be said, and I did not achieve greatly during the first year of my degree. All that changed on the second-year field course, run by Chris Barnard, at Woodchester Park in Gloucestershire. Chris was an inspirational teacher and would become one of my most influential mentors. During that two-week field course,

I wandered the length and breadth of Woodchester Valley, recording bird song and measuring habitat variables, with the aim of determining how habitat might influence song structure. My fellow students worked on vigilance in tufted ducks on the park's lakes, while others did mark–release–recapture studies on the local population of wood mice or observed the fighting behaviour of wolf spiders in the leaf litter beneath the majestic beech trees that line the valley. At night, we observed badgers at the local set. My eyes were opened to the excitement and joy of quantitative field biology, and to the emerging discipline of behavioural ecology. In many ways that field course took me back to my time on the lower site and rekindled in me the passion for biology that was sucked out of me by a rather poor secondary schooling. Later, in his behavioural ecology lectures, Chris introduced me to the work of Geoff Parker, his theoretical modelling of animal behaviour and empirical testing of theory using yellow dung flies. Chris's lectures drove home the value of theoretically driven, experimental behavioural ecology, firmly grounded in field observation of natural behaviour, and made me realise that this was what I wanted to do with my life. Needless to say, when offered a PhD place to work with Chris, I jumped at the chance, choosing to study sexual selection and female choice using field crickets as my model system.

I was to return to Woodchester Park in 1986, invited by Des Thompson to speak at the Liverpool field course and to meet Geoff Parker. Geoff introduced me to his flies as they fought for incoming females on dung pats in the mansion meadow. He pointed out another dung breeding fly, *Mesembrina meridiana*, a large black muscid with distinctive golden wing patches. Geoff mused that the golden patches might mimic the colour of yellow dung fly males, and in so doing deter the mate searching dung flies from persistently pouncing on *Mesembrina* females while they were depositing their larvae. I suggested we could test the idea by blacking out the golden patches with marker pens, and so we spent several afternoons in the sunshine playing with flies around poo like schoolboys. That first meeting with Geoff established a collaboration and friendship that would last a lifetime and would take me to Liverpool as an early career postdoctoral researcher to work with Geoff on sperm competition and optimal copula duration, using his yellow dung flies.

In 1988, I won the opportunity to visit Australia. The Royal Society and the Australian Academy were offering funding for scientific exchanges in celebration of the Australian Bicentenary. I chose to visit the University of Western Australia, where Darryl Gwynne and Win

Bailey had been doing some exciting work on the mating systems of bushcrickets, species of orthopteran insects in which the male provides the female with a large nutrient gift, which forms part of the ejaculate. Some of these species can deliver as much as 30 per cent of their body weight in a single ejaculate, and I was keen to determine whether females might choose among males based on the quality of the nuptial gift. Arriving in Australia as a field biologist was mind-blowing. Everything was different. The plants, birds, mammals, insects, everything I looked at, was different. On my first night, Win took me into Kings Park, armed with a head-torch and a bat detector. Kings Park is a large area of natural bushland, just over 4 km^2, which lies between the city and the university campus. The park consists largely of low banksia woodland, with an understory of shrubs, herbs, grasses and other flowering plants. In August, when I arrived, the red and green kangaroo paws were in full flower, and they were covered in a small stick-like bushcricket on which Win suggested I work. The bat detector was necessary to locate the calling males, whose call is broadcast in the ultrasonic range. I was astonished to discover that the insect was new to science, still as yet undescribed and catalogued as Gen. nov. 22, sp. 1. New to science, right in the middle of a modern city!

Over the next eight weeks I spent my nights in the park, recording the insects' behaviour. Astonished, I watched as females fought over access to males, attempting to dislodge one another from a male offering a nuptial gift. Females were not choosing among males, rather they were fighting for them. It turned out that males were doing the choosing, providing their food gifts only to the heaviest, most fecund females. It was all so exciting. Then one Monday night, I set out with my notebook, head torch and bat detector to find that everything had changed. Females were now completely disinterested in males who competed frantically to gain their attention. What on earth could have happened in just 24 hours? I noticed that the females at my study site had abandoned the kangaroo paws and were now feeding on the 2-m-high flower spike of a grasstree. For several frustrating nights I struggled to observe any matings at all and was about to give up, when it suddenly occurred to me that recording nothing happening was in fact something very significant! The mating system of this little insect had reversed due to the super-abundance of resources offered by the grasstrees. Females no longer needed the nutrients offered by males. This realisation led to a string of field and laboratory experiments that proved Williams and Trivers' long-held theory that the direction of sexual selection depends on the relative parental

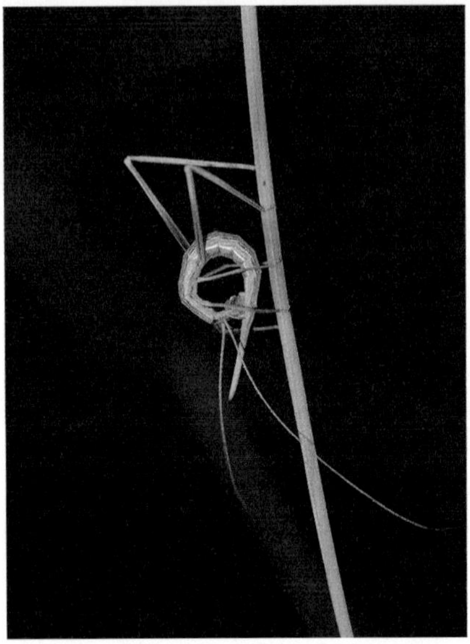

Figure 41.1 A female *Kawanaphila nartee* feasts on the spermatophylax while sperm are transferred from the spermatophore to her sperm storage organs (Leigh Simmons). A black and white version of this figure will appear in some formats. For the colour version, please refer to the plate section.

investments made by males and females (Simmons and Bailey, 1990). Of course, this work took time, and over the years I spent increasingly long periods in Kings Park and less time at home in Liverpool, eventually deciding to move permanently to Perth and my inner-city field site in 1995. The bushcricket is now named *Kawanaphila nartee*, the springtime flower-lover after its pollenivorous diet. And I am fortunate to be able to walk to work through the native bushland that remains my beloved field site. Being in the field, at one with nature, provides the opportunity for deep thought and reflection. It is an endless source of inspiration and has been the foundation of many a sound decision.

Reference

Simmons, L. W. and Bailey W. J. (1990). Resource influenced sex roles of zaprochiline tettigoniids (Orthoptera: Tettigoniidae). *Evolution* 44, 1853–1868.

42 · *Sediment, Wind Turbines and Rhinos: Ah, the Life of a Geographer!*

MIKE SLATTERY

From the moment I set foot in the Namib Desert, I was hooked.

It was January 1988, and my two geography professors at Wits University, Dr Peter Tyson and Dr Justin Wilkinson, were leading a two-week field trip to the Gobabeb Research Station, located right on the Kuiseb River. To the north lay the vastness of the gravel plains of the northern Namib; to the south, the world's tallest linear sand dunes, stretching as far as the eyes could see. In teams of three, we measured local winds, specifically the interaction between the sea breeze and the gradient wind. We did so by releasing weather balloons, every hour, for 10 straight days, which we tracked using rather old, hand-cranked theodolites. At night, we fixed plastic cups to the bottom of the balloons with small candles in each cup and tracked their ascent through several thousand feet, as the sea breeze pushed farther and farther inland. It was an exhausting and utterly exhilarating experience.

In the intervening 30 years, I have experienced many such moments. Some have been incredibly frustrating: losing a $7,000 turbidity probe that I had bolted to a bridge pylon in the lower Trinity River, Texas, neatly removed by someone on their journey downstream. (I am still trying to figure out why anyone would want such a thing displayed above his or her mantelpiece!) Others have been a little terrifying, including an encounter with a group of wild hogs while surveying in the lower Trinity delta using ground-penetrating radar. However, I have conducted all of my fieldwork with one simple goal in mind: to understand a little better the way in which the world works. Moreover, as my PhD supervisor, Tim Burt, once reminded me, never let a good storm get in the way of collecting quality data, even if it occurs on Christmas Eve!

Figure 42.1 The black rhinoceros is now critically endangered (~ 5,500 left in the wild) due to poaching and the illegal trafficking of their horns to South-east Asia. In this photograph, I am assisting in a procedure to change a tracking collar and take DNA samples on young black rhino calf in the Eastern Cape of South Africa. Black rhinos are browsers and generally prefer hilly terrain with thicket and scrub, which means these procedures are almost always done from a helicopter (from the author's photo collection).

In this regard, the morning of 12 July 1996 comes to mind as one of the most memorable field days I have ever experienced. I had only been teaching for about a year at East Carolina University, a large, public institution in North Carolina. My colleague and good friend, Dr Paul Gares, and I had received a multi-year grant from the US Department of Agriculture to document erosion on coastal-plain cropland. As luck would have it, we had only just set up our field site when Hurricane Bertha decided to track directly across eastern Carolina, and over our field site, during daylight hours. There was no decision to be made: we grabbed a departmental van, a velocity probe (that ended up failing), a number of water bottles and a handful of graduate students, and we made a beeline for the tiny community of Clayroot. By mid-afternoon, 66.5 mm of rain had fallen, with winds sustained at 125 km/h, a Category 1 system. We had a brief moment of clarity and suggested the graduate students return to the department, which they did, but Paul and I kept marching across the slopes of this 19.4-ha watershed, trying desperately

Figure 42.2 High-energy, headwater channels in tropical environments are rather poorly understood. Here, I am installing a turbidity probe in a step-pool sequence in a tributary channel to the San Lorencito River in Costa Rica. This channel is located within a 0.3 km^2 watershed at TCU's Tropical Research Station (from the author's photo collection).

to measure velocities of surface runoff in rills on tobacco and cotton fields. With a non-functioning flow meter, we had to improvise and we ended up timing the passage of pieces of wine corks that I had in my rain-jacket pocket (I have no recollection of why they were there), to estimate velocity. In hindsight, I probably should not have parked the van next to the channel (something to do with bankfull discharge and floodplain inundation), but we were able to collect enough data to construct slope hydrographs and link them directly to channel response under torrential rainfall (Slattery *et al.*, 2006). I will never forget one of the reviewers of the manuscript remarking that, 'Despite this paper being a manual on how not to do hydrology [apparently wine corks are not standard hydrologic equipment], the authors should be commended for being able to collect such data under these conditions and make some sense of the hydrologic linkage between slopes and channels.'

Fluvial geomorphology continues to fascinate me, and I have expanded my reach (no pun intended) from coastal-plain rivers in North Carolina and Texas to tropical rivers in Costa Rica. I have also been

Figure 42.3 Hydrological fieldwork means there is no shortcut to getting good data! Here, Dr Paul Gares is about to measure velocity in a rill channel on a tobacco field, during extreme rainfall (Tim Burt). A black and white version of this figure will appear in some formats. For the colour version, please refer to the plate section.

lucky to branch out into other areas of research over the past decade or so. I guess tenure affords one that luxury! I have spent time assessing the socio-economic and environmental impacts of utility-scale wind farms in the Midwest of the United States as well as establishing a rhino conservation programme in the Eastern Cape of South Africa. While I do not pretend to be an expert in either wildlife conservation or renewable energy, what I have found is that my training in both human and physical geography has given me a breadth of knowledge (and an ongoing intellectual curiosity) about human–environment interaction, which seems to have stood me in good stead.

References

Slattery, M. C., Gares, P. A. and Phillips, J. D. (2006). Multiple modes of storm runoff generation in a North Carolina coasteal plain watershed. *Hydrological Processes* 20, 2953–2969.

43 · *Conservation Science: the Need for a New Paradigm Founded on Robust Field Evidence*

WILLIAM J. SUTHERLAND

Fieldwork provides some of my deepest memories, whether exhilarating (being attacked by snowy owls when studying their nests – I still have the multiply ripped jacket), scary (being told the faulty helicopter will probably have to crash land on the pack ice), bizarre (being courted in the dark by a wild, but originally hand-reared, kakapo, who climbed my body and mated with my head), exhilarating (finding the first albatross nest in the North Atlantic – a black-browed albatross courting the gannets in Shetland), satisfying (discovering healthy populations of red-breasted geese when thought to be on the brink of extinction), amusing (being taken for dead by a passing motorist when spread-eagled on a roadside verge, photographing rare plants), disturbing (seeing the all-too-human excitement of chimpanzees nearly killing a red colobus monkey) or breath-taking (the whale shark swimming, mouth-open, directly towards me).

There is also, and especially, the underappreciated emotion of discovery. An observation, such as finding a rare species, is exciting, but to me there is a far higher-level thrill of discovering how something works, and why the world is as it is. Few non-scientists appreciate the sheer unalloyed exhilaration when you realise that probably you alone are the only person with that understanding: I find this to be a deep and profound emotion.

I want to dwell here on how fieldwork relates to conservation. There is terrific field research, described elsewhere in this book, seeking to understand the fundamental scientific problems, such as how populations operate (which of course also underpin conservation science); or identifying problems, such as the impact of organophosphates. But here I want to consider how fieldwork can be part of the solution to dealing with planetary problems.

Mending the world depends on us knowing what to do. Yet, remarkably, despite the huge effort in describing problems, there is surprisingly little experimental or even empirical research examining what actually works. One thing we do not do well, indeed rarely, is reveal what has not worked, or failed. How exactly do we make this planet a better place?

Typically, having identified a problem, we need to test the effectiveness of possible solutions. This appears straightforward, but the challenge of this approach is seen in our review of the global literature, assessing the effectiveness of actions to conserve primates (Petrovan *et al.*, 2018). Few of the studies had the objective of testing specific actions; with almost all papers derived from conservation projects with the admirable objective of benefitting a population, but by using a cocktail of actions (the record was 19!). In such cases we could not adduce which, if any, of the various actions was effective.

A further complication of evaluating the response to conservation actions is that the observed changes in, say, estimated population size may be due to other factors, such as changes in weather. The solution is to do experiments and this is best if a range of experimental options are compared (Smith *et al.*, 2014).

In the United States, there is strong policy support for a move towards 'comparative effectiveness research' (CER) in medicine (Olfson and Marcus, 2013). This came to a head on 17 February 2009, when President Obama signed into law the American Recovery and Reinvestment Act, with just over $1 billion dedicated to CER – essentially, the rigorous evaluation of different treatment outcomes.

I call for the same to apply to conservation. There are some good examples of simple field experiments comparing different methods. Dodds *et al.* (2013) compared five bat box designs for Natterer's bat (*Myotis nattereri*) and brown long-eared bat (*Plecotus auritus*), and they showed that occupancy was much higher for two of the designs. Woltz *et al.* (2008) experimentally compared four different culvert designs for green frogs (*Rana clamitans*): they preferred some substrates (soil or gravel) compared to concrete or PVC (polyvinyl chloride), but there were no preferences for specific width or length of tunnel.

Modifications of conservation fieldwork, with a clear sense of testing particular treatments or practices, could radically improve how conservation operates and make it far more effective. We could enter a new world in which, rather than often having to guess and infer what works, we can build on the collective experience of field practitioners.

My call is for more of the many brilliant minds who study ecology, behaviour and conservation to think about means of improving the effectiveness of conservation actions and find ways of testing what is effective. This promises to combine the innumerable joys of fieldwork with the satisfaction of really making a difference.

References

Dodds, M. and Bilston, H. (2013). A comparison of different bat box types by bat occupancy in deciduous woodland, Buckinghamshire, UK. *Conservation Evidence* 10, 24–28.

Olfson, M. and Marcus, S. C. (2013). Decline in placebo-controlled trial results suggests new directions for comparative effectiveness research. *Health Affairs* 32, 1116–1125.

Petrovan, S. O., Junker, J., Wordley, C. F. R., *et al.* (2018). Evidence-based synopsis of interventions, a new tool in primate conservation and research. *International Journal of Primatology* 39, 1–4.

Smith, R. K., Dicks, L. V., Mitchell, R. and Sutherland, W. J. (2014). Comparative effectiveness research: the missing link in conservation. *Conservation Evidence* 11, 2–6.

Woltz, H. W., Gibbs, J. P. and Ducey, P. K. (2008). Road crossing structures for amphibians and reptiles: informing design through behavioural analysis. *Biological Conservation* 141, 2745–2750.

44 · *The Worst Journey in the World*

DES THOMPSON

> We travelled for Science. Those three small embryos from Cape Crozier, that weight of fossils from Barkley Island, and that mass of material less spectacular but gathered just as carefully hour by hour, in wind and drift, darkness and cold, was striven for in order that the world may have a little more knowledge, that it may build on what it knows instead of on what it thinks.
>
> Cherry-Garrard, A. (1922). *The Worst Journey in the World*

Written by Apsley Cherry-Garrard, this is an exceptional contribution to English literature. As much about exploration, scientific endeavour, human frailties and heroism, the book takes its title from an extraordinary expedition to collect the eggs of emperor penguins (*Aptenodytes forsteri*) in the winter darkness of Antarctica. Its main claim to fame, however, is as a memoir of the tragic quest to reach the South Pole first, on which Captain R. F. Scott and his four companions died on their return.

As a child I had heard of Captain Scott's heroics, and had seen the 1948 film *Scott of the Antarctic*, featuring John Mills, Kenneth More and James Robertson Justice (who lived close to my home village, Culrain, in the north Highlands of Scotland). I found the book inspiring and moving, especially because of what it said about the privations and motivations of expeditionary life and scientific work in the toughest of field situations. As I've grown up with the book, and reread it countless times, it was only after discovering more about the author, especially in Sara Wheeler's revealing biography (Wheeler, 2001), that I came to understand more about this special piece of fieldwork. Ironically, although the expedition was flawed in purpose and outcome, the account of the 'worst journey' is a brilliant exposition on the curious mind.

Observing Waterfowl Feed

Where to begin? In September 1978, on a sweltering hot day, I was studiously watching shelduck (*Tadorna tadorna)* at Edinburgh Zoo. I had just embarked on my final year undergraduate dissertation project at what is now the University of West Scotland, in Paisley, to determine how these waterfowl foraged. At the time it was believed that they fed heavily on the tiny gastropod *Peringia ulvae* (formerly *Hydrobia ulvae*) (Olney, 1965), yet in the upper reaches of the Clyde Estuary these small mudsnails were scarce. The hunch of my supervisor, the renowned environmental educationalist, Professor John Smyth, was that shelduck depended on rich concentrations of tubificid worms in the relatively polluted mudflats of the Clyde (the worms are aptly named 'sludge worms'). Armed with a long descriptive letter from David Bryant at Stirling University, who had published a paper on the birds' feeding habits in the Forth estuary (Bryant and Leng, 1975), I observed them carefully to see how they fed – they scythed, dug, dabbled, head dipped and upended, always carefully filtering the drenched mud through lamellae (comb-like structures along the inside of the bill). We now know that ducks feed using mechanosensitive somatosensory neurons, essentially ever so lightly touching their prey before ingestion (Schneider *et al.*, 2014), but then I simply wanted to see if shelduck could take tiny worms. They did, and I went back to the Clyde to spend a cold winter watching feeding flocks from a portable hide, and, on the basis of what I learnt at the zoo, carefully sampled the mud invertebrates to prove the importance of sludge worms to shelduck (publishing the results in my first two scientific papers; Thompson, 1981, 1982).

Encountering King Penguins Rather Than Emperors at the Zoo

I was very fortunate, for John Smyth's good friend, Professor Roger Wheater, was the chief executive officer of the Royal Zoological Society of Scotland (known as the 'Edinburgh Zoo', and, relieving me from hours spent watching a variety of ducks, took me on a tour of the animals. Built on a hill, it is Britain's only zoo with a Royal Charter, and was inspired by Carl Hagenbeck's 'open zoo' design at Hamburg.

With its spacious enclosures and natural habitats, it was the penguin enclosure that intrigued me. Roger explained that 'Our Zoo' was best known worldwide for its penguins, with its opening year, in 1913, marking the arrival of three king penguins (*Aptenodytes patagonicus)* (from the Christian Salvesen whaling expedition, docked at Leith). These were the first to be seen live outwith the South Atlantic, and as we looked at one feisty bird, Roger explained that he was 'King Nils Olav', named after the then King of Norway and Major Nils Egelien, who had arranged the adoption, six years earlier. Remarkably, this bird is still going strong, and has the role of mascot of the Norwegian Guard, rising from the rank of Corporal in 1982 to Brigadier in 2016, and it was 'knighted' by King Harald V in 2008. But, I asked, what of the world's largest penguin, the emperor (some 30 cm taller than the closely related king)? 'Impossible to keep here!', was Roger's reply, 'So you will have to contend with reading *The Worst Journey in the World* to learn more about them.' And he was right, of course, for only a few establishments have bred them outwith Antarctica (SeaWorld, in San Diego, since 1980; and, more recently, Laohutan Ocean Park, in China, and Wakayama Adventure World, in Japan). I'd not read the book from the perspective of a remarkable piece of fieldwork. But I did so that night, and it changed my outlook!

The 'Worst Journey' to Collect Eggs

On 27 June 1911, Cherry-Garrard, Dr Edward Wilson and 'Birdie' Bowers left the 'Terra Nova' expedition base in Cape Evans for the Cape Crozier headland, 67 miles away, where Scott's earlier 'Discovery' expedition (1901–1904) had found the world's first recorded emperor penguin rookery.

Their mission was to collect the eggs of emperor penguins, for it was believed that the embryos would reveal how birds evolved from reptiles (Haeckel's specious 'recapitulation theory'). Wilson was determinedly sure they could prove this – on the earlier 'Discovery' expedition, he, Scott and Shackleton read aloud from Darwin's *On the Origin of Species* (Darwin, 1859); Wilson was sure that a scientific prize awaited him! The penguin's breeding cycle begins in March, at the onset of the ravaging, dark Antarctic winter, when adults gather in colonies up to 120 km inland from the edge of the pack ice, in temperatures as low as −40 °C. Laid in May, the single, pear-shaped, greenish-white egg, weighing around 460 g, is one of the lightest, relative to its mother's weight, in the bird world (Birkhead, 2016). After 64 days, the chick hatches, and it takes a further 45 or so days before chicks huddle to form a crèche.

An exceptional scientist and artist (and medical doctor), Wilson was unable to join the previous 'Nimrod' Antarctic expedition (1907–1909) led by Ernest Shackleton, as he was heavily involved in the government's Grouse Disease Inquiry (begun in 1905). Wilson led much of the work unravelling the population-crashing impacts of the 'grouse disease' strongylosis (caused by the nematode *Trichostrongylus tenuis*) on economically important red grouse (*Lagopus lagopus scotica*) populations. Published in the two-volume *The Grouse in Health and Disease* (Leslie, 1912), Wilson's contribution was to write a third of the text and extensively illustrate it. This was a seminal study of the drivers of the population biology in birds, and marked out Wilson as an outstanding field biologist. However, such was the pressure on him to join Scott's expedition (as Chief Scientist), which sailed south in June 1910, that he evidently completed one chapter, 'The changes in plumage in the red grouse in health and in disease', whilst on the train to London (and then in a waiting room), and he sent his final instalment from South Africa in August, en route to Antarctica.

Born in Greenock, Bowers was short, stout, dependable, prodigiously hard-working and exceptionally hardy (with exceptional resistance to extremes of cold weather). Little is known about him, with a small display in Rothesay Museum on the Isle of Bute celebrating his 29 years of life.

Cherry-Garrard was one of the youngest expedition members, educated as an Oxford classicist and historian, and so myopic that he should not have been admitted to the expedition. He got to know Wilson, staying at his cousin's house in Kirriemuir, in autumn 1908, having graduated months earlier. He had lost his father the previous November, but now found the father figure who would, in the words of Wheeler (2001), 'tower over the rest of his life like a colossus'. It was to be a short physical influence, but left a life-shattering legacy.

Hauling two sledges with 340 kg of provisions and equipment, the trio was, in the words of Scott, beginning on a 'new and bold venture'. Cherry-Garrard later described it rather differently, as being so awful that 'men do not fear death, they fear the pain of dying'.

It was a horrendous journey. In constant darkness they hauled and trudged. On the second day, Cherry-Garrard got severe frostbite. His spectacles continually froze – in fact their entire bodies were entombed in icy armour, making marching near-impossible, and they were so cold in their frozen sleeping bags that '. . . when your body chatters you may call yourself cold'. It could take five hours to break camp, such was the difficulty of dressing, and then they had to relay-pull each sledge, such were the impediments, adding to the energy-sapping mileage. On the 19th day, they reached Cape Crozier, took three days to build an igloo, and then set out for the rookery – often crawling, such was the lack of light and treacherous terrain of crevasses and deep snow pits.

They were the first people to see incubating penguins, in shafts of twilight, 'a marvel of the natural world'. They took five eggs and three birds (to flense, for their blubber stove) and stumbled back to their igloo (which they almost missed). Blinded and 'already beginning to think of death as a friend', Cherry-Garrard had fallen and broken his two eggs. Seven days after their arrival, in a pernicious blizzard, their tent blew away – without which they would have perished beyond their igloo, now disintegrating and filling with snow. For two days, mummified and without food, they lay, shouted over the howling wind, murmured hymns, prayed and contemplated death. But, in a lull, they got out, and miraculously Bowers found the tent, which he then tied himself to each night. They retreated immediately, and after an awful succession of dreadful nights (the worst part) they made it to Cape Evans on 1 August 1911 – ghostly, wasted and wizened.

Alas, the three eggs had embryos too well developed to test the theory; Wilson deemed the expedition 'fruitless'. But for Cherry-Garrard, it was to determine his fate, for just over a year and three months later, he would discover his two companions and Scott dead in a tent – having made it to the Pole. It was salvation of sorts, for he managed to mine and quarry the experiences of the worst journey, but he was broken.

Tragically, and adding to the poignancy of the worst journey, Cherry-Garrard never forgave himself for the demise of Scott's party. Many travelled south with Scott for the final goal, with groups peeling back as the party ventured further. In the final push for the Pole, just five went on, with the remaining members heading back to base camp, but instructed to return later to meet and support the returning (and hopefully conquering) explorers. It was not to be. Cherry-Garrard was due to meet Scott's party at 'One Ton' depot on their return. Unbeknown to him, as he waited painfully, Scott, Wilson and Bowers lay dying, just twelve and a half miles away. Cherry-Garrard could go no further, with no food for his dogs and a seriously ill companion, and he returned to the safety of Hut Point.

Initially, the media reports on Scott's heroic failure demonised Cherry-Garrard for not going that extra distance to save his companions. Exonerated later, he came to be haunted by depression and despair.

In her biography, Wheeler recounts how, much later, when he was a curmudgeonly landowner, Cherry-Garrard 'chased away boys he caught bird-nesting in the spinneys, took the names of girls who gathered bluebells ... and banned the collection of firewood on the grounds that people were taking too much'. What a curious juxtaposition for someone so driven to collect emperor penguins' eggs in the worst of conditions.

A paper by Li *et al.* (2014), based on the analyses of penguin genomes, proposes that penguins arose around 60 million years ago, coinciding with a period of global warming. Their evolution had nothing to do with reptiles, and their original ancestors probably belonged to the order Procellariiformes (sea-faring albatrosses, fulmars, storm petrels and shearwaters). Cherry-Garrard describes how he arrived at London's Natural History Museum with his precious eggs, only to be treated with indifference – the recapitulation theory had largely been debunked, there was no interest in the penguin embryos. It had evidently been a futile scientific journey.

'Exploration is the Physical Expression of the Intellectual Passion'

What do we take from this? Beyond the brilliance of his writing, and the extraordinary bravery, boundless strength of character, camaraderie and some sheer good luck on the epic journey, I find the tale inspiring. Curiosity and the prospect of a scientific breakthrough motivated the journey. The telling of the story is transfixed in legend, as much for its descriptive verve as for the unbridled tragedy of unfolding events. Had Scott made it back, he would probably have been forgotten. Would the 'worst journey' have featured in our scientific consciousness – I doubt it, for Cherry-Garrard would not have been the tortured soul that emerged. I doubt he would even have written it up, for had Wilson also survived he would have adduced the evidence as a scientific failure. I do wonder if the lure of the egg is what really beguiles, behind the tale – not a chick, not a bird skin, not feathers or bones, but an egg. Men in pursuit of an egg – what does this tell us? Is there something Oedipal, in a Freudian sense, or is it simply a derivative of our hunting instinct? Fieldwork has many motivations, and curiosity and the chance of a discovery are certainly wonderful drivers, but I sense the 'worst journey' has secrets still to tell.

Acknowledgements

I have discussed this book with many friends, to the extent that I cannot resist a liberal sprinkling of quotes, and I thank Tim Burt, Ingvar Byrkjedal, John Matthew Thompson, Roger Wheater and Phil Whitfield for firing my curiosity.

References

Birkhead, T. (2016). *The Most Perfect Thing. Inside (and Outside) a Bird's Egg.* Bloomsbury, London.

Bryant, D. M. and Leng, J. (1975). Feeding distribution and behaviour of shelduck in relation to food supply. *Wildfowl* 26, 20–30.

Cherry-Garrard, A. (1922). *The Worst Journey in the World.* Carrol & Graf, London.

Darwin, C. (1859). *On the Origin of Species.* John Murray, London.

Leslie, A. S., ed. (1912). *The Grouse in Health and Disease. Being the Popular Edition of the Report of the Committee of Inquiry on Grouse Disease.* Smith Elder, London.

Li, C., Zhang, Y., Li, J., *et al.* (2014). Two Antarctic penguin genomes reveal insights into their evolutionary history and molecular changes related to the Antarctic environment. *Gigascience* 4, 3–27.

Olney, P. J. (1965). The food and feeding habits of shelduck Tadorna tadorna. *Ibis* 107, 527–532.

Schneider, E. R., Mastrotto, M., Laursen, W. J., *et al.* (2014). Neuronal mechanism for acute mechanosensitivity in tactile-foraging waterfowl. *Proceedings of the National Academy of Sciences* 111, 14941–14946.

Thompson, D. B. A. (1981). Feeding behaviour of wintering shelduck on the Clyde Estuary. *Wildfowl* 32, 88–89.

Thompson, D. B. A. (1982). The abundance and distribution of intertidal invertebrates, and an estimation of their selection by shelduck. *Wildfowl* 33, 151–158.

Wheeler, S. (2001). *Cherry. A Life of Apsley Cherry-Garrard.* Jonathan Cape, London.

45 · *Field-less Fieldwork in Archaeology's Digital Age*

ANDREW TIBBS

Fieldwork is evolving; no longer is it something which has to be done in the field, or even outdoors. While traditional field techniques such as observations, planning and excavation still have their place today, the development of new technologies means that archaeologists can learn so much about a site before putting a trowel in the earth.

Scotland has a long history of archaeological discovery and exploration, dating back to at least the sixteenth century (Tibbs, 2019). One of the earliest accounts (1564) is by Thomas Randulf, the English Ambassador writing to Queen Elizabeth I, giving an eye witness account of pillars and inscribed stones uncovered at Inveresk, Edinburgh. Shortly after, William Camden published *Britannia* in1586, a detailed study of the upstanding Roman remains in Britain, which is still relevant today as it described many sites which have since been destroyed. However, it was not until the eighteenth century that advances in cartography led to the accurate recording of upstanding Roman remains in Scotland; prior to this, most accounts were based on observation and pacing out the length of sites. As a response to the second Jacobite uprising in 1745, the British government engaged in a programme of forced pacification of the Highlands and deployed military surveyor William Roy to produce the *Great Map of Scotland*, an extensive and detailed snapshot of the country before major agricultural changes began to change the landscape, and in the process destroy many archaeological sites. As part of his survey, Roy recorded more or less all of the visible Roman remains at the time, with the results published as *Military Antiquities of the Romans in North Britain* in 1793.

Figure 45.1 The defensive ditches at Inchtuthil legionary fortress (Andrew Tibbs)

One such site is the Roman legionary fortress of Inchtuthil, a 22-ha fort located in rural Perthshire, and which is bigger than its nearest counterpart at York. Constructed around AD 83, it was abandoned only a few years later and today remains undeveloped, with many Roman features still visible on the ground, such as the ditches that can be seen in Figure 45.1. The peninsula on which it is located also has at least two temporary encampments and two compounds that are possibly associated with the construction of the fortress, parts of which were recorded by Roy in his survey of the site, which also notes additional features such as ditches, roads and evidence of pre- and post-Roman activity.

The 1940s saw the deployment of a new technique, developed during the war as part of intelligence gathering. The method of identifying cropmarks through aerial photography was pioneered by O. G. S. Crawford, and then by J. K. St Joseph, who between them identified at least 80 Roman sites. Aerial photography has remained an important technique, still used to identify such sites. The digitisation of aerial photographs is making it easier to reanalyse images taken over the past 80 years; but it is the development of satellite imagery which is changing the way sites are viewed, owing to high-resolution imagery and laser scanning. Lidar (light detection and radar) is a relatively new remote sensing technique being applied to archaeological sites. By laser scanning

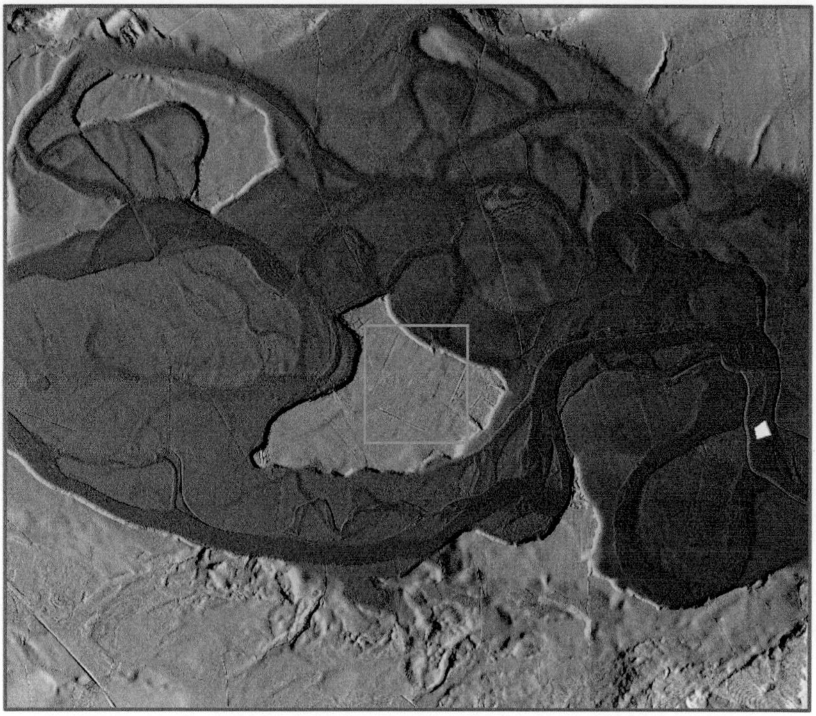

Figure 45.2 Lidar image of the legionary fortress at Inchtuthil. The outline of the ditched defences can be seen in the box. One of the compounds is visible to the right, while the River Tay surrounds the site. (Contains OS data © Crown copyright OS Terrain 50 & Strategi right (2018). Lidar data contains public sector information licensed under the Open Government Licence v3.0.)

a site (from a satellite or drone), it is possible to detect subtle changes in the topsoil, enabling the detection of anomalies under the surface, such as walls and ditches. Figure 45.2 shows the lidar scan of the legionary fortress at Inchtuthil, with the ditches of the fort, camps and compound visible.

Today, the archaeological environment has changed and sites are less likely to be excavated than they were 50 years ago due to funding restrictions and modern, non-invasive techniques, which make it easier to learn about sites without costly excavations. The modern archaeologist has a wide range of technologies available to explore sites; older site surveys now digitised, along with aerial photographs, geophysical surveys (such as ground-penetrating radar, resistivity and magnetometry) can all offer an indication as to what lies underneath the soil. Such technologies have led to the development of a tool which brings together these results in one, geospatial database. Geographic information systems (GIS) are a computer-based framework for gathering, storing and analysing information to enable the digital analysis of sites, taking into account how the topographical and archaeological environment may have changed; of course, all this can be done remotely, without excavating and the potential destruction of evidence.

With the advent of digital technologies and remote sensing techniques, it is no longer necessary to be out in the field investigating sites once the raw data have been gathered. By bringing together multiple data sources as part of GIS, it is often preferable to do the initial processing where there is access to powerful computers and high-speed internet connections (Tibbs, 2018). In an era where time and money for on-site investigations is limited, more can be done efficiently and effectively from an office than being outdoors. Whether or not this is a good thing, when it comes to interpreting sites, is a separate and equally debatable question.

References

Tibbs, A. (2018). Facing the enemy: the orientation of roman forts in northern Britain. In *LIMES XXIII. Proceedings of the 23rd International Congress of Roman Frontier Studies Ingolstadt 2015*, eds. C. S. Sommer and S. Matešić. Nünnerich-Asmus Verlag, Mainz, pp. 1068–1071.

Tibbs, A. (2019). *Beyond the Empire: A Guide to Scotland's Roman Remains*. Robert Hale, Marlborough.

46 · *Reflections on a Career with the Field Studies Council*

SUE TOWNSEND

At School

I was an average student in everything except geography, where I seemed to excel at understanding landforms and developed a thirst for knowledge and a joy of a developing vocabulary. Who wouldn't want to know about castellated ramparts, subduction zones and podsolisation! I struggled with science but felt it was so important to develop and test hypotheses – and biology field studies were a small but joyful part of my school career. I am an English West Country girl and delighted in coastal habitats; on the coast at Studland I identified my first grass – Townsend's Cord grass – of course! To my shame, at that stage in my career, I didn't understand the ethos and discipline of making biological records and I note that this discipline remains absent from our school curricula.

At University

I was determined to further a career somewhere in natural sciences. I wanted agricultural science – but the academic snobbery of the time in my grammar school directed me to the biological sciences – funny how geographical science was 'less academic'. I was undecided on what I wanted to do when I grew up. I still am, so it is good that I remain not quite grown up! My solution was a placement year and I was delighted to be able to split it between animal husbandry research and crop research, with the agriculture seeping in. However, there is only so much joy an undergraduate can get from repeated bacteriology plates from the bovine rumen, or even outside, for weeks on end counting onions.

These experiences, although field-based, steered me away from agriculture and into conservation and my first foray into serious ecology. I was lucky. My university lecturers were inspiring. Cardiff brought me bryologists, freshwater ecologists, conservation, plant and invertebrate specialists and, joy of joys, a field station in north Wales. It was not affiliated to the Field Studies Council (FSC) and was primitive, self-catered, bunk-bedded and redolent of an army camp – but somehow I felt at home. I came back with ringworm, exhaustion and an unintended weight loss, accompanied by an enhanced knowledge and a new vocabulary inspired by my first plant surveys. Principal component analyses, national vegetation classification, *Rhytidiadelphus squarossus* and, joy of joys, my first identifications of freshwater invertebrates – I was hooked.

Manpower Services Commission

Where next? I knew I wanted the great outdoors with opportunities to research and contribute. I am amongst a generation of really fortunate people to be in the right time and place to take advantage of the wonderful Manpower Services Commission 'apprenticeship scheme'. A little like today's conservation organisations exist on short-term grants and legacies, back in the 1980s, this wonderful scheme paid the salaries of keen young graduates to undertake year-long roles within relevant organisations. I was placed in Bristol with the Avon County Community and Environment Services. The scheme worked with the Nature Conservancy Council and County Ecologists to survey sites of botanical interest, to notify potential Sites of Special Scientific Interest (SSSIs) and recommend management. I was learning plant identification on the job and the priority for my year was limestone grasslands. Wow – what an opportunity! The year went too fast and my job hunting was a little disheartening – what could match this?

First Job

Well, the answer was a field study centre in the Lake District - the Brathay Hall Trust. I was expected to teach ecology, and the field courses were for advance-level (A-level) high-school students and were always a week long. The groups came and worked through a progressive and participatory approach, learning skills in identification and field techniques to enable them to develop as field scientists and to lead their own investigations by the end of the week. I saw hundreds of children develop their scientific, interpersonal and communication skills and I felt that the residential field course was a key part in their growth and self-discovery. I remain convinced that residential learning is so effective, enabling people of all ages to gain, practice and develop skills and I have seen so many examples of effective learning across all disciplines. I believe that these experiences should be an entitlement for our schools. With current research and the active health and well-being agendas, and active research into the benefits of both being and learning beyond the constraints of lecture theatres and classrooms, there is no doubt in my mind that they are a huge benefit to all. I also met like-minded individuals from this amazing network of field centres, including many who were to become future colleagues. I was in awe of the great names from that era: John Crothers, Kingsley Iball, Robin Crump, Adrian Bayley and John Bebbington. These people were educators and naturalists par excellence. I also taught at Brockhole in the Lake District, where I met Rob Lucas, at Kindrogan in Scotland and at Castle Head in the Lake District, all before they were FSC centres, connected through the outdoor education network – particularly the embryonic Association of Heads of Outdoor Education Centres (https://ahoec.org/). I remember believing how much I still had to learn and being amazingly keen on plants. Discovering there was more than one species of dandelion led me to a residential identification course aimed at adults at Preston Montford and to my introduction to the Botanical Society of the British Isles, through a field course with the late, great Franklyn Perring. Little did I know how this would shape my career.

Second Job: the FSC

The climate in the early 1980s was buoyant for fieldwork and sharing good practice through the professional networks led me to the National Association of Environmental Education, the National Association of Field Studies Officers, the British Ecological Society and so many others. I couldn't fail to be aware of the FSC, both its centres and its publications (e.g. Berry and Crothers, 1987). The FSC's then Director, Tony Thomas, was keen to build an advisory team to help FSC centres match their field course to curriculum need at a time when the Education Reform Act was taking its toll and school teachers needed to justify field excursions. There was concern across the sector, course lengths were dwindling and schools and universities became more focused on a narrower mix of curricular-focused skills; so we had to sharpen our marketing and demonstrate relevance. I became an FSC Science Adviser and, alongside a geography colleague, was timetabled around our 14 centres to deliver training, fill in for teaching, develop marketing material and become one of the many external faces of this esteemed organisation.

This was a magical five years. I taught at all the FSC centres – and there are not many, even now, who can say that. I even ran a training course at Fort Popton, FSC's then Oil Pollution Research Unit in west Wales. I travelled to European school conferences and presented our work to schools around the country. But I developed a fatigue from the constant moving around and was looking for an opportunity to wake up in the same place for a change.

Then came the lucky break. I became Warden and Director of Studies (what a title!) at Preston Montford field centre in Shropshire, England. The life of a Head of Centre is as diverse as it is exhausting and rewarding. Preston Montford had an even mix of the young school attendees, A-level students, university groups and adult courses. I was inspired by them all in equal measure and, over time, became increasingly aware that the FSC was one of the only national organisations running identification courses, which slowly developed as one of my passions. No one could deny that these courses were not the easiest to maintain. There were no longer Manpower Services Schemes, there was little networking in this sector but a growing need for professional training. There were national bodies, and the British Ecological Society was working with the then embryonic Institute for Environment and Ecological Management (now chartered, and known as CIEEM), the national recording societies such as the Botanical Society of Britain and Ireland (BSBI) and the Dipterists Forum were becoming increasingly aware of the need for training the next generation, and I saw that the FSC might be part of this. I was not alone. The late Sam Berry was key in setting up the National Biodiversity Network; the BSBI was setting up a Training and Education Committee; Alan Stubbs, from what is now Natural England, was noticing the dearth of invertebrate recording groups who offered engagement and training opportunities – hence the birth of Buglife. CIEEM published its skills gap analysis (CIEMM, 2011) and there was a burgeoning recognition that field identification skills were needed to safeguard for the future. There are more examples, too numerous to mention – and with the FSC network of centres in such amazing locations, including now in London with the FSC partnerships in The Royal Parks, the time was ripe for filling that gap.

Biological Recording

Identification was one skill – and I was alerted to the recording networks through a very key individual who has now become a personal friend. Dr Sarah Whild was working in academia and saw the opportunity for accrediting some of the amazing FSC courses that were so often led by the UK's most respected naturalists. With small adjustments in timetabling and opportunities built in for assessment, we began to develop a suite of courses to offer qualifications to people like myself, at the stage I was, back in the 1980s. Recent graduates, young 'apprentice' ecological consultants, county wildlife trust officers and even group leaders from Centre Parks (a commercial holiday-village chain) were looking for training in a re-emerging discipline of biological recording. It was becoming increasingly apparent that UK Wildlife was not only far from safe: there were red lists, action plans for priority species and a real need to record what we had found, preferably with a measure of credibility on the skill and robust nature of the data. Twenty-one years later this suite of courses is still running. It has, of course, undergone changes but it now regularly offers over 20 part-time students, every year, the chance to

Figure 46.1 Vegetation identification on the Long Mynd (Field Studies Council)

Figure 46.2 Earthworm sampling at Preston Montford field centre (Charlie Bell, Field Studies Council)

study biological recording at Masters level with a choice of taxa in which to specialise. The course came under threat when university courses, like schools a decade earlier, were being asked to target their courses on a points-led system, and whole-organism biology was becoming the poor relation. Despite this and with notable support from CIEEM and several national schemes and societies, the course continues to the present day, hosted by Manchester Metropolitan University (MMU; see www2.mmu.ac.uk/sste/courses/short-courses/). I consider this an important part of what the FSC can provide in training future recorders to provide the evidence base for our conservation decisions.

The Future

I have now left Preston Montford in the capable hands of the next Head of Centre and am concentrating on more opportunities to network with partners and support the recording and reporting of our natural heritage. The future is increasingly financially uncertain, commercial viability nestles with grant-funded opportunity and this patchwork of funding will no doubt dictate some of the FSC priorities. However, there are strands of our work which I am convinced will remain. One is the growing need for tried and tested resources, and I applaud the publications department for the increased range of identification guides. The fold-out charts engage so many people and help us to identify our brand across the world. The project work as supported by the marvellous Heritage Lottery Fund (please buy your ticket!) and charitable trusts, such as the Esmée Fairbairn Foundation, remain important to support

Figure 46.3 National Plant Monitoring Scheme training course at Nettlecombe Court field centre (Charlie Bell, Field Studies Council)

Figure 46.4 Freshwater invertebrate identification using Field Studies Council fold-out charts (Field Studies Council). A black and white version of this figure will appear in some formats. For the colour version, please refer to the plate section.

Figure 46.5 Sue Townsend (left) and Dr Sarah Whild (centre) were awarded Honorary Membership of the National Biodiversity Network Trust in 2018 by NBN Chairman Professor Michael Hassell CBE FRS (from the author's photo collection)

volunteers, and the recognition and advice offered by government statutory agencies helps us to integrate our efforts. At Preston Montford, I supported two such projects and felt strategically that there needed to be even more support for specific initiatives to deliver training in biological recording. Out of this commitment, the Biodiversity Training Project (see www.fscbiodiversity.uk), Invertebrate Challenge, Biodiversity Fellows, Tomorrow's Biodiversity and the now prestigious BioLinks all developed and grew, in succession. Innovations have included digital atlases and online identification resources, and it was important to bring conservationists and data providers together. One such innovation for biological recording, in the emerging field of geographic information systems, won the organisation a prestigious CIEEM award in 2017. Engagement to fill the so-called 'nature deficit disorder' is key – and the FSC is responding with family and 'discovery' courses. The terms Bio Blitz and Citizen Science have now come into common parlance within our sector and I might spend just as many of my current field days supporting a recording course as helping family groups spot their first true bug and to make use of the FSC shield bug app. The diversity and importance of this organisation will extend to its next 70-plus years and, as the courses, resources and networks evolve, I believe we can explore even greater connectivity with people, place and landscapes for the future. And me – well, I will be biologically recording with better accuracy, more coverage and a more focused retirement purpose!

References

Berry, R. J. and Crothers, J. (1987). *Nature, Natural History and Ecology*. Linnean Society/Academic Press for the Field Studies Council, London. Reprinted from *Biological Journal of the Linnean Society* 32.

CIEEM (2011). *Ecological Skills: Shaping the Professional for the 21st Century*. See https://cieem.net/resource/closing-the-gap-rebuilding-ecological-skills-in-the-21st-century-ieem-2011/ (accessed July 2019).

47 · *My Love Affair with Rocks That Fizz*

MAURICE TUCKER

For most earth scientists, it all begins in the field ... Many geologists were inspired by something they found in the countryside or on the beach as a youngster, or they were overawed, even puzzled, by the scenery on a family holiday, which led to them wanting to understand what they were looking at – the how and why.

When it comes to current research in the earth sciences, this could also be said to begin in the field – although these days it might not just be the classic, traditional mapping of a specific area or collecting samples/fossils for lab analysis; it could be a cruise to drill into the ocean floor – for sediments, or sampling gases from a belching volcano or taking an ice core. Virtual fieldwork these days is making a significant impact; lidar and photogrammetry of cliffs and mountainsides enable detailed studies of stratal geometries and fractures, for example. And then interpreting satellite images of remote places, or analysing those sent back from outer space, such as those arriving from the Mars rover, Curiosity, as it trundles towards Mount Sharp, are further extensions of fieldwork, producing exciting and amazing results, particularly in the case of Curiosity. But one cannot get a real feel for rocks without being there in the field: observing the exposures, using hammer and lens, appreciating the context (see Tucker, 2011 for a guide to studying sedimentary rocks in the field). An oft-quoted maxim, attributed to H. H. Read (1889–1970, a professor at Imperial College London), is that the best geologist has seen the most rocks.

For me, having spent a career studying limestones in academia (at the universities of Durham (BSc), Reading and Göttingen (PhD), Sierra Leone, Cardiff, Newcastle upon Tyne, California Berkeley, Durham

Figure 47.1 How it all started – my first fossil, found in the back garden at the age of seven: an imprint of a bivalve in a flint pebble derived from the chalk (Upper Cretaceous) (Maurice Tucker). A black and white version of this figure will appear in some formats. For the colour version, please refer to the plate section.

(again, but for 29 years), UWA Perth and now Bristol), but still being fascinated by these intriguing rocks, it all began when I was seven years of age. I was throwing stones into the pond in the back garden (in Newbury Park, Ilford, Essex/Greater London), something I was not supposed to do – and then noticed the imprint of a bivalve on a small pebble of flint (Figure 47.1). I was immediately curious – and the rest is history, as they say. I quickly learnt about fossils, then became more interested in the rocks that contained the fossils – commonly limestones, of course – and so started a long love affair with rocks that fizz.

There is, undoubtedly, an excitement of looking at rocks in the field; you never know what you are going to find – it could be something extra special, like dinosaur footprints (1974, Sully, Glamorgan, Triassic), or amphibian footprints (1993, Howick, Northumberland, mid-Carboniferous) and you end up in the news. In my collecting days (late 1950s to early 1960s), there were no simple ways to identify the more common fossils (the three British Museum books on British fossils had not yet been published). One had to take one's treasured specimens to a museum and compare them with those on display in the show cabinets there. These days, that would not be possible since the majority of museums have removed their fossil displays and replaced them with life-size models of dinosaurs, a few fossils – possibly to handle – and buttons to press. I used to take my best fossils to the Natural History Museum in South Kensington, London, where the palaeontologists working there kindly identified them. In examining my fossils from the Mesozoic Coast (as it should be called!!), Dorset, back in 1962, they

Figure 47.2 In search of viruses – fieldwork on modern microbial mats in Qatar, sampling the mats and their porewaters (Maurice Tucker)

deemed two ever so special that they asked me to donate them. You can imagine how extremely encouraging it was for a young lad to receive a letter thanking me for my 'presentations to the National Collection'.

I have been fortunate and lucky enough to have undertaken fieldwork in some of the most beautiful parts of the world, and often quite remote too. In pursuit of Precambrian carbonates to analyse for information on marine life and seawater chemistry in deep-time, visits were made to Death Valley in California; Glacier National Park in Montana; Porsangerfjord in Arctic Norway; Mussoorie in the Lesser Himalaya and Chhattisgarh in the Eastern Ghats, both in India; Tiout in the Anti-Atlas of Morocco; and the Flinders Ranges, Australia. In trying to unravel the controls on carbonate deposition, i.e. tectonics versus eustasy-orbital forcing (Milankovitch rhythms), fieldwork was conducted in the Great Wall country, northern China (Mesoproterozoic and Cambrian); the karst region of Guilin, southern China (Devonian); the Ardennes, Belgium (Devonian); north-east England (Carboniferous, Permian); the Merida Andes, Venezuela (Permian); Catalan Basin, Spain (Triassic); Yorkshire (Jurassic); the Vercors-Aravis, south-east France

(Cretaceous); and the Zagros Mountains, Iraq (Miocene). Sedimentologists, like myself, often study modern sedimentary environments, which provide analogues and answers to aid interpretation of ancient rocks. Hence, fieldwork has been undertaken in the Bahamas, Belize, Bonaire, Florida, the Pantanal (Brazil) and Western Australia, with recent work focused on microbial mats in Qatar (Figure 47.2), for their hydrocarbon source-rock potential and, really exciting, an assessment of the role of viruses in carbonate precipitation. Viruses are the new frontier in earth sciences!

Reference

Tucker, M. E. (2011) *Sedimentary Rocks in the Field*, 4th Edition, Wiley, Chichester.

48 · *In the Footsteps of John Wesley Powell: Restoring the Sand Bars in the Grand Canyon*

ALAN WERRITTY

On 5 August 1869, John Wesley Powell and his nine companions launched their four boats into the Colorado River near Lees Ferry and soon entered what is now known as the Grand Canyon of the Colorado. Thus began the most hazardous and exciting phase of their exploration of the upper Colorado River, a journey that had begun 74 days earlier, upstream on the Green River. Their aim was to fill in one of the last remaining unmapped regions in the American West by reporting on 300 miles of inaccessible canyon lands. Having entered Marble Canyon, Powell's expedition was immediately confronted with vertical rock walls and death-defying white-water rapids, only some of which could be bypassed by portaging their boats. Sand bars provided temporary campsites, but with food running low and little opportunity for hunting, their emergence mostly safe and sound, 24 days later, was seen as little short of miraculous. Vividly recorded in Powell's *The Exploration of the Colorado River and Its Canyons* (Powell, 1961; first published in 1895), the expedition was deemed an outstanding success, as evidenced by public acclaim back in the east. More significantly, it triggered investment by the Federal government into mapping and developing the resources of this hitherto unknown region in the American West.

In 1993, during research leave from the University of St Andrews, I was invited by Ned Andrews, then of the US Geological Survey in Boulder, to join a group of geomorphologists modelling flow exchange between the main channel and eddies on the Colorado River. Having driven from Boulder to Page Arizona, Ned and I joined the rest of the group where we launched two inflatable rafts into the Colorado River at Lees Ferry, close to where, 124 years earlier, Powell had begun his exploration of the Grand Canyon. Our more modest task was to undertake flow measurements in the vicinity of the sand bars that Powell's expedition had camped upon. These bars are transient alluvial deposits at the margins of the Colorado River, which typically occur where the main flow separates from the river bank and creates an eddy – a recirculating cell within which sediment is deposited. These sand bars provide virtually the only campsites for groups licensed by the national park for white-water boating through the Grand Canyon – an income stream that regularly exceeds that produced by hydropower from the Glen Canyon dam.

Before the Glen Canyon dam was constructed upstream of Lees Ferry, sand bars along the margins of the Colorado River would have formed and re-formed during successive floods. But, following the completion of the dam in 1963, these bars had steadily been encroached upon by thick stands of vegetation and eroded by unregulated spring floods, jointly reducing the availability of campsites and associated tourist income. In 1983, responding to pressure from conservationists and white-water recreation groups, the Bureau of Reclamation initiated the Glen Canyon Environmental Studies project. This in turn led to a proposal for an experimental flood release from the Glen Canyon dam to rebuild and maintain the sand bars in the Grand Canyon.

The design of the hydrograph for what was to prove a successful experimental flood in 1996 was based, in part, on the fieldwork undertaken in August 1993 by Ned Andrews and his group. Given that the rate of sand deposition in an eddy depends in part on the boundary shear stress on the bed, a key component in the numerical model underpinning the specification of the experimental flood was to be derived from velocity profiles in the recirculating cells, adjacent to the sand bars. Accordingly, over a period of 10 days, Ned Andrew's group rafted downstream from sand bar to sand bar, obtaining velocity profiles via stacks of current meters on a rod, whilst at the same time cooking and camping on the bar.

As part of the group, I quickly settled into the daily routine. Waking up was triggered by a rapid sunrise around 6 a.m. Having shaken out my shoes – to check for scorpions – breakfast followed, and the stacks of current meters and their associated loggers were moved into position within the eddy. The daily rhythmic change in water level, driven by the demand for electric power, meant that the size and location of the eddy constantly changed. In turn, the array of current meters was relocated throughout the day so as to obtain as full a record as possible of changes

Figure 48.1 Camping in the Grand Canyon on a Colorado River sand bar (Alan Werritty). A black and white version of this figure will appear in some formats. For the colour version, please refer to the plate section.

in the velocity profile. The cool of the early morning was soon dissipated as the sun rose to its zenith and by the middle of the day I sought what little shade was afforded by the encroaching vegetation. By late afternoon, as the temperature dropped once again, the cook, who doubled up as quartermaster and boatman for one of the large inflatable rafts, began to prepare the evening meal. As the light failed very quickly, conversation and stories took over at the end of each day, often around a campfire built from abundant driftwood rafted down the river. One evening, I produced a long-travelled bottle of Cardhu malt whisky to lubricate the conversation. Sleeping mats were then rolled out and, as the Milky Way slowly rotated over my head, I drifted off to sleep. Some nights, using light from my carefully rationed torch, I attempted to find the same location in Powell's account, over a century earlier, and marveled at his deprivations and our largesse.

Over time, I adjusted to being 1,500 m down in a chasm that at its rim extended over 16 km, and to the fact that for the duration of the trip I was totally isolated from the outside world. Any emergency would have involved climbing to the rim or trying to get the attention of a passing plane by flashing a reflective object towards the sky. One rest day, I did climb part way to the rim and was rewarded with not only a bird's eye view of the river incised into the canyon floor but also the dried 4-m stems of the spectacular Utah agave, which on flowering, every 15–25 years, immediately dies. From this vantage point I could also grasp more fully the bedrock geology, comprising an almost horizontal repeated layering of massive sandstones, limestones and friable shales, resting unconformably on the Precambrian highly resistant schist in the deepest part of the canyon. Powell especially dreaded these reaches in the Inner Gorge as they invariably meant either portaging the boats around the most dangerous rapids or risking disaster by running them. More than a century later, rafting from sand bar to sand bar also involved repeated running of white-water rapids, with the notorious Lava Falls providing the biggest challenge. Ned and some of the others ran several of these in kayaks, emulating Powell and his companions as, bouncing up and down, they skillfully set a course avoiding emergent boulders or being sucked down into threatening turbulent holes. I found running the rapids on an inflatable raft with supplies and all our kit almost as exciting, clutching my camera and soaked to the skin as each time I emerged at the downstream end.

As the days passed and I became more attuned and receptive to the mysteries of the canyon floor, the overwhelming sense was one of contrasts. At dawn and sunset, the colours of the massive cliffs were a kaleidoscope of vermilions, ochre yellows and washed-out greys, thrown into relief by the deep shadows. By noon, as the shadows dwindled to nothing, these colours became muted and distances difficult to judge. Even the apparent aridity was deceptive, with occasional springs in small side canyons yielding rock surfaces locally festooned with brilliant green ferns and bryophytes. Indigenous human activity, although now absent, lingered in petroglyphs on the walls of the cave-like alcoves, carved by repeated floods, and above in the remnants of cliff dwellings, from the long-since vanished Anasazi people. Even the colour of the river was in flux most noticeably where the clear water downstream of the dam was suddenly injected with muddy brown water from the sediment-rich Little Colorado and other large tributaries. Here at the boundary between the two water bodies, the turbulent transfer of energy set off cascades of downstream spinning spirals of the bluish-green and brown waters.

As a scientist I seek to collect data in as objective and dispassionate a manner as possible: as a geographer I am aware that this occurs in a specific place. And place brings with it strong associations which can sometimes be highly evocative. The Grand Canyon is such a place, and since Powell's time has become one of the most iconic and visited places in the American West. But what does the modern tourist really see from the various vantage points along the feted south rim? For Stegner, one of Powell's biographers, the expedition account is uppermost in the mind of the well-read tourist, walking from Powell's statue to the edge of the south rim.

For myself, I'm not so sure. Having experienced the Grand Canyon from the air in a light aircraft in 1968, from a raft on the river in 1993 and from hiking along the rim in 2013, I still struggle to encapsulate it in words. Instead I fall back on the wisdom of Charles Dutton, a geologist who in his *The Tertiary History of the Grand Canyon District* (Dutton, 1882) noted the need for a completely new landscape aesthetic to do justice to the Grand Canyon. Eschewing the canons of landscape beauty appropriate to Europe and the Alps, and straining to find a new vocabulary to describe the Grand Canyon, he wrote 'outlines that at first seemed harsh and trivial have grace and meaning; forms that seemed grotesque are full of dignity; magnitudes which had added enormity to coarseness become replete with strength and even majesty; colours which had been esteemed unrefined, immodest and glaring are as expressive, tender, changeful and, as capacious of effects as any others' (Dutton, 1882,

p. 26). As Stegner (1953, p. 173) observes, 'strange words in a geological monograph' but words that help me make sense of a landscape and a geological phenomenon that is one of the most sublime in the natural world.

Twenty-five years on, this is still my most remarkable fieldwork experience.

References

Dutton, C. E. (1882). *The Tertiary History of the Grand Canyon District: United States Geological Survey Monographs, II.* United States Geological Survey, Washington, DC.

Powell, J. W. (1961). *The Exploration of the Colorado River and Its Canyons.* Dover Publications, New York, NY (republication of *Canyons of the Colorado*, first published in 1895 by Flood and Vincent).

Stegner, W. (1953). *Beyond the Hundredth Meridian: John Wesley Powell and the Second Opening of the West.* University of Nebraska Press, Lincoln, NE.

49 · *Connecting the Next Generation to Their World*

NATALIE WHITE

As a teacher, I am incredibly fortunate to be in a position to offer opportunities for learning outdoors as part of the curriculum. Applying different pedagogical approaches to the outdoor context allows me to engage learners in powerful experiences that encourage and ignite intrinsic motivation, which leads to a journey of discovery. I am privileged to have many wonderful moments with pupils but here are three that have significantly impacted on my teaching.

It's one of those smiles, from ear to ear, and you can't help but smile back. Katie is beaming, the other children all gather around her. Her smile makes every hard day slip into the abyss, it's the *raison d'être* for all educators.

I am lucky enough to be watching a class of five-year-olds on a spring morning, sitting in a local green space close to their school. They are doing a topic based on Jill Tomlinson's book, *The Owl Who Was Afraid of the Dark*, and today they are dissecting owl pellets, pellets that their teacher had strategically placed for them to find at the foot of the trees.

Katie has just identified and matched a mouse jaw bone to the little chart in front of her; her teacher Mrs Baird has confirmed her findings! She is elated, proudly she shows everyone and anyone who will listen, wandering around the group with her bone clenched by the tweezers, her over-sized gloves getting in the way. All that's missing is a little white lab coat. The other children now get excited, they have been spurred on by this discovery; they too want to find out what this owl has been eating for his breakfast.

Viewing this as a teacher I can see deep learning in action, the integration of cognitive, affective and psycho-motor skills; a memorable

Figure 49.1 Nature offers us so many learning opportunities, each one building our connection with the Earth (Natalie White). A black and white version of this figure will appear in some formats. For the colour version, please refer to the plate section.

experience that has involved this pupil using her previous knowledge of what owls eat, her motivation and willingness to engage in the activity and her fine motor skills to dissect the pellet. Katie won't forget this moment and neither will I.

By the time we are heading back into class they have a collection of bones and a plethora of questions. Starting children young, building on their developing sense of discovery, is the perfect place to encourage the enquiring mind. Those 'Why?' questions can be the foundation of so many animated discussions, which, if nurtured, becomes an intrinsic motivation to learn about the world around us. The artefacts they have gathered are given pride of place on the display table and are revisited numerous times over the course of the following weeks. The whole experience has been a catalyst for learning across many subjects – not only science and literacy. The well-employed approach to questioning by the teacher has elicited spontaneous and interesting responses, displaying both depth of knowledge and areas for development. The children have led the learning and are an integral part of the process.

When Thomas, a lively nine-year-old, sprang back from the fence post with a loud 'Wow!', I was again reminded of the power of observing those sparks that ignite the fires in children's minds. After reading Dr Seuss's *Horton Hears a Who*, the children were scattered around the playground, looking at corners of windows, spider holes on the fence, and laying on the grass, inspecting the ground using hand lenses. They were all engaged, intent on finding their own species of 'Who'! For Thomas, that came in the form of a small green aphid and a spider, who were busy having a battle on top of the rotting fence post. His excitement was infectious, rapidly other children were migrating to see what he was shouting about. Thomas gave a running commentary whilst intently watching them through the hand lens, it was likened to a gladiator battle with the aphid, or 'green thing', being the underdog. The incident captured him, he was full of questions, he wanted to know about the insects, he wanted facts, but also his creativity had been sparked. He was animated as he told his friends about the story he was going to write. 'Archie the Aphid' was going to be a great gladiator who was going to rule the fence post for all the other aphids to live safely on. Thomas was excited, excited about learning. Writing his imaginative story wasn't going to be a chore, this was now something he wanted to do, even if it meant missing break time to get it started!

When Amy picked up the binoculars she did so with great care, this new piece of equipment made her nervous, they looked and felt expensive

Figure 49.2 The awe and wonder found in our environment helps to strengthen our connections with nature (Natalie White). A black and white version of this figure will appear in some formats. For the colour version, please refer to the plate section.

and she didn't want to break them. However, the lure of being able to see where the noise was coming from was too great. Watching her, I was overcome with pride; this was a young girl who had run screaming from the classroom cupboard because of the spider. She was a self-declared 'insider'; the outdoors was not for her, it was wet, cold and, worse still, full of bugs and beasties. Yet, in just a short few months, she had become engrossed in our John Muir project. She wanted to learn all about the insects, she had been instrumental in building numerous bug hotels and took real pleasure in sharing her new knowledge with others. On this particular visit to the John Muir project site, we had been erecting bat and bird boxes. We spent time listening to the bird calls and then tried to locate them in the trees. The class were mesmerised and they were surprised that there were so many different birds present. Then Amy found the source of the call through the binoculars, she picked up her identification chart and pointed at what she had seen, a song thrush. I smiled and nodded. Her jaw dropped, she was completely in awe at what she had just done. She won't ever forget the song of the song thrush.

Teaching in the outdoors is a multi-sensory experience, our brain neurons go into overdrive; learning smells, looks, feels and sounds different. Allowing children to participate, explore, play and learn through a variety of hands-on, minds-on activities aids this process. The brain desperately scrambles around making connections with what you already know and creating pathways for the new information. The environment is a stimulating source of interesting inquiry, questions that the children generate through their natural curiosity. The awe and wonder, if successfully captured, can be used to encourage them to explore, to deepen their understanding and appreciation. Katie, Thomas and Amy all know about nature, they have been part of an authentic experience which has helped them to develop values, appreciation and positive attitudes both to learning and our world. They have had a progression of opportunities to develop skills in identifying, recording, analysing and sharing. These skills will be applied again in new contexts as part of life-long learning.

50 · *Beyond the Curriculum: Wider Conceptions of Learning in the Field*

LEWIS WINKS

Introduction

While we often talk about 'the curriculum' – as I have done in the title of this chapter – there is, in fact, no one curriculum, but many, given life by the passions of educators and the places that we visit. In fact, it might even be argued that the most important facets of learning in the field come from the spontaneity and emergent experiences which arise in relationship with the places we visit with our students. While the impetus for fieldwork is provided by a preconceived notion of what is important and what should be learnt (i.e. 'the curriculum'), it is demonstrated that these wider conceptions of learning are also important for learners. Interaction with the environment, with each other, and reflections on personal experiences, can elicit considerable learning and – on occasion – produce transformative, and disruptive, moments, which resonate into young people's lives and which crucially have the power to contribute to a citizenship capable of environmentally and socially responsible decision-making (Selby and Kagawa, 2015; Jickling, 2017). This, it might seem, is an extraordinary ask for fieldwork – but we live in extraordinary times. The generation studying GCSEs and A-levels[1] today face great changes and uncertainty.

This assertion goes hand-in-hand with an important evolution of environmental education throughout the latter part of the twentieth century. At first, environmental education was largely focused on the mantra that

[1] High-school qualifications in the UK: GCSEs for 16-year-olds and A-levels for 18-year-olds.

knowledge of the natural world (and the issues facing it) will bring about a change in various behaviours – towards a pro-environmental stance in learners. Since this time, we have gained considerable understandings of the social psychology of behaviour change, and we now appreciate that this 'knowledge equals action' approach is too simplistic (Hungerford and Volk, 1990). Multiple other factors play into behaviour change – including values, beliefs and, crucially, emotional aspects of learning, triggered by interactions with other learners, with the environment, and prior understandings, beliefs and norms, otherwise known as 'worldviews' (Ajzen, 1991; Hargreaves, 2011; Carmi *et al.*, 2015).

Methods

My work has examined these wider conceptions of learning in the field – seeking to look beyond the taught curriculum and towards the complementary curricula offered by nuanced experiences emerging from the social and environmental settings in which fieldwork occurs. Taking place in 2016, I made use of an immersive ethnographic approach to examine the experiences of students living and learning at Slapton Ley field centre in south Devon, UK. The programmes I worked with were between three and five days in length and encompassed biology and geography A-level courses. Student participants were aged 16-18. As well as observations in the field, the methods included focus groups held at the end of the programme, with photo elicitation – where students each brought one photograph with them to discuss. The excerpts of student discussion, both in the field and during the focus groups, are accompanied here, along with students' own photographs.

It is the purpose of this short chapter to point the reader towards the type of experience which might find greater resonance with the lives of learners, and by doing so, to encourage programme leaders, teachers and educators to further examine the wider and diffuse sets of experiences and curricula which arise during fieldwork.

Vignettes

A Local Bus

In this first vignette, a group of geography students from central London talk about the transport arranged to take them from the field centre to

Figure 50.1 Bus waiting to take students to field site (Lewis Winks)

their field site for the day. It provides much amusement, as well as some more serious observations and reflections on privilege and access.

One student remarked:

> When we first saw this it was quite funny to us. It looked like Stalinist bus that we were getting on, from the Soviet-era.
>
> But then I thought – we are in the countryside we are taking in nature ... with the state of the bus, it's got the mud and the muck and stuff , , , but if this bus was in London, there would be complaints to TfL [Transport for London] straightaway saying why have you got buses that are all dirty ... But here it's like, oh, it's probably just driven through some puddles and stuff.

Another student continues:

> It gives a sense of adventure, all that mud at the bottom ... It's an experience you won't get in London.
>
> It shows us how transport is important for us and how much we take it for granted. Living in London, we see taking the bus as something very easy to do and we have great access to [public transport] and it's widely available ... Whereas in rural areas, in the UK or even abroad, the transport infrastructure

Figure 50.2 Storm damage in Torcross (Lewis Winks)

can often be quite poor ... They often don't have a transport system they can rely on.

I got refunded for a whole journey that was three minutes late! It makes you feel very pompous. We are quite spoilt I think.

Coastal Change

In this next vignette, a group of geography students from the Midlands visit the village of Torcross on the south Devon coastline and witness the immediate aftermath of a storm on the coast, in the month of February. Their initial shock at the damage to the houses is later reflected on in a focus group, eliciting some thoughtful observations regarding living with a changing climate.

Pointing to the houses, a student exclaims:

They all seem to have like, boards up. On the front, downstairs windows. All of them seem to have boards up, like they been smashed or they were protecting them in case there was a storm or something. They all seem to be really cautious about the fact that there was going to be waves coming over the top, like it's happened before.

Later the following day some of the students recount to me how it made them feel to see such damage on the coast: '. . . how the seawall had literally started to fall away, and the massive crack that has appeared ...

You feel sorry for the people that are living or used to live there in the cottages. It's not nice,' one said. Another added:

> [A]nd I think, because we live right in the middle of the country, we don't see these coastal things like, it's not a problem where we live because there is no coast. So . . . seeing all these things, it's mad'.
>
> It's scary thinking about . . . sea level rising at such a rapid rate. It brings home the whole climate change and global warming stuff . . . because we aren't really affected by it at all. If this is going to happen, this is obviously going to have an impact on us, because if sea levels do rise more people have to, I suppose, come into our local area, inland. They will have to come into where we live. . .

Bugs in the Woods

In this final vignette the emphasis shifts away from commonly explored themes which emerge from and connect to the taught curriculum and instead focuses on the more visceral and 'felt' experiences of being in a new, and uncomfortable, environment. In this example, students visited Slapton from Essex, for a biology field trip, in July. On a hot and humid day they visited the woods to carry out leaf-litter sampling.

Once the leaves are placed into the tray, some of the girls don't want to search for insects. A bottle of hand sanitiser is passed round – 'you've

Figure 50.3 Students looking at leaf litter samples in Slapton Woods (Lewis Winks). A black and white version of this figure will appear in some formats. For the colour version, please refer to the plate section.

touched the leaves – do you want to sanitise your hands?' – Another student exclaims – 'this is disgusting'.

The sunshine streams down now through the trees and it is increasingly humid. The midges and flies continue to bother people. I see that one of the students has zipped his waterproof top right up and has the hood up too – people notice and begin to laugh. He responds with a muffled voice – 'The bugs!', he exclaimed – 'I hate them! . . . this is too close to nature for me, Miss'.

Elsewhere, there was a continuing fear and reaction to bugs and insects. One girl screamed at a bug which had landed on her back – 'get it off – f**k!! – get it off!' And another girl shouted: '. . . [T]he bloody ants! And flies! Crawling up you! I just, can't do that. It makes you feel disgusting.'

Later, I talked to some of the students about their experiences in the woods. Speaking about her surprise at the number of insects in the leaf-litter sample, one girl commented:

> . . . I [said], don't [lay the quadrat] there because there's not going to be anything. And then all these things, all these spiders came crawling out and I was like, oh okay! So, yeah, that just showed me that when you're walking through the woods there are so many [insects] . . .

She added, 'I must have killed so many insects by walking in the woods. Poor little things.'

Reflecting on this, another student remarked:

> Normally, when I'm at home I would kill a spider just like that – with my shoe – I wouldn't think twice about it and I [would] be like, oh, it doesn't matter anyway because it's nothing. When we're at home, we're not in the countryside so we don't even have to consider it. You don't really get a lot of bugs where we are, unless you have your window open and light on, and you get moth or something . . .

Another continued, saying:

> If [a moth] was in my room, I would whack it straightaway so that it would be gone. But [here] in the woods . . . I actually feel a bit bad. Like, I've been slapping you with a shoe! I feel quite horrible. And you're looking at them, and [thinking], they're actually living and crawling and they've got everything in their body . . . they've got legs and they move . . . They're actually a bit like us. Do you know what I mean? They are alive. And it's weird to think about it like that.

Conclusion

On consideration of the vignettes, some common conclusions can be drawn. Clearly, to some extent, the experiences of students in each case offered a changed perspective resulting from their time in the field – stepping outside of the normal frames of reference offered by home or school settings. In each of the examples, too, the realisations and discussions emerged surrounding the meaning of 'wider curricular' beyond that which is formally taught in these settings. Finally, it can be said that while *experienced* in the field, it wasn't until later *reflection* in the focus group that some of the deeper and more profound learning emerged in discussion amongst the students.

The vignettes presented in this chapter signpost a wider and more nuanced set of experiences on offer from fieldwork and suggest that we as educators should become more receptive to the beliefs, values and social norms which feed into fieldwork. If we are to step truly beyond the knowledge paradigm and into a more proactive role in encouraging social and environmentally conscious learners, it is important that we also pay attention to the personal and emotional components of learning in the field, as well as the role of challenging and disruptive moments. Becoming aware of the many curricula, and the wider experiences of fieldwork beyond the taught curriculum, is central to fulfilling a contemporary role of environmental education – to transform worldviews and inspire learners to take action for a sustainable world.

References

Ajzen, I. (1991). The theory of planned behavior. *Organizational Behavior and Human Decision Processes* 50, 179–211.

Carmi, N., Arnon, S. and Orion, N. (2015). Transforming environmental knowledge into behavior: the mediating role of environmental emotions. *Journal of Environmental Education* 46, 183–201.

Hargreaves, T. (2011). Practice-ing behaviour change: applying social practice theory to pro-environmental behaviour change. *Journal of Consumer Culture* 11, 79–99.

Hungerford, H. R. and Volk, T. L. (1990). Changing learner behavior through environmental education. *Journal of Environmental Education* 21, 8–21.

Jickling, B. (2017). Education revisited: creating educational experiences that are held, felt, and disruptive. In *Post-Sustainability and Environmental Education*, eds. B. Jickling and S. Sterling. Springer, Berlin, pp. 15–30.

Selby, D. and Kagawa, F. (2015). Drawing threads together: a critical and transformative agenda for sustainability education. In *Sustainability Frontiers*, eds. D. Selby and F. Kagawa. Barbara Budrich, Toronto, pp. 277–280.

Part III

Reflections and Where Next for Field Studies

51 · *Epilogue: Inspiring, Curious and Novel Fieldwork*

TIM BURT AND DES THOMPSON

The Essentials

Arguably, Thursday 1st July 1858 is the most significant date in the annals of field studies, and quite possibly for science. At the Linnean Society of London, in front of an audience of barely 30 people, two papers were read by the Society's Secretary, John Joseph Bennett. One was written by Alfred Russel Wallace, and the other by Charles Darwin; both authors were absent – Darwin at the funeral of his tenth and last child, Charles Waring, and Wallace on an expedition in Borneo.

Wallace's paper was 'On the tendency of varieties to depart indefinitely from the original type', and Darwin's was titled 'Extract from an unpublished work on species' (which was a portion of a chapter lengthily entitled 'On the variation of organic beings in a state of nature; on the natural means of selection; on the comparison of domestic races and true species'), which accompanied an abstract of a letter from him to Professor Asa Gray, dated 5 September 1857. Both papers fundamentally set out the rudiments of evolution through natural selection and speciation and were founded on years of assiduous fieldwork.

Most of the small audience had turned up to hear an obituary notice read by Sir Charles Lyell for the former President of the Society, the Scottish botanist and microscopist, Robert Brown, who had died the previous month. Brown had many claims to fame, notably giving his name to 'Brownian motion', and for reading his own paper to the Society in 1831 (published in 1833) in which he coined the term 'cell nucleus'.

The eminent botanist George Bentham (who three years later became President of the Society) was due to speak first but was persuaded by Sir

Figure 51.1 Charles Darwin - the greatest fieldworker of them all!

Joseph Dalton Hooker to step down to make way for the Darwin and Wallace papers. Hooker, who later became the celebrated Director of the Royal Botanic Gardens, Kew (his tenure was the subject of parliamentary scrutiny, which vindicated him) and President of the Royal Society, was Darwin's closest friend.

The Darwin and Wallace papers were read and followed by six papers covering a variety of botanical and zoological subjects; the meeting ended with no formal discussion. Apparently, afterwards, Hooker commented that there was 'no semblance of discussion', though 'it was talked over with bated breath' at tea afterwards, and many years later he remarked 'the subject [was] too ominous for the old school to enter the lists before armouring'. (This is well described by the historian of science, Janet Browne (2002) from whom the quotations are taken.)

The Darwin and Wallace papers were published the following month in the *Zoological Journal of the Linnean Society* (20 August 1858, pp. 46–50), consisting of a communication letter by Hooker and Lyell, and the above-mentioned papers. Their letter was gracious in intent, stating:

The gentlemen having, independently and unknown to one another, conceived the same very ingenious theory to account for the appearance and perpetuation of varieties and of specific forms on our planet, and both fairly claim the merit of being original thinkers in this important line of inquiry; but neither of them having published his views, though Mr. Darwin has for many years past been repeatedly urged by us to do so, and both authors having now unreservedly placed their papers in our hands, we think it would best promote the interests of science that a selection from them should be laid before the Linnean Society.

A year later, *On the Origin of Species* (strictly, *On the Origin of Species by Means of Natural Selection, or the Preservation of Favoured Races in the Struggle for Life*) was published, on 24 November (Darwin, 1859), with Charles Darwin's name forever associated with the foundation of evolutionary biology. Wallace, too, became a revered figure, albeit much more recently, and in addition to being the co-discoverer of natural selection, was an exceptional zoogeographer and explorer, and one of our earliest environmentalists.

Remarkably, the Linnean Society President of the day, Thomas Bell (a zoologist, who had formally described the reptile and crustacean specimens from Darwin's *Beagle* Expedition, 1831–1836), wrote in his presidential report of May 1859 that 'The year which has passed has not, indeed, been marked by any of those striking discoveries which at once revolutionise, so to speak, the department of science on which they bear.' How wrong he was!

We began this book by considering the origins, history and diversity of field studies. Having delved into these and solicited a wide range of contributions from people active in the field, we are struck by three essential ingredients of field studies. First, fieldwork on nature and the environment has defined the trajectory of some of our greatest thinking and breakthroughs in science. Second, an innate sense of curiosity and passion, often in equal measure, has driven these field-based endeavours. Third, very often what turn out to be breakthroughs are often realised as such only decades after their 'discovery', of which the Darwin–Wallace contribution may be an exemplar.

Touchingly, it is clear from the contributions in this book that field studies in early years have dominated the lives of many of our contributors who have gone on to have busy and increasingly administrative lives, with some only returning to basic fieldwork in so-called retirement. Our younger authors have fewer such impediments to getting outdoors – yet!

Figure 51.2 Children love learning about nature, and these experiences can last a lifetime and cross generations. Here, Tim's grandchildren enjoy discoveries on a rocky seashore (Tim Burt).

Field Studies Are Vital for Society

In Part I, authors reviewed the history of field studies and the reasons why it became so vital to not only the geophysical and ecological sciences but also sections of the humanities and social sciences as well. The caption for Figure 1.2 poses the question: 'Why have we brought you here?' Being in the outdoor world encourages us to be curious, whatever the precise subject matter of interest. For some, the task is to test existing ideas or to place the locality in a theoretical context. Does this place fit in? What elements are different to what we expected? It is these differences between the observed and expected that very often lead to new ideas, sometimes a small embellishment to an existing theory but occasionally something radically different from where we started out.

The Field Studies Council (FSC), a large UK charity of which both the editors are currently trustees, has recently agreed its new strategy to 2025. This is a useful and relevant starting point for us when considering the future of fieldwork. The FSC seeks to inspire everyone to be curious, knowledgeable, caring and passionate about our outdoor environment. The new strapline is 'leading learning outdoors', nicely ambiguous in that the FSC seeks to be a leader in the field of outdoor education but at the same time its main activity is to lead field courses, across the widest range of age groups and subjects. Indeed, the FSC's charitable purposes require it to undertake and encourage field studies in the arts and sciences; to maintain field studies centres; and to promote the benefits of protecting and preserving the outdoor environment. The FSC has three pillars to its 2025 vision:

- get people outdoors;
- provide high-quality outdoor learning; and
- advocate sustainability.

In addition, there is a cross-cutting theme: 'encourage well-being'. For us, these four elements nicely summarise the importance of learning outside the classroom, whether via formal or informal educational opportunities. They will indeed inspire everyone to be curious, knowledgeable, caring and passionate about their world, both the natural and the built environments. It is therefore important to influence national decision-makers in order to develop the wider agenda for outdoor learning and, in particular, to ensure schools and universities recognise its importance and include relevant courses in their curricula. By taking part in out-of-classroom learning, students also will improve their health, happiness, well-being and confidence. There is growing evidence that students learning outside the classroom develop their very capacity and motivation to learn (e.g. chapters by Higgins and Reiss). Figure 51.3 shows a word cloud based on FSC publications, emphasising the range of experience gained on a field course.

The Council for Learning outside the Classroom (CLOtC) points to a wealth of evidence which clearly demonstrates the benefits for young people's learning and personal development outside the classroom. In summary, the CLOtC summarises the benefits of learning outside the classroom as follows:

- It tackles social mobility, giving children new and exciting experiences that inspire them to reach their true potential. These real-world experiences raise aspirations, equipping young people with the skills they need to become active and responsible citizens and shape a fit and motivated workforce.
- It addresses educational inequality, remotivating children who do not thrive in the traditional classroom environment, such as those from disadvantaged backgrounds or with special educational needs. Young people who experience learning outside the classroom as a regular part of their school life benefit from increased self-esteem and become more engaged in their education, both inside and outside the classroom walls.

Figure 51.3 A word cloud based on publications produced by the Field Studies Council, including the acclaimed field guides and charts. (Courtesy of Rebecca-Farley Brown, Field Studies Council.)

- It supports improved standards back inside the classroom, raising attainment, reducing truancy and improving discipline. Learning outside the classroom is known to contribute significantly to raising standards and improving pupils' personal, social and emotional development.

Real-world learning brings the benefits of formal and informal education together and reinforces what good educationalists have always known:

Figure 51.4 Real-world learning can involve the human landscape as well as the natural world. Here, young students are studying 'Key Stage 4' human geography by the River Thames, London. (Courtesy of the Field Studies Council.)

the most meaningful learning occurs through acquiring knowledge and skills through real-life, practical or hands-on activities (www.lotc.org.uk/why). Couple meaningful learning with improved health, well-being and confidence and we have a powerful justification for outdoor education!

Patient Observation

What links our authors in Part II with the likes of Gilbert White and Charles Darwin? It is clear from reading their amazing array of stories that careful, close, sustained observation of the world around them is the answer. In *Crow Country* (Cocker, 2008), Mark Cocker emphasises the abiding importance of 'curiosity' in propelling the work of some of our great naturalists. The value of this is reinforced in Cocker's more recent book *Our Place* (Cocker, 2018), which draws on the work of field naturalists in detailing the state of nature through a thoroughly grounded understanding of its workings. Derek Ratcliffe features prominently in both of Cocker's books as an outstanding naturalist, who quarried his observations of nature to galvanise the nature conservation movement through his writings and wider work (see Thompson *et al.*, 2015). Of course, we must choose what we observe – we cannot observe everything. As the quotation from Mary Midgley points out (in Section 1.5), people are readily drawn to ask questions about a particular thing in the natural world: it may be flowers or stars, or it may be something that other people do not care for at all – toads, beetles, tapeworms.

This is not the place to delve into the psychological reasons for our fascinating choices, but it is clear that there is the widest range of topics available to choose from, from dung flies to urban graffiti. Gilbert White could not record everything, and it is clear where his emphasis lay from the *Naturalists' Calendars* compiled by William Markwick after White's death in 1793 (and reproduced as an appendix in all later editions of White's book): leafing and flowering of plants; harvest times; the first appearance of plants and animals and other aspects of animal life such as bird song. Just as it is impossible to observe everything, so it is impossible to measure everything. White's measurements were simple but accurate – mainly dates with associated accounts – still among the best we possess for the fauna and especially the birds of Great Britain. Of course, White needed extraordinary observational skill to compile such extensive records and we see some of the earliest studies in botany and zoology in his writings (White, 1789).

Charles Darwin was equally a close observer of the world around him. In Chapter III of *The Origin of Species*, Darwin illustrates the breadth and depth of his observations with two examples of the enclosure of heathland. In the first example from Staffordshire, hundreds of acres of barren heath had been enclosed (fenced in) and planted with Scotch fir (*sic*); nothing else had changed, climate or soil, but Darwin notes that 12 species of plants (not counting grasses and sedges) flourished in the plantation that were not found on the heath. He speculates that the effect on insects must have been even greater since six species of insectivorous birds not seen on the heath were present in the woodland, with three distinct species of insectivorous birds on the heath. In the second example from the heathlands of Surrey, Darwin notes that enclosure has led to 'self-sown firs . . . springing up in multitudes so close together that all cannot live' (Darwin, 1872, p. 56). On first sight, Darwin could not see any young firs on the open heath, but close inspection indicated a multitude of seedlings and little trees that were perpetually browsed down by cattle. In one square yard, Darwin counted 32 little trees; 'one of them with twenty-six rings of growth, had during many years tried to raise its head above the stems of the heath, and had failed' (Darwin, 1872, p. 56). The heath appeared barren, but cattle were obviously grazing effectually. Patient observations such as these underpin Darwin's arguments about the struggle for existence and indeed permeate the entire *Origin*. Whatever Darwin chose to study required very patient observation over long periods of time, and careful recording, the basis for his science, such evidence both generating new ideas and being used to test them.

While enthusiasm for observation is undiminished, technology has changed out of all recognition since the 1850s (see our selective list in Section 1.3). There are plenty of examples in Part II of this, for example Ingvar Byrkjedal's ocean voyage with 'sampling, seabed mapping and all manner of recording', Andrew Tibbs's use of lidar to map Roman forts in Scotland, and Kirsty Park's use of bat detectors. Field observation increasingly means field measurement but, as with observation per se, we cannot measure everything. The ability to measure is central to any scientific explanation. The role of a technique (i.e. a measurement procedure) is to allow the variable involved to be quantified in a consistent, rigorous and precise manner. The need to quantify is essential, since a general understanding of any problem is inherently interconnected with one's ability to define the elements contained in that problem. To say that something cannot be measured is tantamount to admitting one's knowledge is limited and that any ideas on the subject must remain vague and confused. The ability to measure helps fix one's general conception of a problem, so that quantity is not regarded as opposite to quality but rather an improvement on it (Anderson and Burt, 1990).

The ever-growing need for numerical measurement has important consequences for fieldwork but it does not undermine or replace the spirit of curiosity, which is the essential prerequisite for any fieldwork. These days, other quantitative approaches will very often complement what we do in the field, including laboratory measurements, computer-based map analysis, remote sensing and computer simulation models. In Chapter 1 we reviewed a wide range of instrumentation, ancient and modern. The message here is, notwithstanding the sophistication of our toolkit, that there is no substitute for being out in the field, curious and alert: that is where new ideas invariably come from, and where ideas generated by other means can be tested and, likely, improved. Not all our authors might regard themselves as quantitative scientists, but patient observation and careful recording is often the start of something significant!

The discoveries and insights which lie in wait must whet the appetite of naturalists – young and old. As this book goes to press, we are intrigued to read in the journal *Science* of the discovery of a fossil find in China from the Cambrian period around 500 million years ago. Palaeontologist Dongjing Fu and colleagues (2019) describe collecting

Figure 51.5 Professors Jeff Warburton and Tim Davies studying the crust of a dried-out bog pool on Bleaklow in the South Pennines, England. Over the past 10 years there have been transformative advances in understanding peat erosion and restoring bogs at a large scale to secure their carbon stocks (Tim Burt). A black and white version of this figure will appear in some formats. For the colour version, please refer to the plate section.

4,351 specimens at a new site, representing 101 different taxa, or groups of organisms. Of the taxa, some 53 per cent have never before been observed. As animal life exploded in diversity and form during the Cambrian, this work will shed new light on biological innovation and diversification during this period (see also Chapter 22).

Coda: Curiosity – the Beginning Not the End

How better to end a book founded on curiosity than to consider the person described by art historian Kenneth Clark as 'the most relentlessly curious man in history': Leonardo da Vinci (Clark, 1939, p. 258). Leonardo's curiosity is perhaps best illustrated in his notebooks, which were crammed full of sketches, notes and calculations. His biographer, Walter Isaacson describes his curiosity as 'delightfully obsessive' (Isaacson, 2017, p. 2): he ranges across the widest range of subject matter, some of which could easily have been included here – eddies of water, plant stems, birds in flight, a

Figure 51.6 Columbine flower (popularly known as granny's bonnet) drawn by Leonardo da Vinci. The flower belongs to the *Aquilegia* genus (which takes its name from the Latin for eagle, *aquila*, as the flower petals resemble an eagle's talon). Columbine refers to the Latin family name for doves, Columbidae, due to the resemblance of the inverted flower to five doves clustered together. (Credit: ilbusca/DigitalVision Vectors/Getty Images.)

woodpecker's tongue – and some well beyond our scope of interest – weapons, flying machines, theatre props. Leonardo knew how to marry science and art, observation with imagination, which made him the consummate innovator in Renaissance art. The important point here is that Leonardo's science was based 'primarily on observation, then discerning patterns, then testing their validity through more observations and experiments' (Isaacson, 2017, p. 174).

More than a century before Galileo, Leonardo's approach anticipated the modern Scientific Revolution: 'his uncanny abilities to engage in the dialogue between experience and theory made him a prime example of how acute observation, fanatic curiosity, experimental testing, a willingness to question dogma, and the ability to discern patterns across disciplines can lead to great leaps in human understanding' (Isaacson, 2017, p. 176).

We cannot hope to match Leonardo's genius, but we can learn from him, aspiring to improve our own skills in being curious about nature, recognising the value of patient, concentrated observation, and always cultivating a sense of awe and wonder in the world around us. Above all, concludes Isaacson (2017, p. 9), 'Leonardo's relentless curiosity and experimentation should remind us of the importance of instilling, in both ourselves and our children, not just received knowledge but a willingness to question it – to be imaginative and, like talented misfits and rebels in any era, to think different.'

Figure 51.7 Patience and deep thought are great virtues in the field. Here, Sally Johnson of Scottish Natural Heritage ponders how best to map arctic-alpine heaths in the western Cairngorms (Des Thompson).

References

Anderson, M. G. and Burt, T. P. (1990). Geomorphological techniques – part one: introduction. In *Geomorphological Techniques*, 2nd edn, ed. A. S. Goudie. Unwin Hyman, London, pp. 1–29.

Browne, E. J. (2002). *Charles Darwin: Volume 2 – The Power of Place*. Jonathan Cape, London.

Clark, K. (1939). *Leonardo da Vinci*. Penguin, London.

Cocker, M. (2008). *Crow Country: A Meditation on Birds, Landscape and Nature*. Jonathan Cape, London.

Cocker, M. (2018). *Our Place. Can We Save Britain's Wildlife before It Is Too Late?* Jonathan Cape, London.

Darwin, C. (1859). *On the Origin of Species by Means of Natural Selection*. John Murray, London.

Darwin, C. (1872). *On the Origin of Species by Means of Natural Selection*, 6th edn. John Murray, London.

Fu, D.,Tong, G., Dai, T., *et al.* (2019). The Qingjiang biota – Burgess Shale – type fossil Lagerstätte from the early Cambrian of south China. *Science* 363, 1338–1342.

Isaacson, W. (2017). *Leonardo da Vinci*. Simon & Schuster, New York, NY.

Thompson, D. B. A., Birks, H. H. and Birks, H. J. B., eds. (2015). *Nature's Conscience: The Life and Legacy of Derek Ratcliffe*. Langford, Peterborough.

White, G. (1789). *The Natural History and Antiquities of Selborne*. Cassell & Company, London (at least 300 editions published).

Contributing Author Biographies

John Archer-Thomson did a botany and zoology degree at Swansea University and a Masters degree in ecology at Durham University, UK. After graduation, he was employed as Conservation Warden for the Dorset Wildlife Trust. A love of conservation and aptitude for communicating its importance led to a job as marine ecology tutor for the Field Studies Council (FSC) at Dale Fort field centre, Pembrokeshire, UK. He went on to become Senior Assistant Head of Centre. From 2012, he has been a freelance coastal ecologist, tutor, writer and photographer. His latest book, co-authored with ex-FSC colleague Julian Cremona, is *Rocky Shores* (British Wildlife Collection Volume 7, Bloomsbury Publishing, 2019).

Arjun Amar is based at the FitzPatrick Institute of African Ornithology, University of Cape Town, South Africa. His research interests are primarily focused on the ecology and conservation of birds of prey. Arjun's passion for raptors was sparked through his boyhood interest in falconry. Arjun's PhD at Aberdeen University, UK, investigated the cause of the hen harrier decline on the Orkney Islands. He then undertook postdoctoral positions with the Game and Wildlife Conservation Trust and the US Fish and Wildlife Service, and then was a Senior Conservation Scientist for the RSPB in the UK for six years before moving to South Africa in 2011.

Alison Averis has spent her entire working life as a field botanist. She studied at Bangor (north Wales) and Edinburgh (Scotland) universities and then worked for the Nature Conservancy Council (latterly Scottish Natural Heritage), first in the Derbyshire Dales and then in Scotland as part of the Upland Survey Project. Since 1993, she has been freelance, and together with her husband Ben she has mapped, surveyed and monitored upland vegetation throughout Great Britain. Her first book, *Springs and Flushes*, was published in Scottish Natural Heritage's 'Scotland's Living Landscapes' series (2003). and she is the first author of *An Illustrated Guide to British Upland Vegetation* (Pelagic Publishing, 2004), which won the Botanical Society of the British Isles/Wild Flower Society Presidents' Prize for best botanical book in 2004.

Hilary Birks is Professor Emerita at the Department of Biological Sciences, University of Bergen, Norway. Her research mainly concerns vegetation, environment and climate change during the transition from the Last-Glacial through to the Holocene. She uses evidence primarily from plant remains (macrofossils) preserved in lake sediments. These are seeds, fruits, leaves, mosses, etc. She has also worked on recent environmental changes in lakes. She is a Foreign Member of the Norwegian Academy of Science and Letters, a Corresponding Fellow of the Royal Society of Edinburgh and a Fellow of the American Association for the Advancement of Science.

John Birks is Professor Emeritus at the University of Bergen, Norway, and University College London, UK. His research is primarily palaeoecology, vegetation history, environmental change and quantitative techniques. His hobby-botany interests are alpine plants, mountain expeditions and photography, and once included bryophytes. He is a Fellow of various learned academies or societies in Norway, Sweden and Scotland and has been awarded academic prizes in Norway, Finland, Poland, UK and Scotland, and from the International Paleolimnology Association.

Chris Burn is Chancellor's Professor of Geography at Carleton University, Ottawa, Canada, whose research focuses on field investigations of permafrost and its response to disturbance, including climate change, in Yukon and western Arctic Canada. He worked with J. R. Mackay from 1986 to 2014, spending many happy weeks with him at Illisarvik each year from 1987 until Ross's last visit for fieldwork in 2011, at the age of 95.

Stephen Burt is a meteorologist from the University of Reading, UK. He is a long-standing committee member of the Royal Meteorological Society's Meteorological Observing Systems Special Interest Group and has been Chairman of the Climatological Observer's Link (COL), the largest organisation of weather observers in the UK and Ireland, since 2011. He is the author of *The Weather Observer's Handbook*, published by Cambridge University Press in 2012. Stephen received the Royal Meteorological Society's Gordon Manley Prize in 2006. His recent book, written with Tim Burt, *Oxford Weather and Climate since 1767*, was published by Oxford University Press in 2019.

Ingvar Byrkjedal retired in 2017 from his position as Curator of Vertebrates at the University Museum of Bergen (formerly Bergen Museum), Norway, and as an associate professor there. His research is partly on bird biology, focusing on breeding waders in mountains and on tundra, on which he has published several books, and on systematics of fish. In recent years he has been amassing reference and research collections for the museum, notably of fish from Greenland, Svalbard and the Arctic Ocean.

Richard Carrick is a Durham University geography graduate, who spent several years at a lecturer in the subject before moving on to a 35-year career working for some of the UK's leading travel businesses. He is now a non-executive director, consultant and investor in the travel and hospitality sector.

Michael Church is Professor Emeritus in Geography at the University of British Columbia, Vancouver, Canada. His research is focused on sediment transport and morphodynamic changes in rivers. He has made long-term field studies of the Peace and Fraser rivers, large rivers in western Canada; the former is regulated for hydropower and the latter presents significant problems of sedimentation. He also studies steep mountain streams with a focus on stability and public safety and has made contributions to regulations for forest practices around streams and to flood management in British Columbia. He is also interested in the methodology of experimental method in geomorphology and in the history of geomorphological thought.

Stuart Corbridge is Vice-Chancellor and Warden of Durham University. From 2013 to 2015, he was Provost and Deputy Director of the London School of Economics (LSE); he was also Professor of International Development there. Stuart studied geography at Cambridge and stayed there for his PhD, which was based on fieldwork in India. He then taught at Huddersfield Polytechnic, Royal Holloway, Syracuse, Cambridge and Miami before going to LSE. Stuart has made several contributions to international development theory and geo-politics, beginning in 1986 with his book-length critique of radical development geography. His major work remains (with Glyn Williams, Manoj Srivastava and Rene Veron), *Seeing the State: Governance and Governmentality in India* (Cambridge University Press, 2005).

Douglas Davies, Professor in the Study of Religion and Director of Durham University's Centre for Death and Life Studies, trained in social anthropology and theology. His monographs include *Meaning and Salvation* (Brill, 1984), *The Mormon Culture of Salvation* (Ashgate, 2000), *Anthropology and Theology* (Berg, 2002), *Theology of Death* (T&T Clark, 2008), *Emotions, Identity and Religion* (Oxford University Press, 2011), *Death Ritual and Belief* (3rd edn, Bloomsbury, 2017) and *Mors Britannica* (Oxford University Press, 2015). An Oxford Doctor of Letters and Honorary Doctor of Theology from Uppsala University, he is a Fellow of the Academy of Social Sciences, The Learned Society of Wales and The British Academy.

Roy Dennis MBE is a field ornithologist and wildlife consultant, living in Moray, Scotland; he has worked in the Highlands and Islands of Scotland since 1959, most notably at Fair Isle Bird Observatory and on the conservation of rare birds, especially osprey and the reintroduction of lost species, including red kite and white-tailed eagle. His Roy Dennis Wildlife Foundation is active on raptor reintroductions and satellite tracking in the UK and mainland Europe. He has long been an advocate for restoring lost mammals to Scotland, particularly beaver and lynx, and has successfully translocated red squirrels. He is a writer, lecturer and broadcaster.

David Evans is a glacial geomorphologist who specialises in glacial landsystems and their applications to reconstructing past glaciations. He was educated at the University of Wales, Lampeter (BA), Memorial University of Newfoundland (MSc) and University of Alberta (PhD) and has worked on glaciers in glaciation in a wide range of locations including the Canadian Arctic, Iceland, Svalbard, New Zealand, Norway, South Georgia and the Canadian Prairies as well as numerous places in the British Isles. After a 14-year spell as lecturer at the University of Glasgow he moved to Durham University in 2004 where he presently works with a large glacial research group. He is the author/co-author of the books *Glaciers and Glaciation* (first published by Hodder Arnold, 1998), *A Practical Guide to Glacial Sediments* (first published by Hodder Arnold, 2004), *Glacial Landsystems* (CRC Press, 2005), *Vatnajokull National Park (South Region): Guide to a Glacial Landscape Legacy* (Durham University, 2016),*Till: A Glacial Process Sedimentology* (Wiley, 2017) and *Glaciation: A Very Short Introduction* (Oxford University Press, 2018). He was awarded the Busk Medal of the Royal Geographical Society in 2017, in recognition of his contribution to glacial geomorphology and research-led teaching in glaciated environments.

Alastair Fitter retired from the University of York in 2013, where he was Professor of Ecology, Head of Biology and Pro-Vice-Chancellor for Research. He worked on the functional biology of root systems and mycorrhizal fungi, and the impacts of climate change on soil ecosystems. He was elected FRS in 2005 and appointed CBE in 2010, and he was President of the British Ecological Society (2004–2005). In addition to scientific papers, he has written numerous popular natural history books, many with his father, the naturalist Richard Fitter (1913–2005).

David Goode has over 50 years' experience as a professional ecologist and an international reputation as a writer on environmental issues. He was appointed peatland specialist in the Nature Conservancy in 1967 and became Assistant Chief Scientist in the Nature Conservancy Council in 1976. As Senior Ecologist at the Greater London Council (1982–1986) and subsequently Director of the London Ecology Unit (1986–2000) he developed a major London-wide programme for biodiversity conservation, which became a model for many other cities. He has written two books on the ecology of cities, *Wild in London* (Michael Joseph, 1986) and a Collins' *New Naturalist* volume, *Nature in Towns and Cities* (2014). David was awarded the Chartered Institute of Ecology and

Environmental Management (CIEEM) Medal in 2015, and he has been an enthusiastic naturalist since the age of 14, with a particular interest in birds.

Andrew Goudie, a geomorphologist, is a graduate of Cambridge University, but has spent all his academic career in Oxford, where he has been Professor of Geography and Master of St. Cross College. He has received the Founder's Medal of the Royal Geographical Society, the Farouk El-Baz medal for desert research, a Gold Medal from the Royal Scottish Geographical Society and the Linton Award of the British Society for Geomorphology. He has been a chairman of the British Geomorphological Research Group, President of the Geographical Association and President of the International Association of Geomorphology. He has undertaken fieldwork in most of the world's hot deserts.

David Harper is Professor of Palaeontology and Principal of Van Mildert College at Durham University. He is best known for his studies on the Cambrian Explosion, Great Ordovician Biodiversification Event and the end-Ordovician extinction, based on extensive fieldwork in many parts of the world. Notably, he has promoted the adoption of more rigorous approaches to palaeontology through the development of user-friendly statistical packages, including PAST. He is author of several textbooks, including *Introduction to Paleobiology and the Fossil Record* (co-authored with Michael J. Benton; Wiley, 2009), which is now widely adopted and leads the field on both sides of the Atlantic.

Heidi Hauffe leads the Conservation Genetics Research Unit and Department of Biodiversity and Molecular Ecology at the Fondazione E. Mach (FEM), northern Italy. She completed her DPhil in evolutionary biology at the University of Oxford in 1993, followed by postdoctoral positions on the role of hybrid zones in speciation at the universities of Pavia and York. She set up the first animal genetics laboratory at FEM in 1997 and has focused on wildlife conservation genomics and zoonotic disease ever since. She applies her research to many different organisms, from viruses to polar bears in collaboration with researchers worldwide.

Peter Higgins is Professor of Outdoor, Environmental and Sustainability Education at the University of Edinburgh, teaching through academic, practical and online approaches. His research is primarily interdisciplinary and across these fields. He is Director of the university's Global Environment and Society Academy, and of the UN Regional Centre of Expertise in Education for Sustainable Development for Scotland. He is a UK Commission for UNESCO 'Expert Panel' member. Since 1999 he has advised the Scottish Government in these fields, chairing the Learning for Sustainability Advisory and Implementation Groups (2012–2016). He is a Fellow of the Royal Geographical Society,

Principal Fellow of the Higher Education Academy, a member of several national advisory boards and a trustee of the Field Studies Council.

Russell Hill is Professor of Evolutionary Anthropology and currently Joint Head of the Department of Anthropology at Durham University. His research focuses on the behavioural ecology of primates and other large mammals and how predation and the landscape of fear shape behaviour. He has conducted and supervised research at a range of field sites in southern Africa and currently runs the Primate and Predator Project in the Soutpansberg Mountains in South Africa. Increasingly, his work investigates human–wildlife conflict from an interdisciplinary perspective that integrates qualitative and quantitative approaches.

Emi Husband graduated from Hatfield College, Durham University, in 2017 after gaining a first-class degree in physical geography. She was also awarded Hatfield College's Mott Fieldwork Prize in 2016. Emi completed an MSc by research at the University of Exeter in 2019. Her research focused on the changing structural complexity of coral reef ecosystems and the development of a coral genus trait-based technique that has the capability to assess and monitor these changes over large spatial scales.

Charles Krebs is an Emeritus Professor of Zoology at the University of British Columbia in Vancouver and Thinker in Residence at the Institute for Applied Ecology, University of Canberra. He has worked on small-mammal population and community dynamics for 61 years in northern Canada, south-eastern Australia and South-east Asia, studying lemmings, snowshoe hares and field voles in North America, house mice in south-eastern Australia and rice-field rat problems in South-east Asia. His major interests are large-scale ecological experimentation and improving ecological methods. He has published 16 books in ecology and rodent biology and 319 papers on pure and applied ecology. He spends his time continuing a 46-year monitoring programme of the boreal forest ecosystem of the southern Yukon, writing books and a blog about how to improve the science of ecology.

Christian Körner was Professor of Botany at the University of Basel from 1989 to 2014. He received his academic degrees at the University of Innsbruck, Austria. His main research fields are high-elevation ecology (alpine and treeline, with books in both fields), forest ecology and global change biology in general. He is one of the authors of the (originally German) standard textbook *Strasburger's Plant Sciences* (Springer, 2013).

Allan Lilly joined the Soil Survey of Scotland in 1980 and has been responsible for systematic soil survey, land evaluation and data collection for the Scottish

Soils Database. Allan has authored over 150 technical reports and maps and has published over 50 scientific papers. He has spent time working with soil scientists in the Netherlands, USA and Australia. He leads work on enhancing and making soils data more widely available and has contributed to the development of Scotland's soils website. Allan is a Fellow of the Institute of Soil Science and curator of the National Soils Archive of Scotland.

Andrés Luque-Ayala is an Associate Professor in the Department of Geography at Durham University. His research focuses on the politics of urban infrastructures in the global South. Currently, he is working on the coupling of digital and material infrastructures as a new security apparatus in the city. He holds first degrees in anthropology and political science, as well as an MSc in city design (LSE), a Masters in environmental management (Yale) and a PhD in geography (Durham).

Peter Marren is a writer on natural history, nature conservation and other subjects. He once worked for the old Nature Conservancy Council as a local officer in Scotland and England, and latterly as its 'author-editor'. He is the author of 22 books including *The New Naturalists* (Harper Collins, 1995), *Nature Conservation* (Collins, 2002), *Britain's Rare Flowers* (Poyser, 2005), *Bugs Britannica* (Chatto and Windus, 2010), *Mushrooms* (British Wildlife, 2012), and *Rainbow Dust* (Blackwell, 2016), a study of the age-old appeal of butterflies. He has also written an autobiography of nature and childhood, *Where the Wild Thyme Blew* (Blackwell, 2016), and his latest is *Chasing the Ghost* (Vintage, 2018), a journey in search of wild flowers and the warm heart of British field botany. In press is a book about the funny names, English and Latin, of moths and butterflies. Some of these books have won awards. He is a regular contributor to *British Wildlife*, including his long-running column, 'Twitcher in the Swamp'.

Rob Marrs retired as the Bulley Professor of Applied Biology at the University of Liverpool in 2018. He is an applied plant ecologist and throughout his career he has worked on everything from the sub-tundra to tropical rainforests. He has focused his research mainly around long-term manipulative experiments, setting up some of his own and working on ones that were set up in the 1950s when he was a babe-in-arms. Rob has worked mainly on the management of heathlands and moorlands with a special interest in the management of bracken and the impact of the managed burning of moorland. He is also interested in the analysis of large country-wide data sets, for example the National Broad-leaved Woodland Survey and Scottish Soft Coasts. He is a Fellow of the Chartered Institute of Ecology & Environmental Management and a Chartered Ecologist.

Jeff McDonnell is Associate Director of the Global Institute for Water Security and Professor of Hydrology at the University of Saskatchewan, Canada. His research focuses on forest hydrology, isotope hydrology and hillslope rainfall-runoff processes. Jeff is the current President of the American Geophysical Union (AGU) Hydrology Section and recent recipient of the International Hydrology Prize (Dooge Medal) from the International Association of Hydrological Sciences, UNESCO and World Meteorological Organization. He has received the Dalton Medal from the European Geosciences Union and is an elected Fellow of the Royal Society of Canada, the AGU and the Geological Society of America.

Stephen Moss is one of Britain's leading nature writers, broadcasters and wildlife television producers. His television credits include the BAFTA-award-winning *Springwatch*, *The Nature of Britain* and *Birds Britannia*, while his books include *Wild Kingdom* (Penguin, 2016), *The Robin: A Biography* (Vintage Publishing, 2017) and *Mrs Moreau's Warbler: How Birds Got Their Names* (Guardian Faber Publishing, 2018). He writes a regular monthly 'Birdwatch' column for the *Guardian* newspaper, is President of the Somerset Wildlife Trust, and teaches an MA in Travel and Nature Writing at Bath Spa University. A life-long naturalist, Stephen has travelled to all seven of the world's continents in search of wildlife.

Stephen Mott has worked in education for most of his life. His first degree is in geology from the University of Durham where he won the Main Prize for Fieldwork. He has taught geology, chemistry and physical geography at secondary schools and has taught in the field with adults. Latterly, he worked in school standards and improvement and informally with university students. He has been interested in birds and wildlife since childhood. He provides his own programme of field studies, writing and research. He enthusiastically promotes fieldwork in all aspects of natural history.

Andrew Nolan is a researcher in soil science, notably soil survey and upland management and is currently Head of the Soil Survey and Land Evaluation Unit at the James Hutton Institute, formerly the Macaulay Institute, in Aberdeen. His expertise ranges from soil and vegetation mapping to assessment of grazing and trampling impacts on upland habitats. He is currently working on a long-term project for the Forestry Commission mapping soil resources at 1:10,000 scale, including training and mentoring a younger generation of soil scientists in soil survey and classification.

Kirsty Park is Professor in Conservation Ecology within Biological and Environmental Sciences at the University of Stirling. After a BSc at Leeds she went on to study bat ecology for a PhD at Bristol. Her research is concerned with the effects of human activity on biodiversity and how best to mitigate the negative impacts, focusing on animal ecology and conservation in heavily managed environments (e.g. urban, agriculture, forestry). She is a Trustee for the UK's Bat Conservation Trust and Chair of the Board for Bats without Borders, a conservation organisation based in southern Africa.

Stuart Rae is an environmental consultant specialising in alpine-arctic ecology in the Scottish Highlands, particularly birds and how they fit in that environment. A firm believer in the value of long-term studies, he has worked on golden eagles for over 40 years and rock ptarmigan for 30 years. Now he lives in Australia and has adapted this approach to tawny frogmouths and, most recently, the initiation of a thorough study of little eagles. Yet he still makes time to return to Scotland every year to continue his original studies, working two sequential breeding seasons.

Jane Reid is Professor of Population and Evolutionary Ecology at the University of Aberdeen. She aims to understand the environmental and genetic causes of individual variation in life histories, and to understand the consequences of such variation for population and evolutionary dynamics. She tackles these aims using long-term data from wild bird populations and has considerable experience as a field ornithologist. Jane was awarded the Zoological Society of London Scientific Medal in 2012, the Marsh Award for Ornithology in 2013 and was elected a Fellow of the Royal Society of Edinburgh in 2017.

Michael Reiss is Professor of Science Education at UCL Institute of Education, University College London, a Fellow of the Academy of Social Sciences and Visiting Professor at the Universities of Kiel and York and the Royal Veterinary College. The former Director of Education at the Royal Society, he has written extensively about curricula, pedagogy and assessment in science education and has directed a large number of research, evaluation and consultancy projects over the past 20 years. His PhD and postdoc were in evolutionary biology and behavioural ecology and he has a long-standing interest in education in field and museum settings.

Lisa Robins has been an Honorary Senior Lecturer at the Australian National University's (ANU's) Fenner School for Environment and Society since 2009. She has 30 years' experience working in natural resource management (NRM) as a researcher, science communicator, facilitator and programme manager. In addition to an extensive record of engagement and publishing in Australia, she has worked in the Mekong Region, Indonesia, Canada and the UK. Lisa holds a PhD on NRM governance and capacity building from the ANU (2005–2009) and was awarded a Master of Science in Environmental Change and Management (1995) from the University of Oxford.

Leigh Simmons is Director of the Centre for Evolutionary Biology and Professor of Evolutionary Biology in the School of Biological Sciences at the University of Western Australia. Leigh studied at the University of Nottingham where he received his PhD in 1986. He held a research fellowship at the University of Liverpool, UK, before moving to Australia in 1995. Leigh's research uses both vertebrates and invertebrates to test the predictions and assumptions of theoretical models of sexual selection and life history evolution. He has published more than 300 papers and articles, has authored books on sexual selection and insect sperm competition, and he has co-edited volumes on dung beetle ecology and evolution, and on insect mating systems. Leigh is Editor-in Chief of *Behavioral Ecology*, and Editor of *Advances in the Study of Behavior*. He held a prestigious Australian Research Council Federation Fellowship from 2005 to 2009, is a recipient of the Zoological Society of London's Scientific Medal and was elected to the Australian Academy in 2009.

Mike Slattery is the Director of the Institute for Environmental Studies at Texas Christian University. His teaching and research focus on human impacts on the environment, including the impact of dams on river systems and the socio-economic and ecological impacts of utility-scale wind farms. He has written more than 85 scientific articles and published five editions of his book *Contemporary Environmental Issues* (Kendall Hunt Publishing). In 2014, he launched the TCU Rhino Initiative, a conservation programme based in the Eastern Cape of South Africa, focused on protecting rhinos in the wild.

William Sutherland is delighted to hold the Miriam Rothschild Chair in Conservation Biology at the Zoology Department in Cambridge. He has a broad interest in global conservation, especially of threatened birds; he has the objective of changing global conservation practice through evidence-based conservation. He has written two books, co-authored another three and edited a further four. He established, and is series editor for, the *Techniques in Ecology and Conservation* series with Oxford University Press. He was awarded the Marsh Award for Ecology, the Scientific Medal of the Zoological Society London and the Marsh Award for Conservation Biology.

Andrew Tibbs undertook his PhD at Durham University where he specialised in the positioning and orientation of Roman fortifications in Scotland. He regularly undertakes fieldwork on some of the most northerly Roman sites in the world, particularly those on the Gask Ridge, Rome's oldest frontier, which is located in Perthshire, Scotland. As well as on-site fieldwork, he is a specialist in the use of remote sensing techniques and the application of modern survey techniques to better understand the Roman occupation of north Britain. He has recently published his first book, *Beyond the Empire: A Guide to the Roman Remains of Scotland* (Robert Hale, 2019).

Sue Townsend has been with the Field Studies Council since 1987; developing partnerships and working in consultation with exam boards, recording societies, national biological recording organisations and, most importantly, with volunteers. As a Chartered Biologist and Fellow of the Linnean Society, her passions are in taxonomic skills, identification and recording. Sue has developed and delivered academic courses, grant-funded projects and enabled field skills to be enhanced across the sector. She is Chair of the Shropshire Environmental Data Network, Secretary to the training and education group for the Botanical Society of Britain and Ireland, and a part-time lecturer with Manchester Metropolitan University.

Steve Trudgill is Emeritus Fellow, Robinson College, Cambridge and an Honorary Vice-President of the Field Studies Council. He grew up in Norfolk where his interests were fostered by the Norfolk and Norwich Naturalists' Society and by studying botany, zoology and geography. He read geography, geology and sociology at Bristol and did his PhD field research on Aldabra Atoll, Indian Ocean. He lectured in Cambridge University Geography Department on Biogeography, Environmental Management and the Social Engagement with Nature. He was a Member of the Executive Committee of the Field Studies Council 1976–2003 and of its Science and Education sub-Committee, which he chaired for two years. He was a member of Slapton Ley National Nature Reserve Management Committee 1990–2005. He is currently Archival and Historical Editorial Advisor to the journal *Progress in Physical Geography*.

Maurice Tucker retired after nearly 30 years at Durham University (1983–2011) in the Earth Sciences Department and as Master of University College (1998–2011) and is now Visiting Professor at the University of Bristol. He has written several textbooks including *Sedimentary Petrology* (Blackwell, 1981) and *Carbonate Sedimentology* (Blackwell, 1990), and received various honours: Major Coke Medal (Geological Society of London, 1993); Percy Allen Award (British Sedimentological Research Group, 2007); Jean Baptiste Lamarck Medal (European Union of Geosciences, 2009); Medaille Andre Dumont (Geological Society of Belgium, 2009); Honorary Fellow, Academy of Sciences and Arts, Croatia (2008); President (1998–2002) and Honorary Member, International Association of Sedimentologists (2010).

Alan Werritty is Professor Emeritus of Physical Geography, University of Dundee and former Research Director of Dundee's UNESCO Centre for Research on Water Law, Policy and Science. Alan's current research focuses on climate change, flooding and societal responses to increased flood risk. He has published over 100 refereed articles and book chapters. Since 2002, Alan has advised the Scottish Government on drafting and subsequently implementing the Flood Risk Management (Scotland) Act 2009. A former member of Scottish

Natural Heritage's Science Advisory Committee, in 2017 Alan was invited to chair a Scottish Government inquiry into grouse shooting, to report in 2019.

Natalie White is the Outdoor Learning Development Officer and Principal Teacher for East Ayrshire Council and an Accredited Practitioner of the Institute for Outdoor Learning. She supports teachers and practitioners to utilise the endless resource which is the outdoors, and to engage our children and young people in learning. Natalie enjoys a variety of activities, both in and outdoors: sailing, walking, climbing and crafts. She is also the Scottish Commissioner for Adventure and Outdoors at Scout Scotland and an Explorer Scout Unit leader. She is passionate about children having the opportunity to experience their own adventures, challenges and have the chance to be 'wowed' by our amazing world.

Lewis Winks is a social science researcher and environmental educator, specialising in sustainability and transformative education, in the Department of Geography at the University of Exeter. He has worked in the environmental sector as both practitioner and researcher with organisations seeking systemic social change. Lewis's research focuses on transformative education and outdoor learning, utilising behaviour change theory to better understand how environmental worldviews and narratives are shaped by our experiences, as well as how we might better equip learners to contribute to sustainable futures. Lewis facilitates teacher training programmes and works regularly with young people in outdoor environments. He is committed to practical action, outdoor adventure and creative cross-disciplinary working.

Index

Printed in the United Kingdom by TJ Clays Ltd.